International Conference on Sustainable Energy Engineering (ICSEE 2017)

IOP Conference Series: Earth and Environmental Science
Volume 73

Perth, Australia
12 – 14 June 2017

ISBN: 978-1-5108-4411-7
ISSN: 1755-1307

Printed from e-media with permission by:

Curran Associates, Inc.
57 Morehouse Lane
Red Hook, NY 12571

CURRAN ASSOCIATES INC. proceedings.com

Some format issues inherent in the e-media version may also appear in this print version.

Printed by Curran Associates, Inc. (2017)

For permission requests, please contact the Institute of Physics
at the address below.

Institute of Physics
Dirac House, Temple Back
Bristol BS1 6BE UK

Phone: 44 1 17 929 7481
Fax: 44 1 17 920 0979

techtracking@iop.org

Additional copies of this publication are available from:

Curran Associates, Inc.
57 Morehouse Lane
Red Hook, NY 12571 USA
Phone: 845-758-0400
Fax: 845-758-2633
Email: curran@proceedings.com
Web: www.proceedings.com

Table of contents

Volume 73

International Conference on Sustainable Energy Engineering
12–14 June 2017, Perth, Australia

Accepted papers received: 3 July 2017
Published online: 10 July 2017

Energy Management and Economic Strategy

Design and Development of New Energy Vehicles

Power and Electrical Engineering

System Modeling and Performance Evaluation

IOP Conf. Series: Earth and Environmental Science **73** (2017) 011001 doi:10.1088/1755-1315/73/1/011001

Preface

It is our great pleasure to welcome you to 2017 International Conference on Sustainable Energy Engineering (ICSEE 2017), which will be held in Perth, Australia during June 12-14, 2017.

The objective of ICSEE 2017 has always been bringing together a large group of researchers, scientists, academics, and engineers in this related area from all over the world, who share much more than technical interests, among other things, culture and history.

This proceedings of ICSEE 2017 presents papers submitted to the conference from universities, research institutes and industries. The papers selected depended on their quality and their relevancy to the conference. The volume tends to present to the readers the recent advances in the field of renewable energy, energy management and economic strategy, energy development and generation system, design and development of new energy vehicles, power and electrical engineering, system modeling and performance evaluation etc.

We would like to thank all the authors who have contributed to this volume and also to reviewers, speakers, chairpersons, sponsors and all the conference participants for their hard work and support to ICSEE 2017.

ICSEE 2017 Organizing Committee
June 5, 2017

IOP Conf. Series: Earth and Environmental Science **73** (2017) 011001 doi:10.1088/1755-1315/73/1/011001

Statement of Peer Review

All papers published in this volume of Earth and Environmental Science have been peer reviewed through processes administered by the proceedings Editors. Reviews were conducted by expert referees to the professional and scientific standards expected of a proceedings journal published by IOP Publishing.

IOP Conf. Series: Earth and Environmental Science **73** (2017) 011002 doi:10.1088/1755-1315/73/1/011002

Conference Committee

General Chair
Tyrone Fernando, University of Western Australia, Australia

Technical Program Chairs
Herbert Lu, University of Western Australia, Australia

Udaya K. Madawala, The University of Auckland, New Zealand

S. M. Muyeen, Curtin University, Australia

Publication Chair
Shu H. Wang, Tongji University, China

Technical Committee Members
Fengji Luo, The University of Sydney, Australia

San Shing Choi, University of Western Australia, Australia

Yasser Alsaleh, Mohammed Bin Rashid School of Government, UAE

Fariha Tariq, University of Management and Technology, Pakistan

Tsung- Mou Huang, National Sun Yat-Sen University, Taiwan

Waleed Al-Saedi, UoT&ECU, Iraq

Peter BRIMBLECOMBE, City University of Hong Kong, Hong Kong

Zhen Li, Beijing Institute of Technology, China

Ben Cheng, The Hong Kong Polytechnic University, Hong Kong

Greg Crebbin, Murdoch University, Australia

Shervin Fani, Western Power Australia

Darren Woodhouse, Safearth, Newcastle, Australia

Sanath Alahakoon, Central Queensland University, Australia

Mohamed Darouach, University of Lorraine, Nancy, France

Dylan Lu, University of Technology Sydney, Australia

Hak-Keung Lam, King's College, London, UK

Steve Ling, University of Technology Sydney, Australia

Bingo Ling, Guangdong University of Technology, China

Dongsheng Yu, University of Mining and Technology, China

Farhad Shahnia, Murdoch University, Australia

Ali Arefi, Murdoch University, Australia

Peer review statement

All papers published in this volume of *IOP Conference Series: Earth and Environmental Science* have been peer reviewed through processes administered by the proceedings Editors. Reviews were conducted by expert referees to the professional and scientific standards expected of a proceedings journal published by IOP Publishing.

IOP Conf. Series: Earth and Environmental Science **73** (2017) 012001 doi:10.1088/1755-1315/73/1/012001

Optimisation and Integration of Hybrid Renewable Energy Storage Systems

E L V Eriksson, E MacA Gray

Queensland Micro- and Nanotechnology Centre, Griffith University, Nathan 4111, Australia

Email: emma.eriksson@griffithuni.edu.au

Abstract. This paper discusses renewable energy system concepts and integration techniques, and reviews modelling and optimization techniques for hybrid renewable energy systems for electricity provision. A proposal to use design criteria that are not limited to performance- and cost-related factors is introduced and forms a background to the following discussion. Optimization techniques in relation to constraints, reliability analysis and algorithms are discussed as well as software tools available for modelling/simulation, component sizing and optimization. The focus is on systems incorporating hydrogen, but the ideas presented have general relevance.

1. Introduction

The faster than projected uptake of distributed photovoltaic systems [1] is one example of the ways in which global energy systems are becoming outdated [2] in terms of technological capability. Furthermore, in the new world of the internationally agreed 2°C climate-change scenario entered in 2016, the approach of designing to meet only goals related to the technology (capacity, availability, reliability) and economics (return on investment, cost to the consumer) must evolve to be able to take into account other goals: environmental goals foremost, especially the sustainability of the entire life cycle of the installation and its constituent materials and waste products, but also those with a social dimension, such as relationship to landscape and noise. Real-world projects are delayed, disrupted or even cancelled owing to non-technical factors. The great majority of optimization studies reviewed address only techno-economic criteria.

Figure 1 illustrates sets of criteria that could be addressed in the optimization of energy systems, with proximity indicating relative weight: technical, based on performance; economic, based on financial factors; environmental, based at minimum on legal requirements for sustainability; and socio-political, addressing public acceptability and facility, for instance. Any of these areas could be emphasized: technical for critical service, environmental for sensitive locations and so on. The virtual exclusion of factors other than technical and economic over many decades, as illustrated in figure 1, has led to the present need for drastic action for more sustainable energy forms. Prioritization of criteria according to the situation and their quantitative inclusion via an algorithm in the optimization methodology can produce well-balanced outcomes.

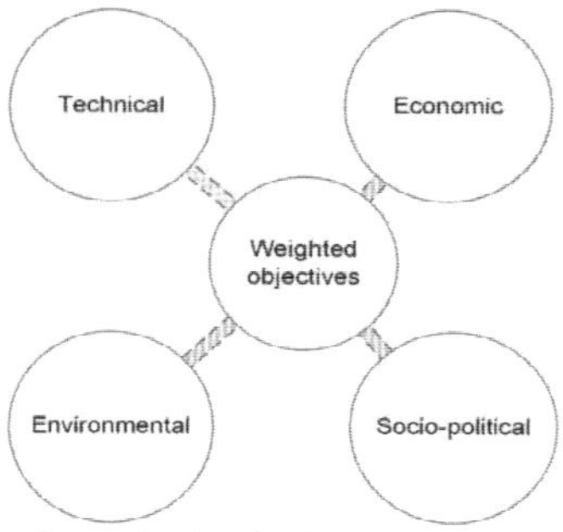

Figure 1. Delineation of criteria for evaluating design outcomes

2. Hybrid energy system integration

Although the load supplied is most likely AC, the generator may be intrinsically AC or DC, and the storage may involve intrinsically AC or DC components. This reality naturally delineates three configurations: DC-coupled, AC-coupled and a hybrid configuration utilizing both DC and AC [3], where "coupled" refers to the electrical bus linking the majority of components. An overarching principle is to minimise the number of energy transformations, since these involve inefficiencies.

In a DC-coupled system the generating components are connected to the bus via DC/DC or AC/DC converters. This configuration can directly feed suitable DC loads, whereas AC loads are fed via a DC/AC converter. Bidirectional converters may be deployed for connected storage components, although batteries may be connected directly to a bus specified for the battery voltage, thus minimising the number of transformations in the energy path. A DC-coupled scheme enables a fairly simple configuration, as synchronization is not required for integration of the various energy sources.

An AC-coupled system integrates generating components such as wind and solar via DC/AC or AC/AC converters. AC loads are fed directly from the bus whereas AC/DC converters feed DC loads. As with the DC-coupled system, a bidirectional inverter may be used for storage systems. Synchronization may be required. In addition, these systems may suffer higher conversion losses, as rectifiers have relatively low efficiency compared to inverters and DC-DC converters [4].

A hybrid DC/AC-coupled configuration has the advantage that it caters for both DC and AC loads, hence eliminating unnecessary energy conversions and increasing the source-to-load efficiency. The downside is that these systems are naturally more complex in terms of control and energy management.

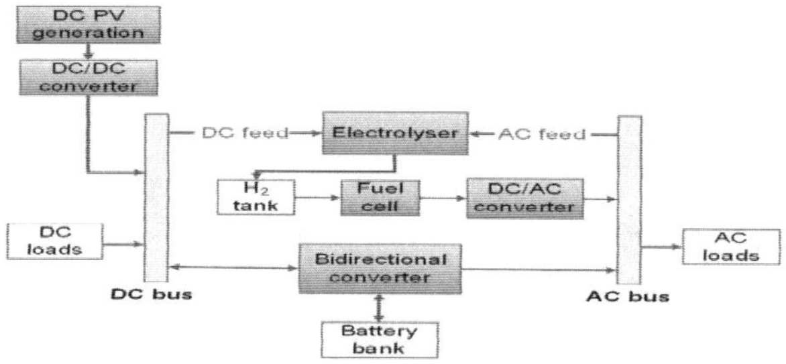

Figure 2. Hybrid bus topology

Individual components may have options for integration, especially electrolysers, which, despite being intrinsically DC, can be purchased as "plug-and-play" AC-powered (fig. 2).

3. Optimisation

Optimization may be seen as getting the best result from the committed resources, including the sources of energy and finance, under constraints derived from the criteria. At the simplest level, the principal output from an optimization process is a specification of the size (energy and/or power capacity) of each component of a system whose configuration was pre-determined. Correct component sizing is the key to achieving a properly functioning system that meets its design goals. At the more sophisticated level, optimization could include an algorithmic approach to settling the system configuration as well, with final parameter values determined by a weighted measure that includes the expanded sets of criteria.

Various configurations of solar/wind hydrogen hybrid energy systems for electricity supply have been studied and reviewed [5], reaching as far back as the 1980's. These energy systems vary in size from 1 kW up to over 600 kW with alkaline and PEM electrolysers with power capacities between 1 and 320 kW, lead-acid and Li-ion batteries ranging from 42 to 1310 kWh, 30 to 120 bar hydrogen storage and PAFC and PEM fuel cells of capacities from 0.5 to 80 kW. Optimization was generally approached via modelling/simulation based on mathematical models describing the physical (electrical, mechanical, thermal) characteristics of individual components, in other words addressing technological criteria, with cost usually addressed separately.

In purely technical terms, accurate simulation of the performance at system level relies on accurate models of the components. In terms of the reliability of the simulation for a given system in a given location, measured or simulated profiles for the energy input and the load determine whether the simulation is usefully predictive:

4. Evaluation constraints

Constraints on the system design function as boundaries and represent parameters for optimization, and so are the key to finding an optimal solution. Typical constraints applied to designs of renewable energy systems reported in the literature are renewable capacity (e.g. insolation, wind capacity), installation size (e.g. PV area, wind area), physical characteristics (e.g. PV panel tilt angle, wind installation height), battery characteristics (e.g. minimum state of charge, maximum charge/discharge rate, cycle life), environmental restrictions (e.g. pollutant emissions, social acceptability) and various grid integration issues (e.g. power fluctuations of injected power and cost of line extension).

4.1. Implementation

Analysis of the reliability of supply plays a vital role when incorporating renewable energy in energy systems, due to the intermittency of the energy source in most cases. Reliability analysis can be carried out in various ways, the most frequently reported method being loss of power supply probability (LPSP), defined as the ratio of the summed energy deficits to the total load energy demand over a selected period, ranging from 0 (load always satisfied) to 1 (load never satisfied. Other less-often adopted methods to measure reliability of supply include loss of load probability (LOLP), loss of load expected (LOLE), loss of load risk (LOLR), deficiency in power supply probability (DPSP), expected energy not supplied (EENS) and level of autonomy (LOA) [6-8]. These measures are interrelated and are all suitable for an evaluation against technical criteria. Frequently adopted economic criteria are annual system cost (ACS) and levelized cost of energy (LCOE), which are calculated by accounting for all of a system's expected lifetime costs (including construction, financing, fuel, maintenance, taxes, insurance and incentives) and in the case of LCOE then divided by the system's lifetime expected energy output. A frequent environmental criterion is to calculate the generated CO2 emissions per generating component in the system. A suitable socio-political criterion could be a risk portfolio evaluation where security of supply is evaluated in line with reliability of renewable and primary energy integration, refer table 1 and table 2. The importance of the latter consideration is illustrated by the severe outage across the Australian state of South Australia in September 2016 [9] when extreme weather disabled a major network at a time when the ~1.5 GW wind power resource was unavailable.

Table 1. Environmental criteria and variables

Environmental criterion	Variables
Emission	Analysis of CO2 emissions as units to be minimized.
Fuel consumption	Minimise the total amount of energy consumption taking into account discount factors and cost of imported fuel.
Life cycle	Analysis of Life cycle (cradle-to-grave) emissions generated during the manufacturing, operation and decommissioning of the system.
Waste management	Analysis of the need to dispose of waste during system operation.
Hazard level	Analysis of type of waste and its severity on the environment.

Table 2. Socio-political criteria and variables

Socio-political criterion	Variables
Land acquisition	Visual impact, shadowing, flicker, eco-system disturbance, acoustic noise, electromagnetic interference.
Political acceptance	Analysis of regulations, policies and subsidies released by energy market commissions effecting the energy sector.
Social acceptance	Review of consensus amongst public for energy initiatives and analysis of negative social effects.
Labour	Analysis of employment opportunities in the energy sector.
Social cost of carbon emissions	Estimation of economic value of impact through carbon emission reduction
Security of supply	Minimisation of exposure to fuel price instability. Analysis of security of supply in the event of severe weather disruption and availability of back up generation and/or grid connection. Analysis of type of waste and its severity on the environment.

5. Algorithms

Numerous approaches and algorithms for optimizing hybrid renewable energy systems have been surveyed [10-12], with a focus on improving the reliability and lowering the computational time of optimization models [13-16]. The most common sizing methodologies reported in the literature are graphical construction methods, probabilistic methods, analytical methods, iterative methods, artificial-intelligence methods and hybrid methods. Probabilistic, analytical and iterative methods require continuity and differentiability of the equations describing the problem, resulting in CPU-intensive calculations, and are not suited to simulating dynamic performance or readily able to account for factors such as thermal constraints and component degradation over the life cycle. If optimization problems have high problem dimensions with several local optima, as is typically the case in a complex energy system, artificial-intelligence approaches such as heuristic optimization techniques are more favourable, as these methods are able to adequately account for dynamic performance and changing system variables with low average error [17]. The selection of an optimization technique ultimately depends on the adopted design criteria, available information, computing resources and the accuracy of the technique.

5.1. Hybrid artificial intelligence methods

In recent years it has become clear that combining an optimization algorithm with other optimization techniques to produce a hybrid (or metaheuristic) algorithm may lead to more efficient behavior and higher flexibility in solving a wide range of optimization problems. This is a complex methodology, but once implemented produces major savings in computation time and has developed into a significant research field [18].

Evolutionary algorithms are the basis of a common meta-heuristic optimization approach where the best solutions are isolated by eliminating the poorest solutions. Meta-heuristic algorithms include: Particle Swarm Optimization (PSO), which uses solutions that move in a design space based on their position [19] and Genetic Algorithm (GA), which uses a fixed linear data structure that allows hierarchical variables [20]. A recent review of optimization studies of renewable energy systems in the period 2013–2015 [21] found that GA (63%) and PSO (30%) were the most-used optimization techniques.

6. Discussion

Of reviewed optimization studies, refer Table 1 Ref. [22], a large majority of optimization assessments assess only techno-economic criteria. There is therefore an opportunity to change the basis of costing to become more comprehensive and thereby more honest, since in all cases the costs are paid somewhere.

Incorporating environmental and socio-political criteria is less straightforward and less certain than a performance- or cost-based assessment. As renewable energy systems become larger, the urgency to accurately predict the outcomes in terms of the selected design criteria will also grow, so that funding, time and public acceptance are not squandered through inappropriate configurations, capacities or locations.

The result of an optimization can only be as good as its inputs, but good inputs will be distorted by poor internal mathematical models. Accurate models of PV arrays are readily available. The stochastic element of wind energy poses a problem, but credible models are readily available. Likewise, battery models are readily available. The development of accurate but tractable electrolyser and fuel-cell models is more recent [13], but these have now reached the point of being both simple enough for incorporation in heuristic optimization and accurate enough to be credible. Hydrogen storage modelling is in a much less advanced state.

7. Conclusion

This paper advocates a more generalised approach to the optimisation of renewable energy storage systems which recognises that performance has different meaning in different contexts and so includes not just technical performance and cost factors, but also environmental and socio-political factors. These can be given weightings to produce an overall performance metric against which the system design is measured during optimisation.

8. Further study

Further studies are recommended into the implementation of a design methodology which takes into account a multi objective criteria whilst carrying out optimization of a system using genetic algorithm. This design methodology would be well placed for analyzing both grid and off grid small and large scale renewable systems.

A survey of modelling/simulation/optimization software packages showed that there is a need for new developments, specifically in modelling hydrogen storage, and generally in adopting a more flexible approach to shaping design criteria additional to cost and technical performance.

References

[1] REN21, Renewables 2016 global status report, in Global Status Report. 2016, Renewable Energy Policy Network for the 21st Centure: Paris.

[2] Bazmi A A and Zahedi G, Sustainable energy systems: role of optimization modeling techniques in power generation and supply—a review. Renewable and Sustainable Energy Reviews, 2011. 15(8): p. 3480-3500.

[3] Farret F A and Simões M G, Integration of alternative sources of energy. 2006: Wiley-IEEE Press.

[4] Starke M, Tolbert M L, and Ozpineci B, AC vs. DC distribution: a loss comparison. IEEE, 2008. 3(8).

[5] Schucan T, Final report of subtask A: case studies of integrated hydrogen energy systems, in International Energy Agency Hydrogenn Implementing Agreement Task 11: Integrated Systems. 2000, Paul Scherrer Institute: Switzerland.

[6] Yang H, Lu L, and Zhou W, A novel optimization sizing model for hybrid solar-wind power generation system. Solar Energy, 2007. 81(1): p. 76-84.

[7] Borowy B and Salameh Z, Methdology for optimally sizing the combination of a battery bank and PV array in a wind/PV hybrid system. IEEE, 1996. 11(2).

[8] Lin X, et al., An improved optimal sizing method for wind-solar-battery hybrid power system. IEEE Transactions on Sustainable Energy, 2013. 4(3): p. 774-785.

[9] Hunger F, South Australia blackout: Once in 50-year storm lashes state, in Sydney Morning Herald. 2016, Sydney Morning Herald: Sydney.

[10] Chauhan A and Saini R P, A review on integrated renewable energy system based power generation for stand-alone applications: configurations, storage options, sizing methodologies and control. Renewable and Sustainable Energy Reviews, 2014. 38(0): p. 99-120.

[11] Evans R, A review of computational optimisation methods applied to sustainable building design. Rnewable and Sustainable Energy Reviews, 2013(22): p. 230-245.

[12] Zhou P, Jin R Y, and Fan L W, Reliability and economic evaluation of power system with renewables: A review. Renewable and Sustainable Energy Reviews, 2016. 58: p. 537-547.

[13] Junghans L and Darde N, Hybrid single objective genetic algorithm coupled with the simulated annealing optimization method for building optimization. Energy and Buildings, 2015. 86: p. 651-662.

[14] Yang C, et al., High throughput computing based distributed genetic algorithm for building energy consumption optimization. Energy and Buildings, 2014. 76: p. 92-101.

[15] Hansen N and Ostermeir A, Adapting arbitrary normal mutation distributions in evolution strategies: the covariance matrix adaption, in Evolutionary Computation. 1996, IEEE International Conference.

[16] Storn R and Price K, Differential Evolution - A simple and efficient heuristic for global optimization over continuous spaces. Journal of Global Optimization, 1997. 11: p. 341-359.

[17] Al Busaidi A S, et al., A review of optimum sizing of hybrid PV–Wind renewable energy systems in oman. Renewable and Sustainable Energy Reviews, 2016. 53: p. 185-193.

[18] Lazaros E M, A review on hybrid optimization algorithms to coalesce computational morphogenesis with interactive energy consumption forecasting. Energy and Buildings, 2015. 106: p. 192-202.

[19] Stoppato A, et al., A PSO (particle swarm optimization)-based model for the optimal management of a small PV (Photovoltaic)-pump hydro energy storage in a rural dry area. Energy, 2014. 76(0): p. 168-174.

[20] Khare V, Nema S, and Baredar P, Solar–wind hybrid renewable energy system: A review. Renewable and Sustainable Energy Reviews, 2016. 58: p. 23-33.

[21] Askarzadeh A, Optimisation of solar and wind energy systems: a survey. International Journal of Ambient Energy, 2016: p. 1-10.

[22] Eriksson L V and Gray E MacA, Optimization and integration of hybrid renewable energy hydrogen fuel cell energy systems – A critical review. Applied Energy, 2016.

IOP Conf. Series: Earth and Environmental Science **73** (2017) 012002 doi:10.1088/1755-1315/73/1/012002

Defining Toll Fee of Wheeling Renewable with Reference to a Gas Pipeline in Indonesia

Amrullah Hakim
Former Member of PMO 35 GW, Student of MBA University of Indonesia, Manager Gas Commercial Medco E&P Natuna

Emails: Amrullah.Hakim@gmail.com

Abstract: Indonesia has a huge number of renewable energy sources (RE) however; the utilization of these is currently very low. The main challenge of power production is its alignment with consumption levels; supply should equal demand at all times. There is a strong initiative from corporations with high energy demand, compared to other sectors, to apply a renewable portfolio standard for their energy input, e.g. 15% of their energy consumption requirement must come from a renewable energy source. To support this initiative, the utilization of power wheeling will help large factories on industrial estates to source firm and steady renewables from remote sites. The wheeling renewable via PLN's transmission line has been regulated under the Ministry Decree in 2015 however; the tariff or toll fee has not yet been defined. The potential project to apply wheeling renewable will obtain power supply from a geothermal power plant, with power demand from the scattered factories under one company. This is the concept driving the application of power wheeling in the effort to push the growth of renewable energy in Indonesia. Given that the capacity of PLN's transmission line are normally large and less congested compared to distribution line, the wheeling renewable can accommodate the scattered factories locations which then results in the cheaper toll fee of the wheeling renewable. Defining the best toll fee is the main topic of this paper with comparison of the toll fee of the gas pipeline infrastructure in Indonesia, so that it can be applied massively to achieve COP21's commitment.

1. Introduction

Indonesia is a vast archipelago comprised of more than 17,000 islands, with over 250 million inhabitants. More than half of inhabitants concentrate in Java Island while the rest spread across Sumatra, Sulawesi, Kalimantan, Nusa Tenggara and Maluku, Papua, and other smaller islands whereas also abundant energy resources are located.

The Government of Indonesia (GoI) through PLN, the state-owned electricity company that is vertically integrated and holds a monopoly over transmission and distribution is responsible for meeting increased electricity demands of its consumers.

With a high number of populations, household has been a major consumer of PLN's amongst three other basic tariff categories on consumer types namely industrial, business and others (public). Per PLN's data, from 2013-2015, average market share measured by electricity sales for household is 42% followed by industry as the second largest market share of 33%. In 2015 electricity sales for household reached 88,682 GWh, industry 64,078 GWh, business 36,978 GWh and 13,106 GWh for others (public) [1]. Industry has always been focus on GoI, in 2015 The President of Republic of Indonesia issued Government Regulation No. 14 Year 2015 concerning Master Plan of National Industry Development year 2015-2033 with its firm vision of "Indonesia as a strong industrial country".

This target would be hard to achieve if energy as one of important backbones for industrial growth cannot be met as needed.

2. Electricity Supply

Supply of electricity has been a domain of GoI. Electricity Law No. 30 Year 2009 stipulates that "*power supplies shall be under the control of the state, which supplies shall be provided by the Government and the regional governments under the principle of regional autonomy*". Supply-demand balance is varying geographically across the country. The generating capacities in Sumatera and East Indonesia (including the smaller islands with smaller electricity systems) are barely sufficient to meet the increasing demand. While Java and Bali are served by generating capacities that are bigger than those are in the rest of other parts of Indonesia combined to meet the power requirement that is mostly supplied by coal-based power generation. Nevertheless, the growth of industry that still concentrates in Java is driving increased electricity consumption. Companies and manufactures are demanding more sustain supply of power from PLN for their production. Alas, PLN's regular maintenance, such as repair at the sub-station, transmission and or distribution lines, major overhaul plus unpredicted event of disruptions or force major could cause reliability of its system where would end up in worse cases with the shutting down of power generations without providing back up with the same amount of power for consumers. To overcome this, some of manufactures install diesel gen-sets at their facilities to prevent the shortage.

Having stable power supply is key for manufactures. With the growth of their production and simultaneously with their environment awareness to take steps toward reducing the environmental impact, manufactures are no longer seeking for power from conventional supplies. They demand of renewable energy is also slowly building up to contribute in their energy mix and to create cleaner productions. Therefore, manufactures are looking for another option to get the power aside from PLN. A few started to install solar PV at their facilities. A small number of them begin to consider getting renewable energy from independent power producers (IPPs) via power wheeling.

3. Electricity Plan

3.1. Implementation

Besides governing power supplies that has to be provided by government, the Electricity Law also allows private sector to take similar part in certain conditions. Electricity Law Num. 30 of 2009 has signalled that PLN is no longer the sole provider of power generation, the private sector who own and manage industrial zones are permitted to build and operate a power plant, along with its transmission lines, for that area.

Investors who generate electricity for their own use rather than for sale to PLN are known as Private Power Utilities (PPUs). A PPU can generate, transmit and distribute electricity for their own use or to their own consumer base (such as Industrial Zone) after getting necessary permits and license to do so.

PT Cikarang Listrindo Tbk is a private power company that provides energy to industrial and residential consumers in Indonesia under the exclusive license [2]. The Company's plants utilize gas turbines that run-on gas or liquid fuel. The license valid for 30 years which has begun since 11 December 2006 as the sole IPP supplying electricity to its tenants of five industrial estates in the Cikarang area which are not served by PLN.

This situation limits manufactures to freely choose source of renewable energy power that they want to have. Because some of them may have a condition where

- They have several facilities located in different areas;
- They are PLN's consumer and only intend to mix their source energy and has no intention to cut 100% power from PLN;
- They plan to their operation with 100% renewable energy source but PLN cannot provide;

3.2. Power Wheeling and Current Regulation

In such cases, power wheeling can be an option for these off-takers. Power wheeling, refers to the practice of a utility that owns a transmission network taking in and passing along electricity produced by another utility or generator. Consequently, besides the availability of electricity, transmission and distribution lines must present to enact power wheeling.

Ministry of Energy and Mineral Resources Regulation No. 01 year 2015 regulates Electricity Supply Cooperation and Joint Utilization of Power Grid. Ideally this should enable companies or manufactures to get electricity supply not solely from PLN but also from IPPs by utilizing the existing transmission and distribution lines.

Unlike any other countries who have successfully implementing power wheeling, there are still a few challenges that need to be addressed by the Regulator and PLN as the owner and operator of transmission and distribution grids, such as:

- The regulation that exists now which only allows power wheeling for IPPs that will supply the power to their own facilities not third parties'. And they cannot supply power to PLN's existing consumers should be reviewed in order to grow private sector interests as well to boost the development of RE;
- The "toll fee" for the usage of transmission and distribution grids and its electricity losses;
- The reliability of the grids;
- Electricity tariff that private sectors pay to IPPs should be competitive with PLN's, in particular, if it comes from renewable energy, and
- The permit for IPPs to have other markets besides PLN should not be impeded.

Private sector who will utilize power wheeling must pay a leasing cost and a reliability cost to PLN which will step in to ensure electricity supply if IPP has a problem.

3.3. Renewable for Electricity Indonesia

With lower demand situation, nowadays, PLN's total power generation capacity is forecast to grow from 50.9 GW in 2015 to 93.6 GW in 2024. Under the assumption that adjustment in the capacity expansion plan from the lower load growth, is limited to deferment of thermal projects, the existing scenario has a 19% share of renewable energy by 2024, assuming all hydro and geothermal projects in the 2015 power supply business plan are implemented according to the plan. To reach the Government target of 25% by 2025, or 24% by 2024, would require an additional 22.3 TWh of renewable energy capacity which would be met by a combination of additional wind, PV, small hydro and geothermal of the total 3,935 MW. Only the geothermal capacity is assumed to have a full capacity credit and would displace coal.

3.4. Transmission and Distribution CAPEX

Transmission CAPEX from the 2015 power supply business plan divided by two phases, the expenditures in the years 2015-2018 are dominated by the completion of the 500kV system, at around $4 billion per year, then fall sharply to around $1 billion by 2021 and thereafter. At the first phase, the transmission CAPEX per kW is $1,300 while the distribution CAPEX per kW is $460, resulting the total transmission and distribution CAPEX per kW at $1,760 [3].

4. Toll Fee Determination in Gas Transmission Pipeline

In order to support the need to expand its pipeline network to raise gas penetration rates and cut oil dependency, Government of Indonesia has introduced the concept of open access to gas pipelines [4]. Through the regulatory agency, BPH Migas, that was established on 30 December 2002 has function to assume state oil and gas company Pertamina's regulatory roles in relation to downstream activities (Articles 46 and 47 of Law No.22). BPH Migas is charged with assuring sufficient natural gas and domestic fuel supplies and the safe operation of refining, storage, transportation and distribution of gas and petroleum products via business licences.

Toll fee for gas pipeline must consider the economics of pipeline owner and interest of the shippers and consumers. In order to calculate the toll fee, BPH Migas follows the following steps:

- Determine the base parameters
  - Base Asset Value = 151 US$ mio
  - Capital Loan = 18 US$ mio
  - Equity Loan = 133 US$ mio
  - Length of Project = 8years
  - Risk Free Rate = 2.55%
  - Mature Equity Market Premium = 5.26%
  - Indonesia Country Risk Premium = 3.7%
  - Beta = 1.018
  - Cost of Equity = 11.67%
  - Incentive IRR on Equity = 0%
  - Loan Capital Interest = 6.67%
  - Corporate Income Tax = 25%
  - Cost of Debt = 5%
  - WACC (Capital Loan/Base Asset Value) * Cost of Debt + Equity Loan/Base Asset Value) * Cost of Equity = 10.89%
  - Inflation rate in USA = 2%
- Operational Parameters
  - Volume Forecasting and Volume Annual Gas Flow in mmscfd
  - Gas Losses
  - Operational Cost
- Income Calculation and Levy to BPHMigas (3%)
- Depreciation Calculation
- Cost of Service Calculation

Grissik-Duri pipeline (Duri line) is the milestone for onshore indonesian gas transmission which has been completed in 2001. It is 536 km long with 28 inch diameter pipeline from Grissik to Duri pipeline. There are facilities along Grissik to Duri pipeline, i.e. 22 Sectional Valves, 5 Pig Trap Stations, and 2 Compressor Stations which located at Sakernan and Belilas, with total capacity of 430 mmscfd.

Following the previous calculation steps, BPH Migas through the BPH Migas Regulation No. 01/2015 decided that the tariff transportation of natural gas for Duri line which defined as total Cost of Service divided by Annual Gas Flow shall be 0.466 US$/mscf, equivalent to 0.49 US$/MMBtu. This amount of toll fee provides an IRR to the owner of gas pipeline at 11%.

According to PGN (a state owned gas company), the toll fee for South Sumatera- West Java line (SSWJ) I with 400km length that built in 2007, has been regulated by BPH Migas to be 1.55 $/mscf while for SSWJ II with 600 km length is 1.47 $/mscf.

5. Converting Toll Fee of Gas to Electricity

If we take the reference of pipeline gas tariff calculation above, we can either count the conversion of 1 kWh to mmbtu directly for power wheeling toll fee on an open access transmission system. It is known that 1 kWh equal to 0.0034 MMBtu, so the toll fee in gas pipelines by 0466 US$/mscf will generate power wheeling rates at 0.17 US cents/kWh.

Meanwhile, if we calculate the total investment required for the construction of transmission and distribution principle using rate of return regulation of an electricity transmission company, then it is computed in three steps [5] as below:

IOP Conf. Series: Earth and Environmental Science **73** (2017) 012002 doi:10.1088/1755-1315/73/1/012002

- Estimate the amount of electricity (kWh) the transmission system will be delivering over a given year
- Estimate how much it will cost the transmission company to transmit that quantity of kWh
- Divide the second number by the first to compute a transmission price in cents/kWh

In setting prices for the transmission industry, policymakers must take into account special technological circumstances of electricity transmission: load balancing, line losses, loop flow, and user-specific lines. Transmission and distribution CAPEX per kW is known in the above at $1,760. Assuming that the transmission and distribution will be utilized for 20 years, then the cost of transmission and distribution which will be assumed as a power wheeling charge is as follow:
$1,760,000 / (20 years x 365 days x 24 hours) = 10$/MWh.

Further, from the data compiled by the Project Management Office of 35GW of Government of Indonesia, the evaluation of transmission CAPEX and distribution CAPEX are as follows:

- 2016: 1285 $/kW and 457 $/kW
- 2017: 1111 $/kW and 455 $/kW
- 2018: 927 $/kW and 406 $/kW
- 2019: 607 $/kW and 401 $/kW
- 2020: 271 $/kW and 370 $/kW
- 2021: 186 $/kW and 333 $/kW
- 2022: 159 $/kW and 313 $/kW
- 2023: 123 $/kW and 278 $/kW
- 2024: 33 $/kW and 231 $/kW

Respectively above numbers reflected to a power wheeling charge as described in Figure 1 below:

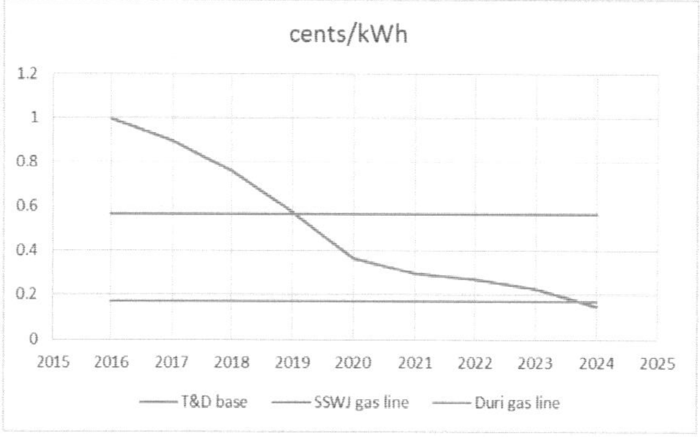

Figure 1. Indicative Toll Fee for Power Wheeling has range between 0.2 – 0.6 cents $/kWh as a baseline

The best approach to define the toll fee for renewable wheeling in Indonesia to have competitiveness with the gas transmission pipeline from Figure 1 is at 0.6 cents $/kWh.

6. Conclusion

Indicative Toll Fee for Power Wheeling has a range between 0.2 - 0.6 cents $ / kWh as a baseline can be used by the Government of Indonesia as a prefix to the implementation of power wheeling, for example, for a 5 MW geothermal IPP for later evaluation if required. If this approach is successful, this could be a pioneer in introducing the application of power wheeling in the effort to push the growth of renewables in Indonesia and expectantly could be followed by other companies or manufactures. In addition, the geothermal power plant company will have a new perspective to explore the development of small-scale geothermal projects (3/5/10 MW scale) which will also increase the exploration activities.

The increased number of geothermal power to be used to provide electricity for factories manufacture in Indonesia will support the utilization of renewable energy so that Indonesia's commitments in COP21 can be met.

Acknowledgments
Authors wishing to acknowledge assistance or encouragement from Mr Nur Pamudji, former CEO of PT PLN (Persero) and former Chairman of PMO 35 GW, and Ms Gina Lisdiani from Allotrope Partners.

References

[1] PLN *RUPTL 2015*

[2] PWC November 2016 *Power in Indonesia, "Investment & Taxation Guide"* 59

[3] ADB February 2016 *PLN Corporate Options Study 2015-2024: Financial Analysis* 29

[4] PWC May 2014 *Oil and Gas in Indonesia, "Investment & Taxation Guide"* 29

[5] Timothy J. Brennan, Karen L. Palmer, Raymond J. Kopp, Alan J. Krupnick, Vito Staglino and Dallas Burtraw 1996 *A Shock to the System – Restructuring America's Electricity Industry* 67

IOP Conf. Series: Earth and Environmental Science **73** (2017) 012003 doi:10.1088/1755-1315/73/1/012003

Bioethanol Production from Waste Potatoes as a Sustainable Waste-to-energy Resource via Enzymatic Hydrolysis

A A Memon, F A Shah, N Kumar

Department of Chemical engineering, Mehran University of Engineering and Technology, Jamshoro, Pakistan

Email: aziza.bano@faculty.muet.edu.pk;

Abstract. Ever increasing demand of energy and corresponding looming depletion of fossil fuels have transpired into a burning need of time to vie for alternative energy resources before the traditional energy sources are completely exhausted. Scientists are continuously working on sustainable energy production as an alternate source of energy to meet the present and future requirements. This research deals with conversion of the starch to fermentable carbon source (sugars) by fermentation through liquefaction by using yeast and alpha- amylase. The results show that the significant bioethanol production was achieved while using the parameters like temperature (30 ^{0}C) pH (6) and incubation time of 84 hrs. About 90 ml of bioethanol was produced from potato intake of 800 g. Pakistan being an agricultural country is rich in potato crop and this research bodes well to open new vistas to arrest the energy shortage in this part of the world

1. Introduction

Keeping global energy demand pathetic environment quality in view, it is dire need of renewable and sustainable liquid fuel and that can be achieved after the production of bioethanol. It will make future commitment to the energy solution [1].

Potato requirements of the Pakistan are estimated to be around 3 million tons per annum. This year the country is expected to have a record production of over 3.7 million tons. The hoarders have offloaded an estimated 300,000 tons of the last year's stocked potato in the wholesale market. [Source: Ministry of National Food Security and Research]

Pakistan may export 120,000 to 125,000 metric tons of potato to Russia as the demand for potato in Russia is increasing, said Harvest Trading's CEO Ahmad Jawad. Pakistan is making efforts to export the surplus to Sri Lanka, Turkey and Russia. It is important to mention that Pakistan lacked cold storage facilities in the past. The rising exports of citrus have helped develop a good number of storages in the agriculture heartland. The same facility is now being utilized for the storage of vegetables and fruits. http://www.pakistantoday.com.pk/2016/01/22/pakistan-to-export-500000-tonnes-of-potato

Globally interest is growing for production of ecological sustainable bio fuels. The target in the European Union is to adjust 5.75% of the fossil fuels which is used by traffic with biomass-based fuel from potatoes and corn as an ecologically sustainable bio-fuels by the year 2010 and 20% by the year

2020 [2]. Ethanol is a significant product of 21st century with its versatile usages and widely consumption across the globe.
The U.S. Energy Information Administration's recently released *International Energy Outlook 2016* (IEO2016) projects that world energy consumption will grow by 48% between 2012 and 2040. Most of this growth will come from countries that are not in the Organization for Economic Cooperation and Development (OECD), including countries where demand is driven by strong economic growth, particularly in Asia. Non-OECD Asia, including China and India, accounts for more than half of the world's total increase in energy consumption over the projection period. OECD-FAO, Agricultural outlook 2011–2020.
Pakistan exiting capacity of fuel grade ethanol is 270000 tons per year, while it has a potential of 400000 tons per year. (Environment.gov.pk). Ethanol production from potatoes have enormous potential for industrial production which is environmentally friendly, highly productive, economic with low cost, and can be easily manipulated [3].
There is dire need to investigate the economical ethanol production from waste potato by applying the optimized fermentation process [4].

2. Material and Methods

In order to attain maximum fermentable sugar conversion optimum parameters are applied. Temperature: 30 ^{0}C pH: 6 and Incubation Time 84 hrs. Ethanol is mostly produced depends on quality of raw materials and the optimum conditions [5]. Alcohol from potatoes obtained from two enzymatic methods respectively [6]. Using Yeast only 2. Using Alpha amylase and yeast.
First of all raw potatoes of market are being washed properly as contamination along with layers could be removed. After that mash of potatoes were prepared as smooth as silk with the help of grinder or juicer by adding proper amount of distilled water. Homogeneously of solution is made which is after wards processed to bio chemical or enzymatic reaction.

2.1. Yeast addition

The most common and simple method of forming alcohol from potatoes are yeast reaction to potato mash. First of all mash of potatoes are being prepared via heating the flour along water made up by grinder. Solution of water to potatoes is maintained at ratio of 2:1 (400 ml to 200 g of potatoes). Mash is afterwards heated to 30 to 60 minutes during that time temperature approximately climbs to 50 to 60 ^{0}C. After that solution is cooled down when it reaches to 30 to 35 ^{0}C. Normally the yeast (Rossmoor Food Product) added is 2 gms in diluted mash prepared [7].
Fermentation is something where microorganism comes in to play to let reaction happens. In that reaction ethanol along with CO_2 is formed. Fermentation includes solution which is yeast added is prepared than fermentation process takes place. Solution is kept at shaker for fermentation cause. Usually retention time for fermentation is 84 hours at room temperature as shown in figure 1 (A).

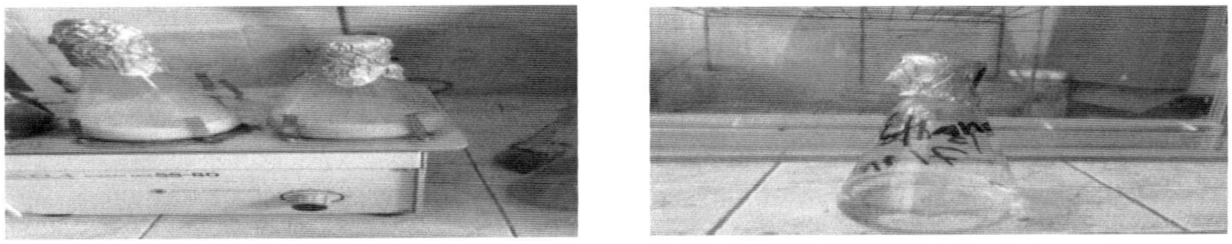

A **B**

Figure 1. Shake Flask Fermentation (A), Distilled Ethanol (B)

IOP Conf. Series: Earth and Environmental Science **73** (2017) 012003 doi:10.1088/1755-1315/73/1/012003

Distillation could be called as most important step of ethanol formation from potatoes using yeast. Boiling point of ethanol is 78 ^{0}C and distillation of fermented shake is maintained at 78 ^{0}C along with continuous monitoring as shown in figure 1 (B) as distilled ethanol.

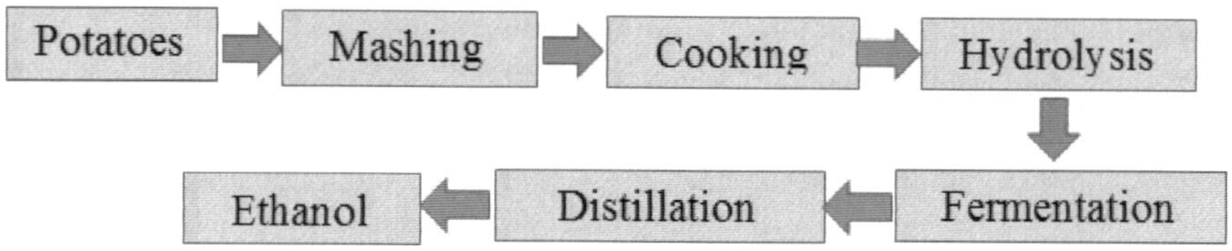

Figure 2. Ethanol Production by Yeast

2.2. Alpha Amylase Addition

This is process as compared to process described above. In this process first of all starch is being extracted from potatoes, which is nothing but the isolation process of starch.
If starch is used as raw material, potato mashing is not needed. Alpha amylase (AVONCHEM limited) enzyme is inserted. About 5 ml for mashing water and potatoes should be in 2:1. Than mash is cooked and stirred well almost for an hour. It is obvious that during that one hour temperature of mash will jump above 80 ^{0}C centigrade. Now that mash is cooled down until 60 ^{0}C centigrade and at this point remaining alpha amalyse is added and mash is allowed to cooled down until 30 ^{0}C centigrade where 2 grams of yeast is added and fermentation is followed afterwards [8]. Generally, applied methodology step wise as shown in figure 2.

3. Result & Discussion

The study to produce alcohol from waste potatoes due as an inexpensive and cheap feed stock in Pakistan as an agricultural country is selected and it was first attempt.

Table 1. Survey of Jamshoro City

Area In Jamshoro	Process able Amount(kg)	Spoiled Amount(kg)
Phatak	10000-12000	700-900
Society	2000-3000	300-400
Wapda Colony	1200-1500	100-150

Table 2. Material Balance

SUBSTANCE	WEIGHT
WASTE POTATOES	800 g
MASH SUBSTANCE	200 g
YEAST	2 g
DISTILLED WATER	400 ml
ALCOHOL FORMED	90 ml

The survey has been done by collecting data from different areas of Jamshoro City, Pakistan. It was observed that large amount is left as waste potato shown in table 1, can be recycled for ethanol production.

About 800 g of potatoes were used and significant quantity about 90 ml of alcohol was produced from waste potatoes in Chemical Engineering Department of Mehran University of Engineering and Technology Jamshoro as shown in table 2.

Optimum parameters were selected as pH. 6, temperature 30 ^{0}C and incubation time 84 hrs

Finally the produced ethanol was analyzed and obtained satisfactory results as shown in figure 5.

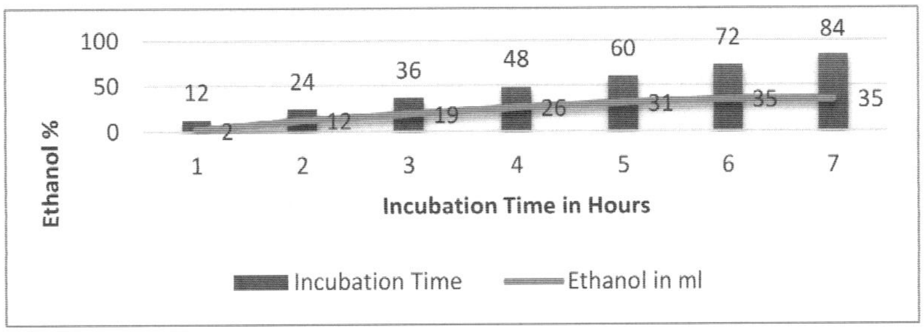

Figure 3. Ethanol % by Yeast

Concentration of alcohol determined from calibration curve, also known as a standard curve.

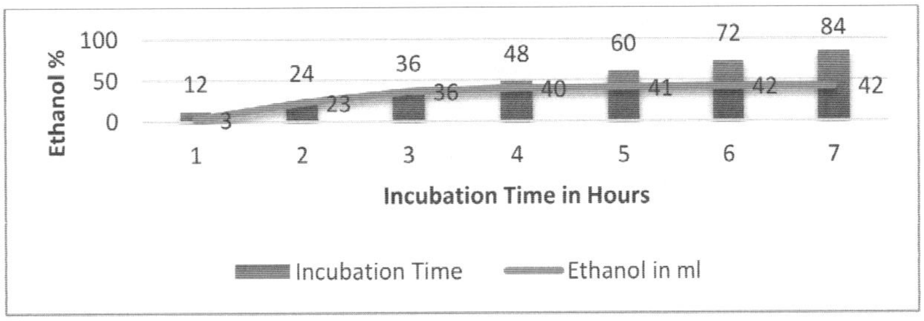

Figure 4. Ethanol % for Alpha Amylase

It has been analyzed that increased quantity around 10% was obtained by using alpha amylase as compare to yeast due to liquification by Alpha Amaylase as shown in figure 4 and 5.

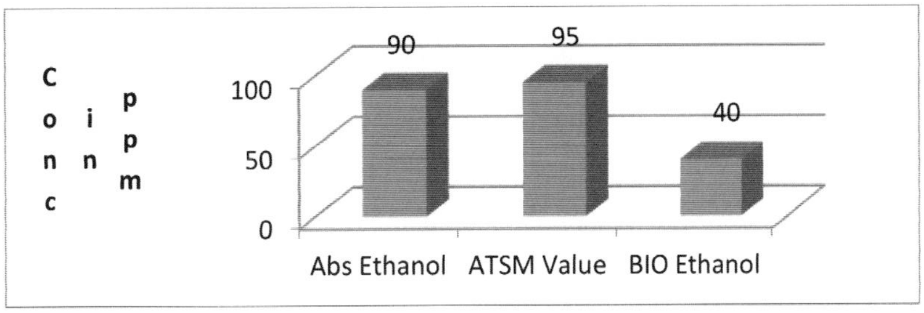

Figure 5. Net Concentration graph

According to ASTM and Absolute standards, the properties of alcohol were compared and found satisfactory as shown in figure 5.

4. Conclusion

Initially liquefication was performed before fermentation process by adding first yeast then secondly yeast and Alpha Amylase. It was observed that after 72 hrs incubation period by yeast maximum conversion of fermentable sugars to alcohal obtained but in case of Alpha Amylase and yeast, the maximum conversion of fermentable sugars to alcohol is after 48 hrs, so it is concluded that waste potato contain great amount of starch that can produce significant quantity of bioethanol by enzymatic hydrolysis as alternative fuel. By yeast it is very low cost and cheap to produce bioethanol as by alpha amylase enzyme product is expensive. Therefore, it is still need to study/investigate the economical production of ethanol as a common source of energy by achieving optimized fermentation process.

5. Recommendations

It is feasible to establish bio fuel plant and would be economical and contributing to overcome energy crisis we are facing now a days in Pakistan surely address the Global warming issues. Feasibility report for Plant Cost estimation will be calculated.

References

[1] Khan R A, Nawaz A, Ahmed M, Khan M R, Azam F D, Ullah S, Sadullah F, Ahmad A, Shah M S, Khan N. *Production of bioethanol through enzymatic hydrolysis of potato*. Afr. J. Biotechnol. 11: 6739–6743 (2012) Cross Ref Google Scholar

[2] Kilpimaa S, Kuokkanen T, Lassi U: *Ethanol Production from Waste Potatoes 2010,* University of Oulu, Department of Chemistry, FI-90014 University of Oulu, ENERGY RESEARCH 21 at the University of Oulu

[3] Wang F, Jiang Y, Guo W, Niu K, Zhang R, Hou S, Wang M, Yi Y, Zhu C, Jia C, Fan X: *An environmentally friendly and productive process for bioethanol production from potato waste 2016.* Biotechnology for Biofuels2016 **9**:50

[4] Gulten Izmirlioglu (2010) *Ethanol production from waste potato mash using saccharomyces cerevisaiae*, The Pennsylvania state university, (Thesis Report)

[5] Abul Kalam Azad, Nilufa Yesmin1, Shanjit Kumar Sarker1, Abdus Sattar, Rezaul Karim (2014): *Optimum Conditions for Bioethanol Production from Potato of Bangladesh*. Advances in Bioscience and Biotechnology, 2014, 5, 501-507

[6] D. Arapoglou, Th. Varzakas, A. Vlyssides, C. Israilides: *Ethanol production from potato peel waste (PPW) (2014).* National Technical University of Athens, Department of Chemical Engineering, 9, Heroon Polytechniou St., Zographou 157 70, Greece

[7] Sankaranarayanan M (2014): *Studies on alcohol production from sweet potato*. Institut Supérieur d'Agriculture et d' Elevage, ISAE, Busogo

[8] Sanat Rath, Ajay Kumar Singh, Harison Masih, Yashab Kumar, Jyotsna (2014): *Bioethanol production from waste potatoes as an environmental waste*. International Journal of Advanced Research, 2 (4), 553-563, 2014, International Journal of Scientific Engineering and Technology Research

IOP Conf. Series: Earth and Environmental Science **73** (2017) 012004 doi:10.1088/1755-1315/73/1/012004

Performance Simulation of an Air-Heated Humidification Dehumidification Desalination System

W F He, and D Han
Nanjing University of Aeronautics and Astronautics, Nanjing, Jiangsu Province, China

Email: wfhe@nuaa.edu.cn; handong@nuaa.edu.cn;

Abstract. This paper focus on an air-heated humidification dehumidification (HDH) desalination system driven by low grade waste heat. The working processes within the air-heated HDH desalination system are demonstrated, and the performance of the HDH desalination unit is gained and analyzed on the basis of the building mathematical models. In addition of the energy analysis, the entropy generation of the desalination system is also obtained to validate the energy efficiency. The simulation results show that the performance of the HDH desalination system is especially sensitive to the mass flow rate ratio between the working seawater and humid air. It was found that the actual maximum value of gained output ratio (GOR) is GOR=1.80 at the case of m_{sw}/m_{da}=0.68. Furthermore, the maximum value of GOR is consistent with the minimum entropy generation within the desalination, which validate the performance analysis in some extent. The obtained results provide some significant references for the design and optimization of the HDH desalination system.

1. Introduction

In the light of the water shortage all over the world, extensive attentions were paid on the desalination methods to produce freshwater. However, traditional systems [1, 2], were not suitable for small scale requirements of freshwater because of the relevant low efficiency. Recent years, a high-performance desalination method, called HDH desalination method, was proposed [3], and investigations have been achieved on the system constitution and the relevant performance analysis.

Al-Sulaiman [4] gave out a parabolic trough solar collector to power the open air, open water desalination system, and the related thermodynamic performance of two different specific configurations were simulated, acquiring the influences from the structures of the solar heater on the performance of the HDH system. It was discovered that the solar collector was very suitable for the air heated HDH systems in the places with high radiation, and the configuration with the air heater between the humidifier and the dehumidifier is more efficient with a higher freshwater production.

A theoretical simulation model of a water-heated solar desalination unit was also suggested by Hamed [5], and energy equations of each component within such unit considered were solved to evaluate the relevant performance as well as the freshwater yield. Two periods were designated to calculate the freshwater production of the proposed system. The final simulated results told that the highest freshwater production was found in the second period, after preheating before entering the humidifier at 13 pm to 17 pm. Furthermore, according to validate the accuracy of the theoretical results for the proposed desalination system, the relevant experimental platform was built to test the heat and mass transfer process within the system, and a good agreement of the comparison between experimental and theoretical results presented the significance of the new water-heated solar desalination system.

In addition of the energy analysis of the HDH desalination system, entropy generation was also calculated to optimize the system characteristics for better operation parameters [6-8]. Mistry [9] assessed the entropy generation with the HDH desalination system both with humid air and water heated, was accomplished. The mathematical models to obtain entropy generation and exegetic were built. In view of the results both from the energy and entropy generation analysis, it was validated that the minimization of the entropy generation was necessary and significant to fixed the components and the relevant operating conditions in order to get an optimal performance.

In the paper, the waste heat powered air-heated HDH desalination unit is proposed. The corresponding working principles of the air-heated HDH system are illustrated, and the performance for the HDH desalination unit is calculated and analyzed in view of the mathematical models with the analysis of energy and entropy generation. The research method as well as the corresponding obtained results provides the principles to define the configuration of the coupled HDH desalination system, and the entropy analysis also gives out the optimization orientation to reduce the final investments.

2. Numerical model and computing method

2.1. System description

Configurations of the air-heated HDH desalination system are presented in Fig. 1. It is found that the desalination system is mainly made up of the direct contact humidifier, the dehumidifier and the heat exchanger, and there existing two cycles, including the open seawater cycle and the closed humid air cycle. The initial seawater first enters the dehumidifier to cool the hot humid air, and the steam in the air condenses to freshwater. As a result, the initial seawater is heated to a higher temperature, and then the solvent of the hot seawater is carried by the dehumidified air in the humidifier, and the obtained concentrated seawater is exhausted to the environment direct while the humidified air flows into the heat exchanger to close the air-cycle.

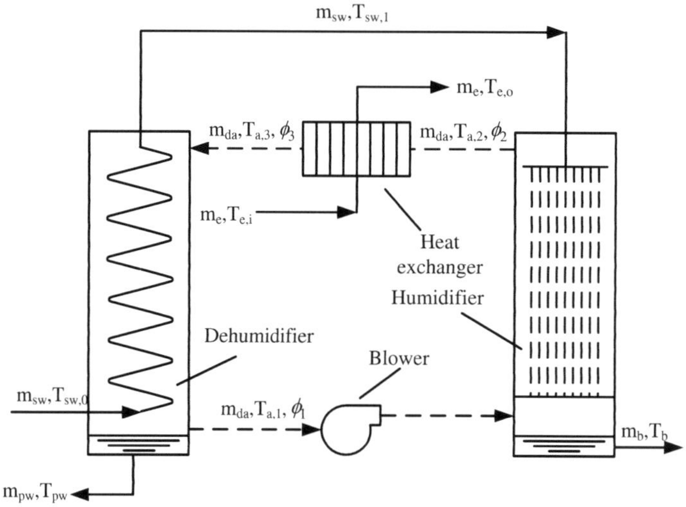

Figure 1. Configurations of the air-heated HDH desalination system

2.2. Mathematical models

In the heat exchangers, the saturated humid air out of the humidifier absorbs the released heat from the waste heat sources with a counter flow pattern. The relevant energy balance and entropy generation can be expressed:

$$Q_r = m_e \times (h_{e,i} - h_{e,o}) = m_{da} \times (h_{a,3} - h_{a,2}) \tag{1}$$

$$S_{gen,p} = m_{da}(s_{a,3} - s_{a,2}) + m_e(s_{e,o} - s_{e,i}) \tag{2}$$

For the dehumidifier process, the related conservation equations and entropy generation in the surface heat exchanger are demonstrated as:

$$m_w = m_{da}(w_3 - w_1) \tag{3}$$

$$m_{sw}(h_{sw,1}-h_{sw,0})+m_w h_w=m_{da}(h_{a,3}-h_{a,1}) \tag{4}$$

$$S_{gen,d}=m_{sw}(s_{sw,1}-s_{sw,0})+m_{da}(s_{a,1}-s_{a,2})+m_w s_w \tag{5}$$

For the humidification process, the corresponding equations are respectively expressed as:

$$m_{sw}-m_b=m_{da}(w_2-w_1) \tag{6}$$

$$m_{sw}h_{sw,1}-m_b h_b=m_{da}(h_{a,2}-h_{a,1}) \tag{7}$$

$$S_{gen,h}=m_b s_b-m_{sw}s_{sw,1}+m_{da}(s_{a,2}-s_{a,1}) \tag{8}$$

2.3. Assessment

In the HDH desalination system, freshwater is separated from the humid air continuously with a mass flow rate of m_w. As a result, the definition of gained-output-ratio (GOR) is defined to indicate the efficiency of the energy utilization.

$$GOR=\frac{m_{pw}\gamma}{m_e\cdot(h_{e,i}-h_{e,o})} \tag{9}$$

Furthermore, specific generation, $s_{gen,t}$ is applied to illustrate the irreversible loss within the desalination thermal cycle.

$$s_{gen,t}=\frac{S_{gen,t}}{m_w} \tag{10}$$

3. results and discussion

The investigated HDH desalination system is powered at the air aspect by the waste exhaust through the heat exchangers, and the corresponding performance is simulated and presented. The thermal parameters of the waste exhaust from a furnace, are listed in Table 1. Furthermore, for the humidification and dehumidification processes, the definition of the effectiveness, ε, and modified heat capacity ratio, HCR, are used to characterize the relevant performance, and all the designed parameters of the HDH desalination system during the simulation are listed in Table 2.

After the energy analysis of the HDH desalination system, the freshwater production and the humidity ratio of the humid air with the mass flow rate ratio from m_{sw}/m_{da}=0.68 to m_{sw}/m_{da}=1.68 are presented in Fig. 2. For the air-heated HDH desalination system, it is apparent the seawater solvent is first carried by the air during the humidification process. Thus, the freshwater is acquired after the carried steam condenses under the cooling effect of the feed seawater in the dehumidifier. It is seen that corresponding to the change principles of the humid air temperature, the difference of the humidity ratio between the inlet and outlet of the humidifier decreases continuously from w=0.031kgkg^{-1} to w=0.006kgkg^{-1} with the increase of the mass flow rate ratio, m_{sw}/m_{da}. According to the calculation method of freshwater production, the final flow rate of the freshwater production drops from m_w=73.56kgh^{-1} to m_w=12.42kgh^{-1}. It was verified that the performance of the HDH desalination will alternate at the balance condition of the dehumidifier, HCR=1[10-12]. Hence, it is shown that the balance condition of the humidifier, HCR_h=1, changes the variation trend for the freshwater production, which drops more sharply after the balance point.

Table 1. Parameters of the low grade heat source

Term	Unit	Value
$T_{e,i}$	K	413.15
m_e	kg/s	0.4
x		
CO_2	%	31.2
N_2	%	68.8

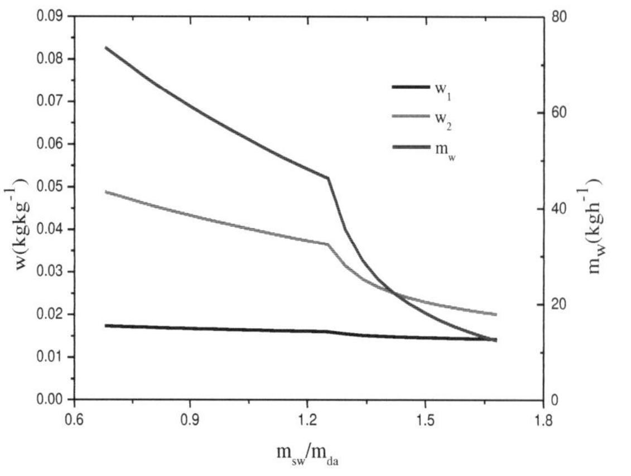

Figure 2. Variation trend of the water production in the HDH desalination system

Table 2. Typical parameters within the HDH desalination system

S (g/kg)	ε_h	ε_d	φ_1	φ_2	$T_{sw,0}$ (K)	$T_{a,3}$ (K)
35	0.85	0.85	1	0.9	288.15	353.15

As previous statement, gained-output-ratio is considered to characterize the energy conversion efficiency within the air-heated HDH desalination system, and the relate values with the elevation of the mass flow rate ratio are demonstrated in Fig. 3. It is obvious that the peak value, GOR=1.88, emerges at the balance condition of the dehumidifier, HCR_d=1, and it is inferred that the characteristics of the desalination system is more sensitive to the performance of the dehumidification process. Nevertheless, due to the negative entropy generation of the dehumidification process at the case of HCR_d=1, the correspond working point is impossible to arrive. Hence, the actual peak value, GOR=1.80, is realized at m_{sw}/m_{da}=0.68. In spite of the smaller top value of GOR, the proposed air-heated HDH desalination still shows great advantages compared to the traditional methods with lower energy consumption in response of the fixed water production. It is evident that the driving energy from the heat exchangers is first obtained through the seawater flow during the dehumidification process. Then the heated seawater releases the obtained energy to heat the humid air. Therefore, the thermal cycle in the HDH desalination system is accomplished with a GOR value farther larger than 1.

As the scale of the irreversibility loss within the HDH desalination system, the total specific entropy generation is also simulated and presented in Fig. 3. Obviously, it seems that the values of the total specific entropy generation for all the cases are positive. Actually, the negative region for the specific entropy generation for of the dehumidifier exists. Furthermore, It is gained that the variation trend of GOR is completely consistent with that of the total specific entropy generation. Taking the balance condition of the dehumidifier, $HCRd$=1, for example, a minimum irreversibility loss within the HDH desalination system arises when the bottom value of the total specific entropy generation is $s_{gen,t}$=1.21kJkg^{-1}K^{-1}, corresponding to the maximum value of GOR, GOR=1.88 at HCR_d=1, and the actual maximum GOR case has the value, $s_{gen,t}$=1.26kJkg^{-1}K^{-1}, of the total specific entropy generation.

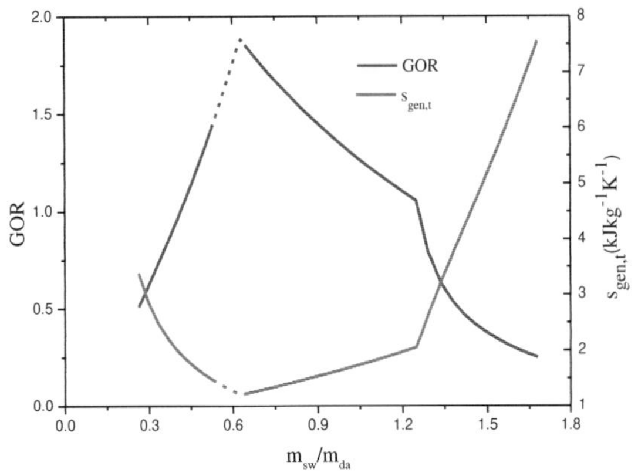

Figure 3. Performance of the HDH desalination system

4. Conclusions

In this paper, the mathematical models of the air-heated HDH desalination system powered by waste heat is established, and the corresponding performance is iteratively calculated in the platform of Matlab. The simulation results show that the analysis based on the first and second law should be simultaneously completed to verify the practical operation cases. A actual top value of GOR=1.80 at m_{sw}/m_{da}=0.68 appears despite the impossible maximum value, GOR=1.88, at the balance condition, $HCRd$=1, during the dehumidification process. The variation laws of GOR are completely consistent with the total specific entropy generation of the HDH desalination system. The obtained results and the relevant research method give some significant references for the further design of the HDH desalination system. Moreover, the thermo-economic and relevant optimization analysis should be achieved in the future.

Acknowledgments

W F He thanks the support by the National Natural Science Foundation of China (Grant No. 51406081) and China Postdoctoral Science Foundation (Grant No. 2016M601801).

Reference

[1] M. A. Al-Weshahi, A. Anderson, G. H. Tian. *Organic Rankine Cycle recovering stage heat from MSF desalination distillate water*. Applied Energy, 130 (2014): 738-747.

[2] S. Lu, Y.H. Zhou, M.S. Bi, J.J. Ren, Z. Cai. *An integrated optimization model and application of MEE-TVC desalination system*. Desalination, 371 (2015): 1-8.

[3] V. G. Gude, *Energy storage for desalination process powered by renewable energy and waste heat sources*. Applied Energy, 137 (2015): 877-898.

[4] F. A. Al-Sulaiman, M.I. Zubair, M. Atif, P. Gandhidasan, S.A. Al-Dini, M.A. Antar, *Humidification dehumidification desalination system using parabolic trough solar air collector*. Applied Thermal Engineering, 75 (2015): 809-816.

[5] M. H. Hamed, A.E. Kabeel, Z.M. Omara, S.W. Sharshir, *Mathematical and experimental investigation of a solar humidification-dehumidification desalination unit*. Desalination, 358 (2015): 9-17.

[6] C. Muthusamy, K. Srithar, E*nergy and exergy analysis for a humidification dehumidification desalination system integrated with multiple inserts*. Desalination, 367 (2015): 49-59.

[7] F. A. Al-Sulaiman, G. P. Narayan, J. H. Lienhard, *Exergy analysis of a high-temperature-steam-driven, varied-pressure, humidification-dehumidification system coupled with reverse osmosis*. Applied Energy, 103 (2013): 552-561.

[8] K. H. Mistry, *Second law analysis and optimization of humidification-dehumidification desalination cycles*. Master Thesis, Massachusetts Institute of Technology, 2010.

[9] K. H. Mistry, J. H. Lienhard, S. M. Zubair, *Effect of entropy generation on the performance of humidification-dehumidification desalination cycles*. International Journal of Thermal Sciences, 49 (2010): 1837-1847.

[10] G. P. Narayan, K.H. Mistry, M. H. Sharqawy, S. M. Zubair, J. H. Lienhard, *Energy effectiveness of simultaneous heat and mass exchange devices*. Frontiers in Heat Mass Transfer, 023001 (2010): 1-13.

[11] W. F. He, D. Han, L. N. Xu, C. Yue, and W. H. Pu, *Performance investigation of a novel water-power cogeneration plant (WPCP) based on humidification dehumidification (HDH) method*. Energy Conversion and Management, 110 (2016): 184-191.

[12] W. F. He, D. Han, C. Yue, and W. H. Pu, *A parametric study of a humidification dehumidification (HDH) desalination system using low grade heat sources*. Energy Conversion and Management, 105 (2015): 929-937

IOP Conf. Series: Earth and Environmental Science **73** (2017) 012005 doi:10.1088/1755-1315/73/1/012005

Natural Aggregation Approach based Home Energy Manage System with User Satisfaction Modelling

F J Luo[1,2], G Ranzi[1], Z Y Dong[3], J Murata[4]
[1] School of Civil Engineering, University of Sydney, Sydney, NSW 2006, Australia
[2] State Key Laboratory of Power Transmission Equipment & System Security and New Technology, Chongqing University, China
[3] School of Electrical Engineering and Telecommunications, University of New South Wales, Sydney, NSW 2052, Australia
[4] Department of Electrical Engineering, Faculty of Information Science and Electrical Engineering, Kyushu University, Fukuoka 812-8581, Japan

Email: tracyluofengji@gmail.com; gianluca.ranzi@sydney.edu.au; joe.dong@sydney.edu.au; murata@cig.ees.kyushu-u.ac.jp;

Abstract. With the prevalence of advanced sensing and two-way communication technologies, Home Energy Management System (HEMS) has attracted lots of attentions in recent years. This paper proposes a HEMS that optimally schedules the controllable Residential Energy Resources (RERs) in a Time-of-Use (TOU) pricing and high solar power penetrated environment. The HEMS aims to minimize the overall operational cost of the home, and the user's satisfactions and requirements on the operation of different household appliances are modelled and considered in the HEMS. Further, a new biological self-aggregation intelligence based optimization technique previously proposed by the authors, i.e., Natural Aggregation Algorithm (NAA), is applied to solve the proposed HEMS optimization model. Simulations are conducted to validate the proposed method.

1. Introduction

Driven by the developments of two-way communication infrastructure and building automation technologies, Home Energy Management System (HEMS) has attracted lots of attentions in both academia and energy industry. HEMS is considered as a kind of delegation of end users to cope with the complexity of varying electricity tariff and automatically manage the home energy resources.

Many works have been conducted on the design and development of HEMS. In our previous works [1-3], we designed HEMSs for managing the residential air conditioning systems; [4] optimally schedules the appliances under the forecasted real-time electricity pricings; [5] coordinately scheduled the Battery Energy Storage System (BESS) and household appliances with high solar penetrations; in [6], a load commitment framework was proposed to minimize the household operation costs; in [7], a HEMS was designed to schedule the appliances in each dwelling unit, and based on it the demand of whole community was forecasted and reported to the utility. In [8], a two-stage (day-ahead and real-time stages) HEMS was designed to manage the residential BESS with solar penetrations. In our recent work [9], we discussed the vision of household recommender systems based on service recommendation techniques.

Existing HEMSs schedule and control the household appliances by using the user pre-specified allowable operation time ranges [4-6] as constraints. However, this method ignores the fact that in practical situations, the user often has different tolerances on the operation time of different appliances. For example, for a rice cooker, the user could strictly restrict its task termination to occur before 6pm,

IOP Conf. Series: Earth and Environmental Science **73** (2017) 012005 doi:10.1088/1755-1315/73/1/012005

as the user does not want to have dinner too late. In this paper, we refer such requirement as the "rigid restriction". For some appliances such as washing machine, the user could have a preferred operation time range (e.g., 10am to 8pm), but the user could also tolerate running it out of the preferred operation time range if a considerable amount of electricity cost can be saved. In this paper, we refer such requirement as the "elastic restriction". The major contribution of this paper is to propose a new HEMS by considering above different kinds of operational requirements of the user on the household appliances. The proposed HEMS aims to coordinately schedule the residential energy resources (a residential BESS and multiple controllable appliances) to accommodate the residential photovoltaic solar output and minimize the 1-day energy cost, while respecting the user's satisfactions and life habit. Further, a metaheuristic optimization technique recently proposed by the authors, i.e., Natural Aggregation Algorithm (NAA), is applied to solve the proposed optimization model.

This paper is organized as follows. Section 2 formulated the modelling method for the user's psychological satisfaction on the appliance operation; Section 3 presents the HEMS model formulation; Section 4 presents the solving approach for the proposed HEMS model; Section 5 discusses the simulation results; conclusions and future works are drawn in Section 6.

2. User Satisfaction Modelling

As discussed in the introduction part, the user's requirements on the appliance's operation can be categorized as rigid restriction and elastic restriction, respectively. For the appliances with the rigid restriction, they must be restricted to be operated within the user-specified preferred time ranges; for the appliances with the elastic restriction, they are allowed to be operated out of the user's preferred time ranges. But in this case, the psychological satisfaction of the user on the appliance operation needs to be modelled.

In [4], the authors assumed the user always preferred to have the appliance finish its work as earliest as possible, and measured the disturbance of an appliance schedule to the user by the 'delay time rate (DTR)' metric, in which the later the appliance finishes its work, the larger of the DTR value. However, this strategy in [4] may not truly reflect a practical user's willingness. To explain this, simply imaging a user who prefers to have the clothes dryer run between the period of 9am to 5pm (after he/she leaves home to work in the morning and backs home after work), then the clothes dryer finishes the work on 11am or 2pm would probably has no difference to the user. What actually disturbs the user is to run the clothes dryer out of the user's preferred time range (e.g., run from 4pm to 5:30pm). Based on this realization, in this paper we propose the metric of "User Disturbance Value (UDV)" to measure the satisfaction degree of an appliance schedule to the user. The smaller value of UDV, the more the user satisfies with the appliance schedule.

The UDV of the operation of *i*th controllable appliance is calculated as:

$$UDV_i = \frac{\sum_{t=1}^{T} s_i(t)\Delta t \delta_t}{d_i} \tag{1}$$

$$\delta_t = \begin{cases} 1 & \text{if } t < l_i^{begin} \text{ or } t > l_i^{end} \\ 0 & \text{if } l_i^{begin} \le t \le l_i^{end} \end{cases} \tag{2}$$

where T is the total number of time intervals over the whole scheduling horizon; Δt is the duration of each time interval (hour); d_i is the duration of a full operation cycle of the *i*th controllable appliance (hour); l_i^{begin} and l_i^{end} are the beginning and end time of the user's preferred operation time ranges of the *i*th appliance; $s_i(t)$ is the state of the *i*th controllable appliance (0-OFF, 1-ON). The meaning of models (1) and (2) is that if a controllable appliance is scheduled within the user's preferred operation time range, then it is considered that the user is fully satisfied with the schedule; otherwise, the psychological disturbance degree of the appliance schedule to the user is proportion to the operation time outside the user's preferred time range. Models (1) and (2) will be applied on the appliances with the elastic restrictions, and will be incorporated into the HEMS model presented in the next section.

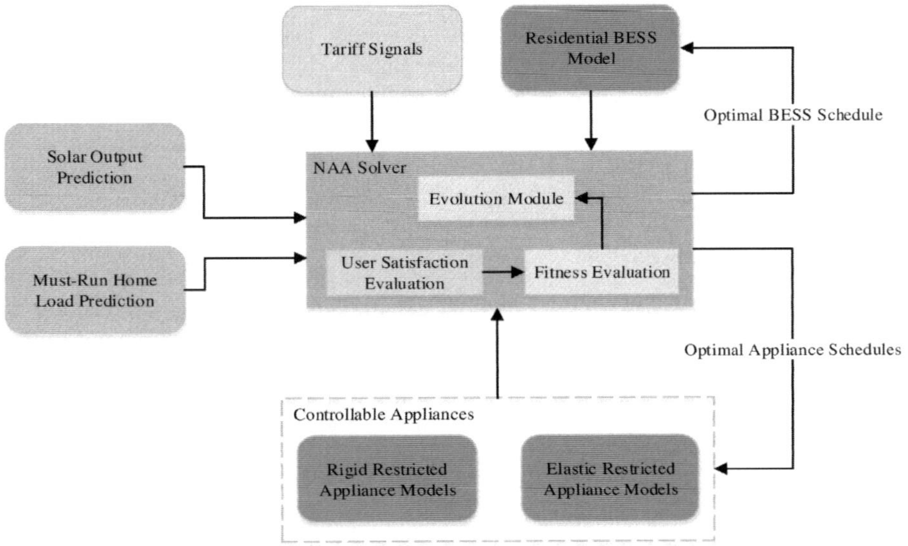

Figure 1. Schematic of the proposed HEMS

3. HEMS Model Formulation

In this paper, we study a smart home environment with roof photovoltaic source installed. The Time-of-Use (TOU) tariff is considered. A residential BESS is considered to accommodate the intermittent solar power output. The residential energy resources (RERs) considered in this study include the BESS and multiple controllable appliances. The schematic of the smart home can be illustrated in Figure 1, and the HEMS model is formulated as below.

$$\min F = C_{cost} + \eta \cdot \sum_{i=1}^{N} UDV_i \tag{3}$$

$$C_{cost} = C_{tariff} + C_{bess} + C_{pv} \tag{4}$$

where N represents the number of controllable appliances; C_{cost} represents the overall home energy cost; η is the penalty factor of the user's dissatisfaction. UDV_i is calculated as equation (1). C_{cost} consists of three items: electricity tariff cost (C_{tariff}), BESS depreciation cost (C_{bess}), and the discounted roof solar panel installation cost (C_{pv}). C_{tariff} is calculated as:

$$C_{tariff} = \sum_{t=1}^{T} \left(pr(t)\Delta t \cdot \max(P^{net}(t),0) \right) \tag{5}$$

where $pr(t)$ is the electricity price at time t (\$/kWh); $P^{net}(t)$ indicates the net load of the house at time t, calculated as:

$$P^{net}(t) = \sum_{i=1}^{N} s_i(t)P_i + L(t) + P^{bess}(t) - P^{pv}(t) \tag{6}$$

where P_i is the rated power of the ith controllable appliance (kW); $L(t)$ is the overall power consumption of the non-controllable home energy facilities at time t (kW); $P^{bess}(t)$ is the charging/discharging power of the BESS at time t (negative-discharging, positive-charging); $P^{pv}(t)$ is the forecasted solar power at time t (kW).

The operational cost of the BESS is calculated as:

$$C_{bess} = \gamma \cdot \left| P^{bess}(t) \right| \cdot \Delta t \tag{7}$$

$$E^{bess}(t+1) = E^{bess}(t) + \Delta t \cdot P^{bess}(t) \tag{8}$$

where γ is the cost factor of the BESS (\$/kWh); $E^{bess}(t)$ is the energy stored in the BESS at time t (kWh).

The daily discounted photovoltaic source investment cost is calculated by its installation fee paid by the user and the guarantee years provided by the vendor:

$$C_{pv} = \frac{C_{install}}{N^{GY} \cdot 365} \tag{9}$$

where $C_{install}$ denotes the installation fee of the solar panel (\$); N^{GY} is the number of the guaranteed year of the solar panel.

The decision variables of the HEMS model include the commitments of the controllable appliances ($s_i(t)$) and the charging/discharging power of the BESS ($P^{bess}(t)$). And the model is subjected to following operation constraints:

(a) Residential BESS operation constraints. In this study, we assume that power output of the residential BESS is only allowed to serve the local load, and not allowed to inject into the main grid. This restriction is based on the secure and reliable operation considerations of the grid, and is applied in some countries such as Japan. The residential BESS operation is therefore subjected to following constraints:

$$\left|P^{bess}(t)\right| \le P^{rbess} \qquad t = 1:T \tag{10}$$

$$-P^{bess}(t) \le P^{net}(t) \qquad \text{if } P^{bess}(t) < 0 \text{ and } P^{net}(t) > 0 \tag{11}$$

$$SOC^{lower} \le SOC(t) \le SOC^{upper} \tag{12}$$

$$SOC(t) = E^{bess}(t) / E^{rbess} \tag{13}$$

$$SOC(T) \ge SOC^{init} \tag{14}$$

where SOC^{lower} and SOC^{upper} are allowable lower and upper SOC limits of the BESS; SOC^{init} is the initial SOC of the BESS at the beginning of the scheduling horizon; $SOC(t)$ is the state-of-charge of the BESS at time t; P^{rbess} is the rated power capacity of the BESS (kW); E^{rbess} is the rated energy capacity of the BESS (kWh). The meaning of constraint (14) is to ensure the BESS have sufficient energy at the end of the scheduling horizon, so as to continuously provide services for the house in the incoming day.

(b) Operation cycle constraint of the controllable appliance. The time duration of the operation cycle of the controllable appliances must be ensured:

$$\sum_{t=1}^{T} (s_i(t) \cdot \Delta t) = d_i \qquad i = 1:N \tag{15}$$

(c) Operation constraint of the non-interruptible controllable appliances. Controllable appliances in a smart home can be categories as non-interruptible appliance (NIAs) and interruptible appliances (IAs). The operation of NIAs cannot be interrupted until it finishes its work:

$$\sum_{t=t_i^*}^{t_i^* + d_i/\Delta t} s_i(t) = 1 \qquad i = 1:N^{NIA} \tag{16}$$

where t_i^* represents the time interval when the ith NIA is first time to be turned on; N^{NIA} is the number of NIAs.

(d) Minimum online/offline time constraints of IAs. For the IAs, the minimum online and offline time constraint is applied to protect their mechanical devices, shown as:

$$\begin{cases} \tau_i^{on}(t) \ge \tau_{\min,i}^{on} & s_i(t) == 0, i = 1:N^{IA} \\ \tau_i^{off}(t) \ge \tau_{\min,i}^{off} & s_i(t) == 1, i = 1:N^{IA} \end{cases} \tag{17}$$

where $\tau_i^{on}(t)$ and $\tau_i^{off}(t)$ are accumulated online and offline time of ith IA at time t (hour); N^{IA} is the number of IAs; $\tau_{\min,i}^{on}$ and $\tau_{\min,i}^{off}$ are minimum online and offline duration limits of the ith IA (hour). There is $N = N^{NIA} + N^{IA}$.

(e) Operation time constraint of the controllable appliances with rigid restrictions:

$$s_i(t) = 0, \quad \text{if } t < l_i^{begin} \text{ or } t > l_i^{end}, \forall i = 1 : N^{rg} \tag{18}$$

where N^{rg} is the number of rigid restricted controllable appliances.

4. Solving Approach

The proposed HEMS model (3) is a mix-integer, combinatorial optimization problem over a finite horizon. In this paper, a new metaheuristic optimization algorithm previously invented by the authors, i.e., NAA [10], [11], is applied to solve the proposed model.

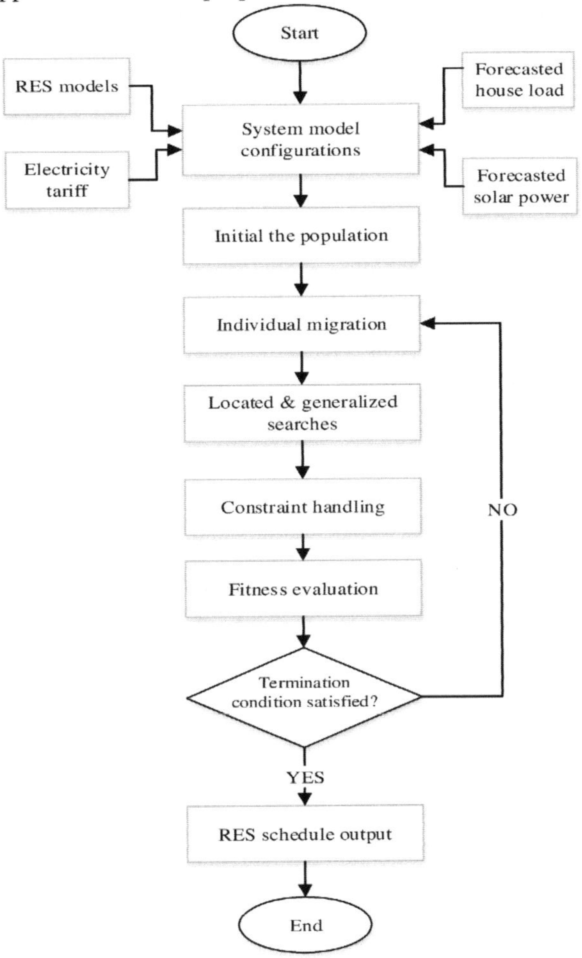

Figure 2. Workflow of the NAA-based solving approach

NAA is a biological intelligence inspired optimization algorithm. Specifically, NAA is motivated by the self-aggregation intelligence of the group living animals. In NAA, the whole population of the individuals are organized as multiple sub-populations, and a stochastic migration model is utilized to migrate the individuals among the sub-populations. Meanwhile, the located and generalized search strategies are used in NAA at each generation. Due to the page limitation, we do not give detailed introduction for NAA in this paper, which can be found in [10], [11].

4.1. Encoding scheme
By applying NAA on the proposed HEMS model, each individual is encoded as a potential RER scheduling solution, with the dimension of $(N+1)\cdot T$. The first $N\cdot T$ dimensions are binary variables, representing the commitment status of the N controllable appliances, where every T dimensions represent the status of an appliance over the T time intervals. The last T dimensions are continuous variables, representing the charging/discharging power of the BESS.

4.2. Solving procedures
The overall workflow of the NAA-based HEMS model solving approach is shown in Figure 2. Firstly, the system models are inputted, which include the profiles of forecasted 1-day solar output and non-controllable home energy consumptions, and the operational models of the RERs. Then, the individuals are encoded in NAA and the evolution process is started. In each generation, for each individual, the constraint handling is performed to make it feasible, and the individual's fitness value is evaluated by calculating equation (3). Finally, the scheduling results are outputted.

5. Simulation study
In this section, simulations are conducted to validate the proposed HEMS model. All programs are implemented on Matlab and executed on a DELL workstation with 128-Gegabyte memory and 2 Intel Xeon processors.

5.1. Simulation setup
We simulate a smart home with five kinds of controllable household appliances and one residential BESS. The scheduling interval is set to 15 minutes, the configurations of the controllable appliances and BESS are shown in tables 1 and 2, respectively. 1-day Australian solar power data with 1-minite sampling frequency is used, shown in Figure 3. The data is then averaged in 15-minute basis for simulation. One-day non-controllable house energy consumption profile is generated from the Australian "Smart Grid, Smart City" data [12], shown in Figure 4. The TOU tariff published by the Energy Australia is used, shown in table 3. It is assumed that the critical peak time price is activated. The values of both $\tau_{\min,i}^{on}$ and $\tau_{\min,i}^{off}$ are set to 15 minutes. The parameter settings of NAA are as follows: population size is set to be 200, maximum generation time is set to be 600, $N^S=8$, $Cp^S=25$, $\delta=2$, $Cr_{local}=0.8$, $\alpha=1.5$, and $Cr_{global}=0.2$.

Table 1. Residential BESS Configurations

Power Capacity	3kW
Energy Capacity	6kWh
SOC Upper Limit	90%
SOC Lower Limit	10%
Initial SOC	30%

Table 2. TOU Tariff Structure

Time-of-Use	**Rate ($/kWh)**
Peak: 2pm-8pm	0.3564
Shoulder: 7am-2pm, 8pm-10pm	0.1408
Off-peak: 10pm-7am	0.0814
Critical Peak Price	**Rate ($/kWh)**
5pm-8pm	2.000

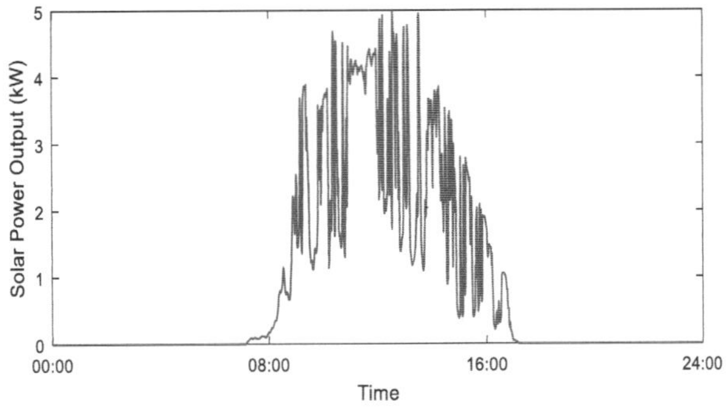

Figure 3. 1-Day solar power output

Table 3. Controllable Appliance Configurations

Appliance	Operation Cycle Duration	Interruptible	Restriction	Preferred/Required Operation Time Range	Rated Power
Dish Washer	2 hours	NO	Rigid	[7:30pm, 9am]	0.8kW
Rice Cooker1	0.75 hours	NO	Rigid	[10:30am, 12am]	0.5kW
Rice Cooker2	0.75 hours	NO	Rigid	[5pm, 6:30pm]	0.5kW
Washing Machine	1 hour	NO	Elastic	[5pm, 7pm]	1.0kW
Clothes Dryer	1.5 hour	YES	Elastic	[7:30pm, 10pm]	1.0kW

[a] Note: the superscripts of rice cooker mean its first and second time running.

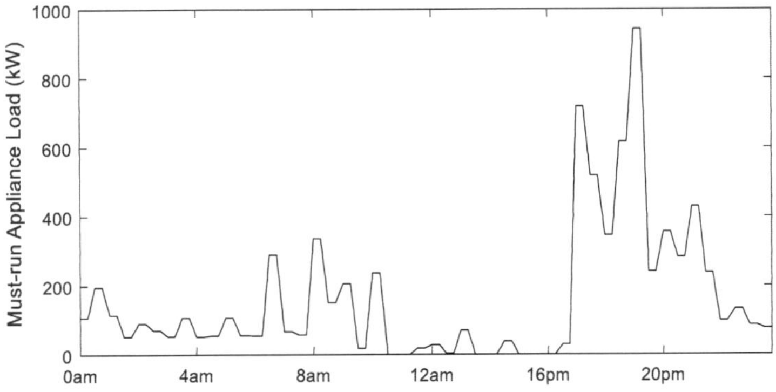

Figure 4. Profile of 1-day must-run load of the house

5.2. Residential energy resource scheduling results

We set the value of the user satisfaction penalty factor η to be 0.4 and run the simulation. And we consider two cases: in case 1, the cost factor of the BESS (γ) is set to a low value, i.e., 0.06\$/kWh. This means the battery operation cost is lower than the electricity price even in the off-peak period. In case 2, γ is set to a high value, i.e., 0.15\$/kWh, which means the battery operation cost is higher than the electricity price in off-peak and shoulder periods, but lower in the peak and critical peak time periods. We then investigate the RER scheduling results under the two different BESS operational cost settings as below.

5.2.1. Case 1: Low BESS Cost Factor

Figure 5 shows the final scheduling of the five kinds of controllable appliances in case 1. The TOU signals are also plotted in Figure 5. It can be clearly seen that most of the appliances are well scheduled to avoid the critical peak period (5pm-8pm). The two rigid restricted appliances, i.e., dish washer and rice cooker, are scheduled to operate within the user-specified time ranges. It is also noticeable that although the preferred operation time range of the user on washing machine is [5pm, 7pm], it is scheduled to operate in the period of [11:am, 12:00am], so as to avoid the peak and critical peak pricings.

Figure 6 shows the optimization results of the residential BESS. The profiles of the solar power output and total power load of the house are also plotted. The negative value of the BESS output indicates discharging while the positive value indicates charging. The results clearly show that the BESS charging/discharging profile follows a desired shape. That is, during the period of [0am, 8am] when there is no solar power, the BESS is discharged to serve the house load; at the noon time, the solar power is surplus, and the BESS is therefore charged by the solar power. In the evening time, the BESS is discharged again to fully serve the house load. Figure 7 shows the corresponding SOC profile of the BESS under the optimized schedule. It can be clearly seen that the BESS is well controlled within its safe SOC lower and upper limits. And at the end of the scheduling horizon, the BESS is charged back to larger than its initial SOC level (30%), which makes it can continuously serve the house load in the incoming day. Table 4 summarizes the cost items and the UDV value under the final schedule.

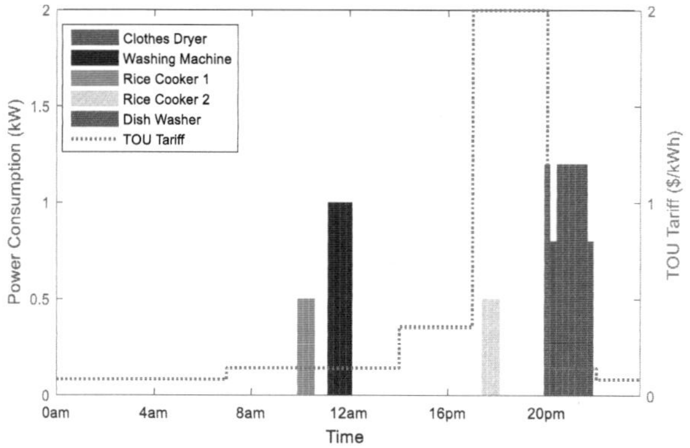

Figure 5. Controllable appliance scheduling results of case 1

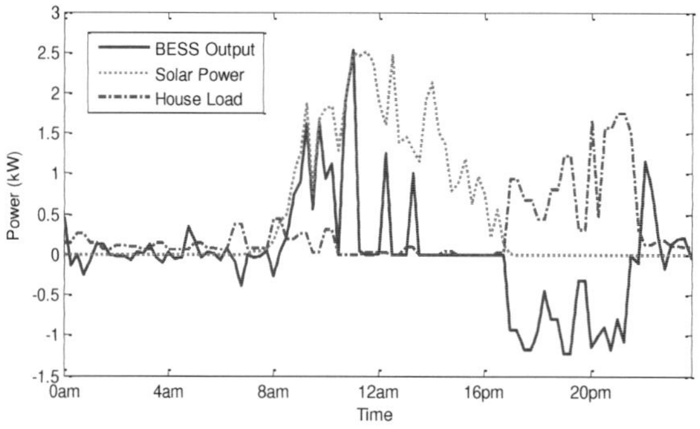

Figure 6. Residential BESS scheduling results of case 1

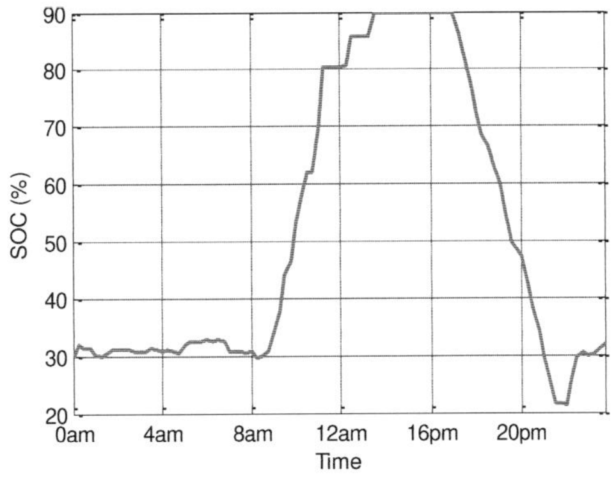

Figure 7. SOC profile of the residential BESS of case 1

Table 4. Cost Items and UDV of Case 1

Item	Value
Electricity Cost	\$0.54
BESS Operation Cost	\$0.28
Discounted Daily Solar Generation Cost	\$0.66
Total Cost	\$1.48
Total UDV	1.0

5.2.2. Case 2: High BESS Cost Factor

The scheduling results of the controllable appliances are shown in Figure 8. The appliance schedules follow the similar trend with those of case 1. That is, all the appliances with elastic restrictions are properly scheduled to avoid peak or critical peak periods. The residential BESS scheduling results are shown in Figure 9 and Figure 10. Comparing with the BESS scheduling results of case 1, it can be clearly seen that in the critical peak period, the BESS is well discharged to serve the house load. However, after 20pm in the evening, when the house load is high and electricity price is low, the BESS is not discharged to serve the load. This is because the cost factor of the BESS in case 2 is higher than the electricity price of the off-peak period.

Table 5 reports the house operation cost and total UDV of case 2. With the same setting of the penalty factor η (0.4), the total UDV of both cases 1 and 2 are the same, which is equal to 1.0. Meanwhile, both the electricity cost and BESS operation cost of case 2 are larger than those of case 1.

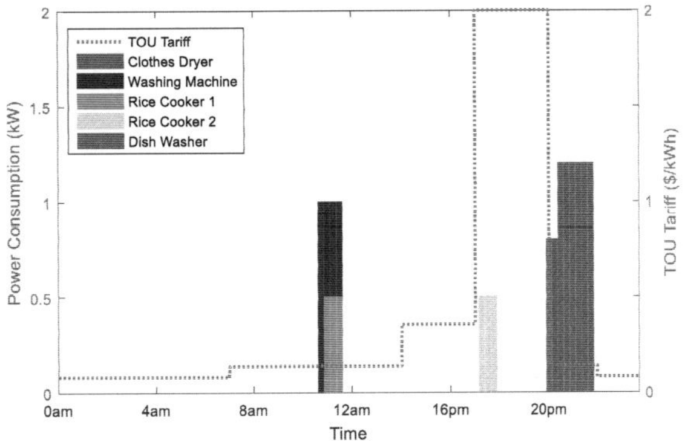

Figure 8. Controllable appliance scheduling results of case 2

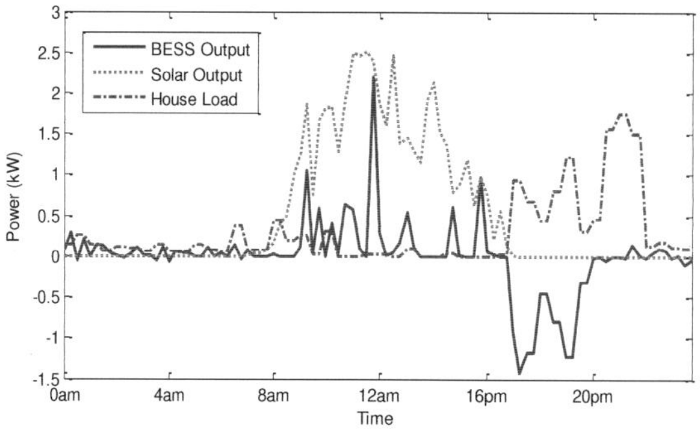

Figure 9. Residential BESS scheduling results of case 2

5.3. *Sensitivity Study of Penalty Factor* η

Finally, we conduct the sensitivity study for the penalty factor η, so as to invistigate the relationship of user satisfaction and the house operation cost. For both cases, we run the simulations under different settings of η, and the results are shown in Figure 11 and Figure 12. With the increase of the value of η, the UDV decreases, and the house operation cost increases. Therefore, different settings of η represent different compromises of the user satisfication and household energy cost.

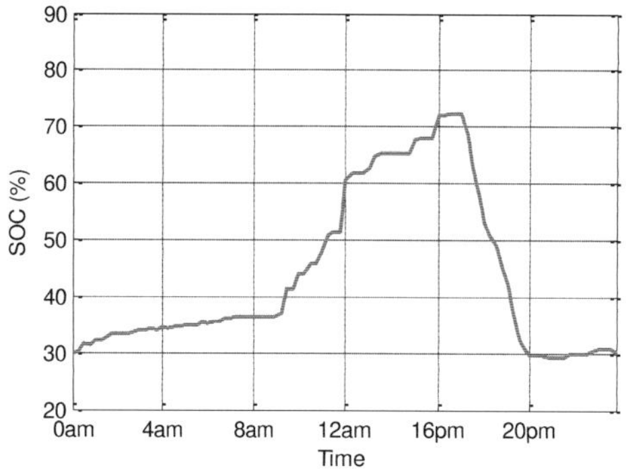

Figure 10. SOC profile of the residential BESS of case 2

Table 5. Cost Items and UDV of Case 2

Item	Value
Electricity Cost	\$0.75
BESS Operation Cost	\$0.81
Discounted Daily Solar Generation Cost	\$0.66
Total Cost	\$2.22
Total UDV	1.0

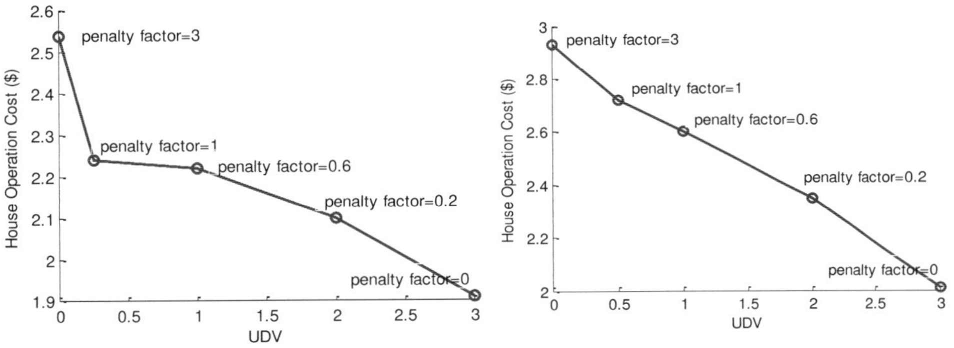

(a) case 1 (b) case 2

Figure 11. Sensitivity study of penalty factor of case 1 & 2

6. Conclusions and future works

In this paper, we propose a new HEMS with the considerations of different operational requirements of the user on different kinds of appliances. The penetrations of residential photovoltaic source and BESS are simulated. And a recently proposed metaheuristic optimization algorithm, i.e., NAA, is applied to solve the proposed HEMS model. The results clearly show that the controllable appliances and BESS can be coordinately scheduled by taking account into the user satisfaction. The simulations also show that different operation cost factors of the residential BESS will significantly affect the decision-making of the HEMS.

Future researches can be conducted in different directions. The stochastic nature of the renewable energy output can be modelled and incorporated into the HEMS. The vehicle-to-home (V2H) technology can be also integrated into the smart home energy management. While many existing HEMS researches assume the renewable energy source is private and can only serve the load of a single house, how to coordinately perform the energy management for multiple units in a shared renewable energy environment is also a problem deserving of study.

Acknowledgment

This work is supported in part by the Australian Research Council through its Future Fellowship scheme (FT140100130), in part by the Visiting Scholarship of State Key Laboratory of Power Transmission Equipment & System Security and New Technology (Chongqing University, China) (2007DA10512716401), and in part by the Early Career Research Program of Faculty of Engineering and Information Technology, The University of Sydney, Australia.

References

[1] Luo F, Zhao J, Dong Z, Tong X, Chen Y, Yang H and Zhang H. “Optimal dispatch of air conditioner loads in southern China region by direct load control,” *IEEE Transactions on Smart Grid*, vol. 7, no. **1**, 2016. 439-51

[2] Wang H, Meng K, Dong Z, Xu Z, Luo F and Wong K. “Efficient real-time residential energy management through MILP based rolling horizon optimization,” in *Proc. IEEE Power and Energy Society General Meeting*, 2015. 1-6

[3] Wang H, Meng K, Luo F, Dong Z, Verbic G, Xu Z and Wong K. “Demand response through smart home energy management using thermal inertia,” in *Proc. Australasian Universities Power Engineering Conference*, 2013. 1-6

[4] Zhao Z, Lee W, Shin Y and Song K. “An optimal power scheduling method for demand response in home energy management system,” *IEEE Transactions on Smart Grid*, vol. 4, no. **3**, 2013. 1391-1400

[5] Pedrasa M A, Spooner T and MacGill I. “Coordinated scheduling of residential distributed energy resources to optimize smart home energy services,” *IEEE Transactions on Smart Grid*, vol. 1, no. **2**, 2010. 134-143

[6] Rastergar M, Fotuhi-Firuzabad M and Aminifar F. "Load commitment in a smart home," *Applied Energy*, vol. 96, 2012. 45-54
[7] Ozturk Y, Senthilkumar D, Kumar S and Lee G. "An intelligent home energy management system to improve demand response," *IEEE Transactions on Smart Grid*, vol. 4, no. **2**, 2013. 694-701
[8] Iwafune Y, Ikegami T, Oozeki T and Ogimoto K, "Cooperative home energy management using batteries for a photovoltaic system considering the diversity of households," *Energy Conversion and Management*, vol. 96, 2015. 322-29
[9] Luo F, Ranzi G, Wang X and Dong Z. "Service recommendation in smart grid: vision, technologies, and applications," in *Proc. 9th International Conference on Service Science*, 2016.
[10] Luo F, Zhao J, and Dong Z. "A new metaheuristic algorithm for real-parameter optimization: Natural aggregation algorithm," in *Proc. IEEE Congress on Evolutionary Computation*, 2016. 94-103
[11] Luo F, Dong Z, Chen Y and Zhao J., "Natural aggregation algorithm: a new effcient metaheuristic tool for power system optimizations," in *Proc. IEEE International Conference on Smart Grid Communications*, Sydney, Nov. 2016.
[12] Smart Grid, Smart City [Online]. Available at: http://www.industry.gov.au/ENERGY/PROGRAMMES/SMARTGRIDSMARTCITY/Pages/default.aspx

IOP Conf. Series: Earth and Environmental Science **73** (2017) 012006 doi:10.1088/1755-1315/73/1/012006

Challenges and Strength of Current Industrial Energy Efficiency Management Practices in Steam Industries

S. B. Nkosi[1] **and J. H. C. Pretorius**[1, 2]
University of Johannesburg, South Africa

Email: sbnkosi444@gmail.com;

Abstract. The aim of this study is to achieve greater output by examining the existing way of coordinating the determined attempts of Steam Industries in South Africa to successfully reach a sustainable industrial development by using energy source adequately in a more competent way. Furthermore into the study we look at obstacles that prevent and those that leads to maximum utilization of energy management measures and also highlights the effects of implementing cheap available energy source in South Africa. The investigation and analysis have shown that energy is not well managed in Steam Industries and that the use of energy is minimized and not fully utilized due to poor management and lack of knowledge. Another detection was that lack of government structured and strategic measures of implementing and motivating the use of energy effectively. The effective and rational use of available power by Steam Industries in South Africa is a key player in developing a sustainable industrial development. The use of energy efficiency management strategies has contributed an increase in economic and improve environmentally friendly in the industrial sector. The slow pace adoption of energy saving and cost effective management programmes are negatively impacting on the benefits to Steam Industries in South Africa. In conclusion the study finds that the economy can be boosted by implementing energy efficiency management programmes and environmentally friendly. These will also stabilize the negative impact of energy raising prices.

1. Introduction

Energy is inseparable from matter. All material phenomena are associated with energy. Energy is therefore essential to life and the quality of energy consumed per capita can be used as one indicator of a country's development level. No thought was given to this in the past but today it is being included in energy efficiency management activities. Energy supply and demand must be treated with both a macro-economic and a micro-economic approach. There are different aspects in planning energy demand, whether for country, a factory or a household, but each must still be investigated individually given the particular characteristics involved [1]. Energy efficiency management is by definition the most cost effective, efficient use of energy. The term Energy Efficiency Management is relatively new and is a sign of our time. Until recently the efficient use of energy in all its forms was the tacit responsibility of many people in an organization. Now, in many companies, the responsibility for the overall management of energy and its widest contexts is focused formally on the Energy Manager. It is the Energy Manager's duty to optimize energy usage not only in individual items of equipment but within the total limits of his specified area [1]. If this area is a net exporter or importer of energy he will have to look beyond the horizons of his own area to investigate the optimization of energy at a company or even inter-company level. Thus a hot effluent stream from an adjacent area may be harnessed to provide process, factory or office heating.

2. Background

The backbone of the country's economy is the extraction and processing of minerals. To date an estimation of 93% of South Africa's power is generated from coal – fired power stations, while 5% is nuclear and hydro - electric, whereas 2% is for pumped storage and gas turbines [2]. Current statistics indicate that South Africa is lagging behind in the implementation of energy efficiency management systems.

2.1. Energy use and Market in Steam Industries in South Africa

Depending on the size and variables of the Steam Industry; the energy cost and utilities consumes between 3% and 8% of a Steam Industry's general budget. As it can be seen in figure 1, that Steam Industries consumes much of thermal and electric energy. Thermal energy is used to heat water in boiler house and in steam turbines. The process of steam system is typically the largest consumer of electric energy but the boiler-house, turbines and waste water treatment plant can account for substantial electricity demand [2].

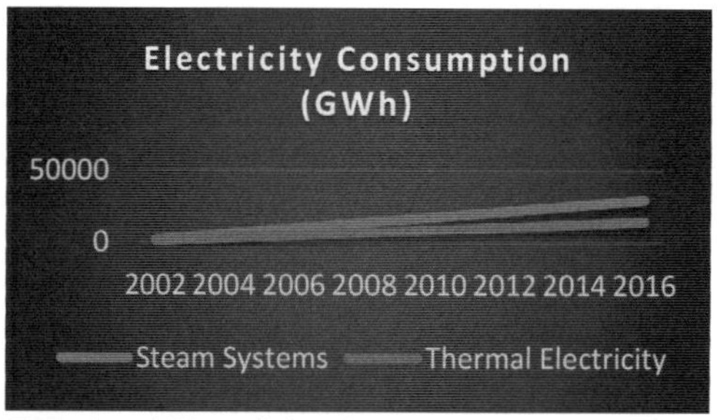

Figure 1. Trends of Electricity Consumption by Steam Industries

Energy is the most essential resource to South Africa's economy as to the rest of the world. Success of systems and strategies in the organizational and sectorial levels will be more successful if they be embedded in a broad based and long term national environmental programme. More especially granting energy prices to take after external environmental costs will prove to be more energy efficiency and cost efficient and promote awareness among higher administration [3]. In South Africa similar policies introduce preservation of the ecological tax reform. The increment of tax rate on fuels and electricity diminishes financial risk linked with investments in energy utilization and grants long-term planning [3]

2.2. Challenges facing the Steam Industries in South Africa

Issues of energy security constantly threaten South Africa's economy; these issues stem from challenges facing the steam industries of South Africa. With rising energy costs and prices around the world, the Steam Industry continues to seek innovative solutions to these problems. The demand for Steam worldwide continues to rise faster than supply, and the shortage has created a demand for Steam and that will require Steam producers to be more efficient [4].

3. Industrial Energy Management

3.1. *Effective features of energy efficiency management systems*

Energy management made based on Plan-Do-Check Act model, which serve as an organizational culture of stable advancement in energy efficiency. The culture of stable advancement assures that the set of objectives are obtained in a progressive manner. It also assures that those set of objectives are practical, obtainable and accommodates the available resources of the firm.

3.1.1. *Plan Phase*

Planning has been defined as (1) outlining a course of action to achieve an objective, (2) devising a method to attain a defined goal, and (3) developing a specific way to obtain a desired result. Each definition contains two elements: Knowing what is wanted to be accomplished and deciding how to accomplish it. The energy planning process consists of three steps:

- deciding what is wanted to be accomplished, or setting objectives
- deciding how to accomplish the objectives, or outlining procedures
- deciding how to apply the skills needed to meet objectives, or assigning responsibilities. This step is necessary because the manager does not carry out plan singlehandedly. It is the manager's job to get other people to do this.

3.1.2. *Do phase*

To gain the maximal conceivable savings capability, the investigated action plan should be prioritized and restated into a work plan. And also involve the accountabilities and timeframe for different processes, the work plan must very time contain the essential resources. Highlight the energy aims and short-comings, but it will only be successful if there is enough funds and availability of resources. In addition, Energy Manager must note the achieved targets and analysis the benefits in terms of cost saving. The achieved cost saving and reduction in environment pollution should be demonstrate to motivate the employees.

3.1.3. *Check Phase*

Check phase aims to supervise the performance (by handling energy audits) in the expressions of energy saving and analyse the set targets. Whenever encounter any loss, it is then important that the root causes are investigated and determined to form correction in order to recognize set objectives. Hence, it is necessary that set objectives are significant to promote the appraisal development.

3.1.4. *Act Phase*

This phase mostly involves management analysis of audit, internal and external reports referring to the achievement of the energy management program. This information plays a huge part for the organization to spot any loss and act up on it to provide continuous progress.

4. Environmental, Economic And Social Benefits Of Industrial Energy Efficiency

4.1. Environmental
The use of environmentally damaging fossil fuels as a primary energy source has provided to be a difficult issue for the Steam Industry to address. As known that, oil-based fuels have some built-in convenience, density, and portability features that make them particularly easy to transport and especially suitable as a form of energy to power transportation activities. The Steam Industry has taken some important measures to minimize the impact of fossil fuel energy on the environment.

4.2. Economic
The principal aim of any energy conservation project is to save money using limited energy or using energy more efficiently. The economic assessment of energy projects can often be complicated since it involves the engineering techniques needed to classify the magnitude of energy savings, together with the economic principles associated in checking in case that capital investment in the project is supported. The utilization of the energy plant is important because if it is pushed beyond the rated flow-sheet capacity, the variable costs may rise rapidly, because of inadequate time for preventive maintenance, resulting in excessive wear or breakdowns and high labour costs for weekend and holiday maintenance work. These are called super-production costs.

5. Energy Efficiency Gap
One of the major problems that Steam Industry faces in relation to energy is an inability to balance energy demand with energy supply. The nation will not get very far with its efforts to converse fuel and improve energy efficiency as long as demand for energy continues to increase and even surpass energy supply. The real problem with energy thus reaches beyond oil reserves, gasoline prices, fuel efficiency, and similar issues, it is a problem based in a way of life, one in which convenience, speed, and technologies have driven Steam Industry attitudes about energy and lifestyle.

6. Economic Barriers Non-Market Failure Related

6.1. Hidden costs
Hidden costs are the most essential and dominant statement for the 'efficiency gap'. "The engineering economic studies fail to stand for their reduction in utility connected with energy efficient technologies or the additional costs connected with their use" [5].

6.2. Organizational Barriers
Organizational barriers in this concept refers to those organizational factors that can withhold a range of possible energy efficiency investments. This factors include power, culture and awareness of the firm. These factors are related to the structure, size and available infrastructure of the organization [5].

6.3. Culture and Awareness
The standard of energy awareness among employees was found to be very poor. All grounds had used different awareness campaigns – like posters and information stands – but all reported that these were inefficient. Awareness rising could potentially be helping through responsibility for energy costs, using the environment impact of energy use to motivate staff, and moves towards 'team – working'

and higher accountability for energy costs centre performance. But all grounds lacked a high standard champion for energy efficiency and all reported that awareness of energy efficiency was low within the 'Steam Industry Culture' [5].

6.4. Credibility and Trust
The credibility of expertise and trustworthiness are the core significant aspect of information that interfere with the energy efficiency improvement. Hence the successful circulation and absorption of information rely on the trustworthiness as a source. Whenever the credibility of the information source is doubtful, firms may hesitate to invest in energy efficiency based on that information.

6.5. Values
In additional to the financial incentive for improving energy efficiency, individuals and Steam industry may be motivated by a desire to improve the Steam Industry's environmental performance. In some case, this may have (or be expected to have) an influence on shareholder value.
Steam Industry must have a robustly driven culture to environmental values, either on the part of respondents or of senior management. Respondents must implement significant energy efficiency measures that exhibited a higher level of environmental concern and emphasized the importance of management commitment. The MD must raise the profile of energy in the organization, through emphatically for cost-savings reasons, and the new emphasis on the environmental agenda. However, 'energy is not seen to be as significant as the environment', apparently, so that intention turns to be directed more towards broad-based environmental improvements rather than energy efficiency.

6.6. Advantages of an energy efficiency management (EnEM)
An explanation to energy efficiency management needs systematization on energy saving methodologies. In the long run, this consequences in certain size of energy savings and cost – related improvements, as well as efficiency in process. At the lower standard, it widely procurer and the supplier, an energy efficiency management [5-6]. The main reasons are discussed below:

6.6.1. Cost reduction
EnEM's introduction can save up to 10% of energy costs in the first years of implementation by systematically identifying the weak points of your energy usage and addressing them with basic measures. Investments in compressed air, refrigeration, ventilation systems, pump systems, materials handling technology will see a 5 – 50% reduction in power usage, and payback time of less than two years [6-7].

6.6.2. Environmental protection
Climate change is one other major catastrophe and not overlooking its effect of on the environment. The result of increasing temperature gives rise to flooding of Coastal regions and low lying Island nations; increasing desert areas and melting glaciers. White climate change is a universal challenge stern measures have to be taken to reduce the green- house effect. An efficient energy management is therefore an important element as it can contribute considerably to reducing greenhouse gas emissions [8].

6.6.3. Improvement of public image

IOP Conf. Series: Earth and Environmental Science **73** (2017) 012006 doi:10.1088/1755-1315/73/1/012006

An ISO certification, grants your company credibility to public and that you are operating in respect to energy efficiency and hence, taking good care of the environment. Environmental requirements are progressively a significant aspect in public proportions in South Africa, inclusive of atmospheric-friendly purchasing. Both from the perspective of the procurer and the supplier, an energy efficient management support the measurement of CO2-emissions.

6.6.4. Use of financial incentives

Lack of energy saving seems to be problematic impending the development and success of Steam Industries in South Africa. Consequently, the Power Generation Sector (Eskom) deficient the capacity to sufficient invest to meet the increasing demand of energy in the country; this challenge is caused by the energy pricing, since the sector is not cost effective. Again South African energy service organizations have a very poor financial scheme when it comes to the energy market to make profits [8].

7. Results

In general, energy efficiency management had a very low status compared to research and teaching. However, the study results suggested a correlation between the statuses granted for implementation of energy efficiency management and its activities. An assessed status was based on the subjective judgment of the people interviewed, the extent to which top administration were interested in energy issues and whether were environmental or energy guidelines. According to this ranking, Power Generation sector was the only organization on energy efficiency management status rated relatively high. At this organization, more weight was attached to energy efficiency management when it came to prioritizing investment decisions within the given budgets. A general lack of awareness of energy performance and energy costs was prevalent in all Steam Industries and awareness raising activities were limited and uncoordinated. Where energy savings programmes did exist, energy efficiency management staff tended to abstain because they had 'better things to do'. Motivating staff was also difficult because they were only temporary part of the system and had little incentive to get involved. While reward schemes for staff were not seen as an efficient means to generate new ideas, they could play some role in raising awareness.

7.1. Current energy efficient technology adoption in Steam Industries in South Africa

Improvements in technology adoption in steam generator efficiency result primarily from reductions in waste heat energy losses in the stack gases and expelled water. Procedures that reduce the mass flow and energy content of these flow streams directly benefit unit performance. Other losses occur from surface heat transfer to the atmosphere and incomplete combustion of the fuel.

7.2. Barriers to energy efficiency improvement in Steam Industries in South Africa

Basically at the industry level, incentive to save energy costs was not allowed to be used for savings due to the reductions in future budgets. And the basic cause of that is public accounting which limits the transferability of funds. The unspent money budgeted for a certain cannot either be used on other task and not either hand – over to budgeting periods. For example, savings in energy costs may not be used on investment in energy efficiency, steam systems maintenance or heating equipment. Previously, this was the instance in all Steam Industries. At the time the interviews were conducted, budgeting principles were in the process of being changed in most federal states. However, the policies and

IOP Conf. Series: Earth and Environmental Science **73** (2017) 012006 doi:10.1088/1755-1315/73/1/012006

future plans on how to allocate savings differed across organizations. For most Steam Industries, limited negotiability of retained funds to other functions with the same budgeting period was actual planned for the upcoming connection with the initiation of 'global budgeting' [9]. Similarly, for savings at the industrial level, some type of cost sharing arrangements with the organization administration were being planned, but not all the saving made in the past will remain at the Steam Industries for future periods. Thus, incentives for the industry to save energy costs are likely to improve, but remain constrained.

7.3. Benefits of energy efficiency measures in Steam Industries in South Africa

Energy is a direct reflection of cost, the more energy is saved the more cost is reduced. The first stage of raising energy saving awareness can be achieved without financial capital. On the other hand equipment repairs like broken insulation and conductors can save energy. Maintaining proper combustion, conditions of boilers and furnaces and equipping the operators with skills and knowledge will contribute in energy saving in industries [10].

Energy efficiency management is an effective way of saving the in short supply of energy and material resources. Take a brief on boiler operations, a 3mm diameter hole on a pipeline carrying 7 kg/cm2 steam would waste 32, 650 L of Fuel oil per year. A good housekeeping measure that fixes the hole saves that amount of resources [11-12]. The ability to implement energy saving measures in industrial and residential sectors is a cause for concern. The lack of awareness to start by no cost awareness and followed by cost saving low cost awareness measures of preserving and saving energy has been largely ignored by managers.

8. Discussion of Findings

8.1. Challenges of current industrial energy efficiency management practices in Steam Industries in South Africa

Currently there are policy instruments opposing the development of industrial energy efficiency management in steam industries in South Africa. South Africa's energy market is characterized as a regulated market, hence, government is trying to ease the heavy load on energy prices (certainly with electricity and petroleum products) by subsidizing energy. Such subsidies misrepresent the actually cost of energy supplied to industries and inefficient send the energy prices signals to the industries.

8.1.1. Lack of policy framework

The ignorance of policy and regulatory plays an important role in poor improvement of energy efficiency measures in steam industries in South Africa. In the occasion of policies, these both cover the national and local government policies. In a number of steam industries, predominantly in South Africa, there is directly none policy or, if there is, it can be unconcerned about the energy efficiency. Regulations that hold up indecorous tariffs can resist the curiosity in energy efficiency. For instance, it is standard to see tariffs that provide for diminishing energy prices for incremental energy consumption by considerable customers. This in turn reacts as a discouragement for such customers that engage or commit themselves to the activity of energy efficiency. Empowering policy and regulatory environments for energy efficiency comprise with setting targets, either binding or discretionary should be considered, from which strategies, for giving support towards increasing levels of energy efficiency, can be developed. The legislation must be enacted within the framework of a national energy policy that has continuity and is not allowed to become a football for party politics. This consistency must apply to any Steam Industry measure that has a potential effect on energy.

Legislation concerning energy must be both self-consistent and more readily comprehensible than in the past.

8.1.2. Lack of access to Capital

Lack of financial injection for investment was a significant barrier to upgrading energy efficiency at the Steam Industries. Access to capital proved to be a problem at the industrial level as well as within individual Steam Industries. While business risk and low profitability proved to be the major reasons for lack of capital in the Steam Industries in South Africa. Hence, the monetary support for digger investment is provided by the state, of which is very rigid. The circumstance is really heavy due to the fact that, previously, Steam Industries were not allowed to lend on the capital market. As indicated that taking up a credit for investment in energy efficiency has not been a solution, however profitable may the investment be.

8.1.3. Lack of management awareness

In the work place energy saving between staff and management was found to be very poor. As a consequence, the general level of housekeeping (for example, turning lights and computers off) was also poor. Factors contributing to this included the relative insignificance of energy costs, the lack of responsibility for energy costs and the dominance of production and other priorities. While a number of Steam Industries had undertaken limited awareness campaigns (for example, stickers, exhibitions) these were generally considered to be ineffective. The environmental consequences of energy consumption were considered to be of little help in motivating staff, owing to their relative invisibility compared to problems such as waste. Partly as a consequence of this, estates departments preferred to concentrate on improving energy efficiency through better controls, rather than improved housekeeping.

8.2 Strength of current industrial energy efficiency management practices in Steam Industries in South Africa

Disregarding the present odds of energy efficiency management rules in steam industries within South Africa. There is recent feasible information aimed for the supplementation of energy efficient practices within Steam Industries, involving the manner of industrial framework for activity development, promotion and coordination. This plan of action provides for the urgent supplementation of low – cost and no – cost interventions, and again to those higher – cost measures with deficient payback periods. These with be undertaken by medium – term and long – term investment opportunities in energy efficiency. The action plan recognizes that there exists indicative possibility for energy efficiency improvements across all sectors of Steam Industries.

8.3 Bridging the industrial energy efficiency gap in Steam Industries in South Africa

Political intervention is of great significance to energy efficiency improvement and this has been provided with the evidence as a proof for its existence. The political impact will subsequently take part in the bridge of the industrial energy efficiency gap steam industries among South Africa. Based on the results found, the management of steam industries in South Africa has the responsibility to establish a wide ranging substructure to accelerate the refinement of industrial energy efficiency. The drive of the substructure has to aim its focus in modifying the inefficient energy use among South African Steam Industries, again establish political instruments of market economy that will express correct energy fare significance to industrial customers. The application of the discretionary industrial energy efficiency management plan with irresistible financial advantages can also be a useful mechanism in developing industrial energy efficiency management in Steam Industries in South Africa. South African Steam Industries can prescribe energy efficiency in their territory by developing the coherence towards departmental facilities and operations, and sustaining energy efficiency developments for its employees, equipment and Industrial machines.

9. Conclusion

The conclusions of this research cover a generic analysis of the energy efficiency management practices in the Steam Industries in South Africa. South Africa‘s energy system can be defined as consolidated system with Steam Industries as the dominant steering body. The review has acknowledged that, the Steam Industries over the years have made powerful efforts to upgrade energy efficiency management in South Africa; by developing policy instruments and commencing energy efficiency systems and curriculums. Nonetheless, an extensive efficiency gap still stands in the Power Generation Sector and the logic behind is that, government‘s efforts to develop energy efficiency has been aimed towards residential and commercial sectors of the economy.

References

[1] Anderson, S.T., and R.G. Newell (2014), Information Programmes for Technology Adoption: The Case of Energy – Efficiency Audits, Resource and Energy Economics, Vol. 26, pp 27-50

[2] Armah, B. Retrieved July 2, 2013 from www.unmillenniumproject.org/documents/south africa_energy.doc

[3] Barber, W.J., Cochrane, J. L., Marchi, N., Yager, A. J., Graufurd, D. G. (2013). Energy Policy in Perspective – today’s problems, yesterday’s solutions, N.W., Washington D.C. 20036.

[4] Caffal, C. (2015), Energy Management in Industry, Centre for the analysis and Dissemination of Demonstrated Energy Technologies (CAD-DET), Eskom

[5] Christoffersen, L.B., A. Larsen, and M. Togeby (2014), Empirical Analysis of Energy Management in Steam Industry, Journal of Cleaner Production, Vol. 14, No.5, pp 516-26

[6] Cooke, P. W. C. (2015), Energy Saving in Distribution. Pitman Press, Bath.

[7] Dow.RetrievedJuly22http://www.dow.com/energy/perspectives/efficiency.htm

[8] Energy Commission South Africa, (2016a), Strategic National Energy Plan 2010 – 2030, Main report.

[9] Energy Commission South Africa (2016b) Strategic National Energy Plan 2010 – 2030, Annex Four of Four. Energy Supply of the Economy Wood fuels and Renewables.

[10] Energy Commission South Africa (2016c) Strategic National Energy Plan 2010 – 2030, Annex One of Four. Energy Demand Sectors of the Economy.

[11] Ernst, F.O. (2014), Integrating Energy Efficiency Performance in Production Management – Gap Analysis Between Industrial Needs and Scientific Literature, Journal of Cleaner Production, Vol.19, No. 6-7, pp.667-79

[12] Goodman, G.T., Rowe, W.D. (2012). Energy Risk Management, Harcourt Brace Jaranovich, Publishers.

IOP Conf. Series: Earth and Environmental Science **73** (2017) 012007 doi:10.1088/1755-1315/73/1/012007

Investigation of Sustainable Energy Policy: Nairobi Case Study

Y Shengyuan[1, a], J F L Habiyaremye[1, 2, b,*], and W Yingying[1, c]
[1]School of Mechanical and Electrical Engineering, Harbin Engineering University, Nantong Street, NO.145-1, PO.BOX 150001 Harbin City-China
[2]School of Science and Technology, University of Rwanda, PO.BOX: 3900 Kigali-Rwanda.

E-mail: [a]yanshengyuan@hrbeu.edu.cn, [b,]*habijealuc@yahoo.fr, [c]weiyingying2007@126.com

Abstract. A plan for actively achieving green energy obligation is a strategic tool for policies that point forward the diminution of the fossil fuel consumption and greenhouse gas (GHG) in conformity with the Paris environment-friendly accords (COP21) and updates of other eco-system agreements. To achieve the concrete implementation of the sustainable energy strategy (SES) and to accomplish its objectives, an investigation is a critical factor. SES investigation has to consider both the advancement of each particular action and its wide-ranging green effect, which necessitates multiple levels of improvement. In this study, a consolidated eco strategy for evaluating, monitoring and handling the SES via investigation and execution process is established. The city of Nairobi was used as one of the geographical positions to test the effectiveness of this approach and to investigate its robust and weak points. Specifically, benefit-cost analysis, reliability, peer review and general level of participation were renowned as vital tools for attaining a functional SES investigation and for then drafting successful energy guidelines. Some suggestions were put forward to highlight the research and execution methods and to draw a road map of how SES can be strategically placed into practice.

1. Introduction

For many years ago, the SES has turned out to be a central tool for establishing regional energy policies and the initiative program covered all 17 municipalities in Nairobi. All partners have subsequently believed in a progressive role to reach the Kyoto Protocol goals and, now of the Paris climate change accord. The positive achievements of this scheme and its continuing advancement called National Climate Change Action Plan (NCCAP) do not take away the difficult coupled with SES execution. The SES works as a vital tool to be improved and performed with maximum efficiency based on gained results of compliance with the GHG reduction strategy. In agreement with the methodological perspective, the SES builds on the outcome of the inventory on gas emissions that quantify the consumed energy and emitted CO_2 of the urban area and categorizes various actions to be taken in priority locations. The SES can achieve its goals through proper planning, execution, and monitoring with coordinated efforts at the municipal level bringing together public institutions, the private sector, and local communities. Throughout the implementation, the SES has to deal with changing needs and confront evolving situations. Industrial expansion, public policies, economic position and regulatory mechanism continually changes, hence demanding an equivalent evolution of the SES [1]

Therefore, the monitoring tasks are a vital opportunity to investigate the implementation level and assess the planning process. According to **Kenya** Vision 2030, throughout the SES application, the local counties try to enhance the quality of the environment through a continuous improvement process. This way of viewing things is described as investigation and monitoring (I & M) that, adequately represents the ideal process[2, 3]. A smooth implementation of the SES solidly depends on this repetition process of continually checking and improving environmental quality. This process is not easy; various researchers have shown that SES I& M necessitates much attention because of the important challenges that frequently hold back an efficient execution of this process [4].

In this case, starting with the viewpoints suggested in the NCCRS about monitoring, assessment, and case study know-how skills, the researchers answer a few major questions. Preferred actions should be able to achieve an anticipated GHG emission diminution, but the pathway of their execution and updates do not suffice. The first point was to know if there was any other step forward to test whether the actions taken are adequate. The second aspect was to know how monitoring can develop into an adequate tool to contribute to the overall goal. This study determines the strategic plan for SES follow-up and uses the case of Nairobi SES for reflections on the ground experience. Subsequently, after discussing the questions mentioned earlier, some recommendations were put forward.

2. Research background

The implementation of an SES has to take into considerations the needs of updating and changing, the know-how skills and related political will; simultaneously, the economic and regulatory mechanism and the counties' feedback need to be considered too. Therefore, this paper has developed valuable tools for the execution of actions, with regards to repeated process illustrated in Figure 1.

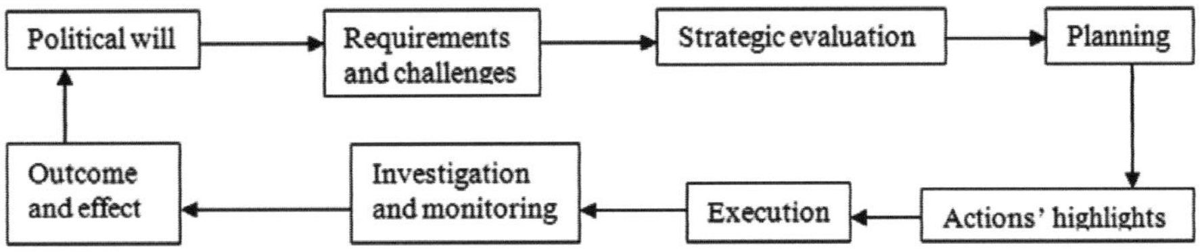

Figure 1. SES monitoring practice loop.

After the submission of the SES, all municipalities and counties were committed to compiling data and producing biennial reports. The first report quantifies the information on efforts to assess, monitor and confirm the implementation and the effect of the SES progress. The second report provides updated data on CO2 emission situation that is compared against NCCRS inventory. This study embraced methodologies based on multiple conflicting criteria, strictly for SES planning stage[5]. A multi-criteria decision analysis (MCDA) helped establish the appropriate combination of energy scheme and techniques for undertaking investments with maximum effectiveness. Nevertheless, they have primarily been used for examining the efficiency of one particular source of energy or technical solution[6].Although monitoring tool shows a positive outcome, it cannot guarantee the accuracy of final results alone by a simple tracking. There must be a need to collect and assess feedbacks from the field to orient decisions in the adequate direction.

2.1. The monitoring of the SES

As a result of the embracing the environment-friendly energy plan and Kyoto agreement on climate change, the Kenyan Government initiated a countrywide climate change response strategy (KNCCRS 2010) and climate change action plan (KNCCAP, 2013). These initiatives aimed at kick-starting the development, the actualization of energy and climate change policies in compliance with global eco-friendly accords [2, 7]. The KNCCRS had objectives of raising awareness and educating people about the reality of climate change in Kenya. The KNCCAP launched on 27th March 2013 and the execution tool it upholds, the SES are established within this agenda and encourage the implementation of Kenya's commitments to the Kyoto Protocol with an independent and intentional contribution of all partners [2]This article has analyzed the techniques presently used to monitor the SES, and possible execution of this course of action was assessed. Therefore, the present method for this issue must be demonstrated, and its strong points and limitations will be drawn to attention. The next section explains an approach, puts in place a benchmark and establishes an environment for a critical analysis that uncovers much more useful practice. The reporting template in the SES monitoring is divided into three major subsections:

- Course of action to evaluate the SES implications;
- Adjusting finances and providing updates to stakeholders and
- Inventory for GHG emissions.

Monitoring and updating the measuring system in the SES progress comprises four stages:

- Identification of the persons in charge of SES on municipal level;
- Organizing general meetings and setting up relevant timetable;
- Holding sessions of professionals to share expertise and updates
- Examining qualitative study process and exploring quantitative research blueprint in SES evolution

The NCCRS gives a control template for the SES execution in which each measure shows new area to be filled in such as personnel qualifications, expenditures and some challenges compromising the SES performance. The SES monitoring took into considerations the sectors recommended in the NCCRS such as houses (HSs), lights (LTs), transportation (TRP), municipal planning (MP), public procurement (PRO), power production (PPD), air conditioning & refrigeration (ACR) and miscellaneous (MLS).Although it was not technically easy to quantify, the Nairobi city council was committed to investing in untapped green energy resources and purchasing power saving products. On the other hand, the ecosystem monitoring was focusing on power savings and equivalent carbon emission reductions in every area of application in 2015 (see Figure 2).

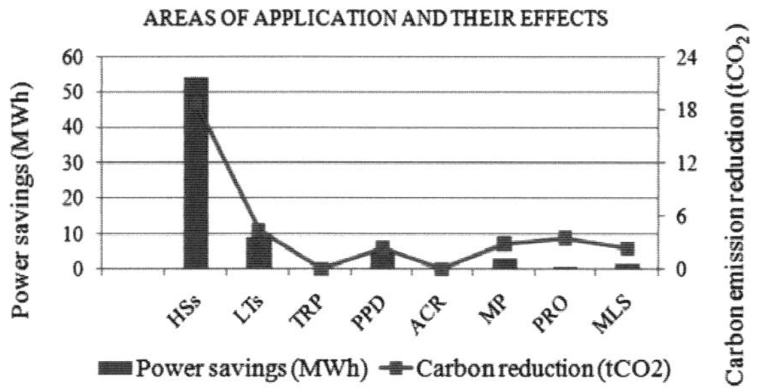

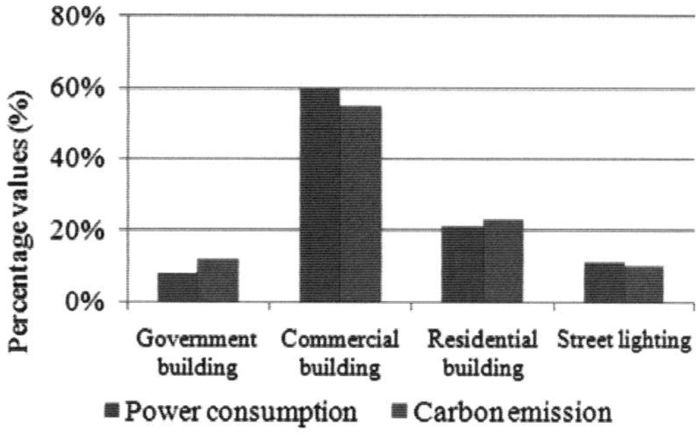

Figure 2. Areas of application & power savings, 2015. **Figure 3.** Power consumption and GHG, 2010.

It is clearly noticed that the amount of reduced carbon emission varied from case to case (see Figure.2), this gives details of fundamental challenges affecting the results. Furthermore, the comparison between intervention action advancement and the equivalent ecosystem improvement provides comprehensive knowledge of the critical challenges. With regards to the progress of SES corrective steps, for some 176 intervention works, some have already started, and others are under development process. Throughout the SES monitoring, critical challenges related to intervention actions are no more valid just because the administration has updated policies or because incredibly complex intervention actions had to be deleted. Figure.3 presents the status of power consumption and GHG emission in 2010.The business buildings consume most of the power generated and subsequently emit a high level of carbon emission, followed by residential building.

The comparison of NCCRS inventory (2010) and SES-GHG inventory (2015) with regards to the type of fuel (see Figure.4) shows a significant improvement with all types of fuel. The liquid gas is leading the trend with a potential reduction in GHG emission going to 31.3%, fossil fuel to 24.3 %, biomass to 18%, natural gas to 14.4 % and electric power to 7.63%.Figure.5 compares CO_2 between the year 2010 and 2015, the reduction of CO_2 amounted to 20.9% in business buildings and 6.9% in public buildings. Although there was an improvement, the SES execution faces some challenges: a) difficulties in retrieving data, b) financial constraints and market dynamicity, c) long process for technological implementation and d) inadequate procedural methods. Also, the cost-benefit evaluation must be used in a simplified manner since the primary objectives are reducing carbon dioxide gas emission. Cost estimates help the operators update and identify the most beneficial and priority projects. Subsequently, the MCDA techniques seem to be suitable for the SES monitoring advancement since they consider cost as a key criterion[8].

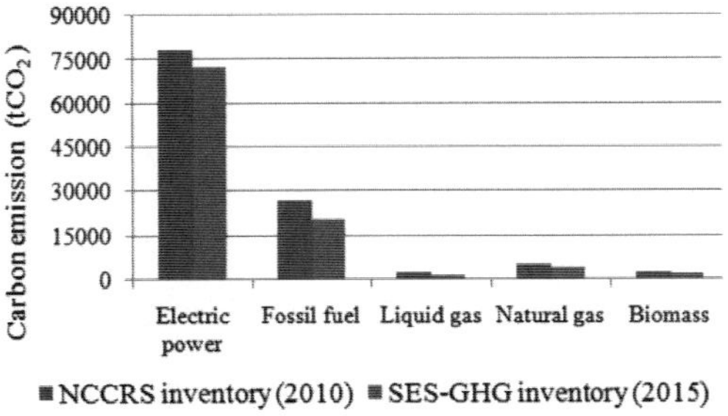

Figure 4. Types of fuel and CO_2.

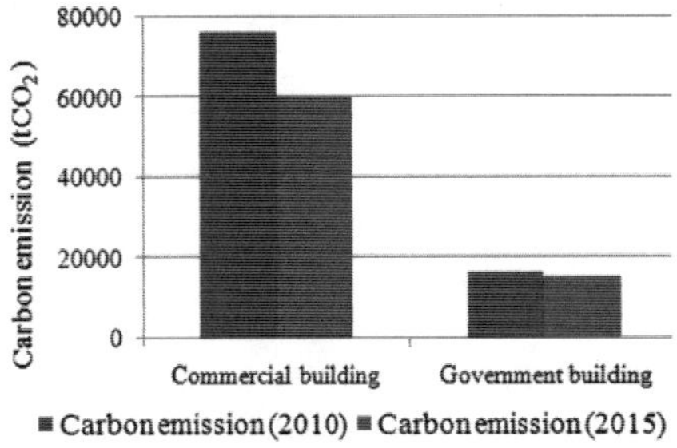

Figure 5. $CO_2$and some areas of application.

2.2. *SES advancement*

The SES monitoring necessitates being regularly updating and informing to the relevant authorities for the examination of the GHG emission reduction and the discussion on qualitative and quantitative advancement. The SES progress report shows in Figure.6 that the vision 2030 climate change target ranges between 50% and 70 % carbon reduction, whereas the for the last seven years the highest carbon reduction level was 33%. And Figure.7 shows the CO2 emission status at the launch in 2010, the current situation in 2017 and the CO2 emission target in 2030 for various areas of application.

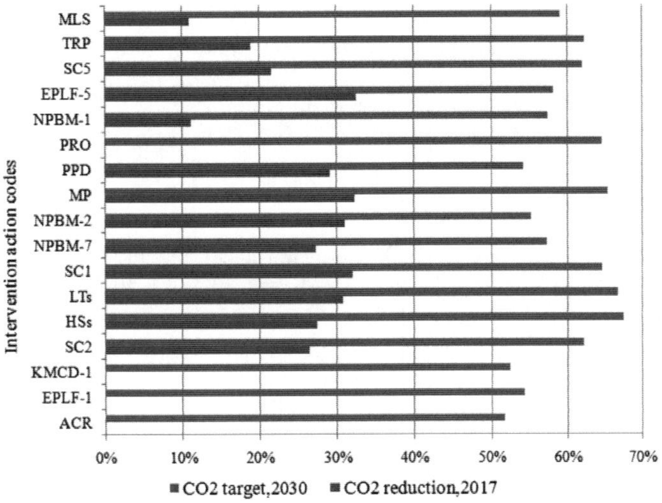

Figure 6. SES advancement report.

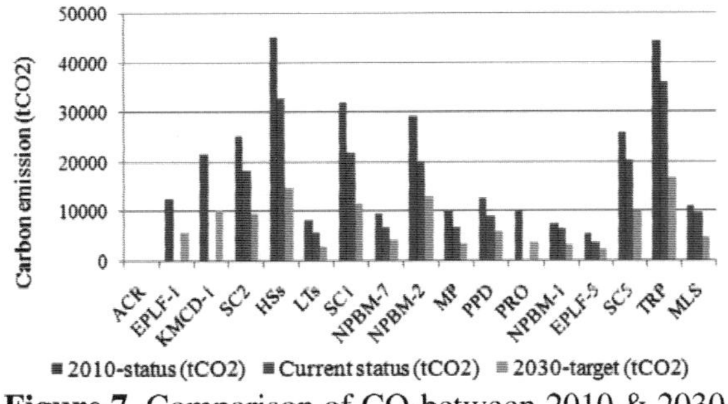

Figure 7. Comparison of CO_2between 2010 & 2030.

3. Conclusion

The SES must be continually updated in content and methodologies for better evaluation progress and policy making. Besides, the SES monitoring should consider the advancement of every action and its global ecological effect since it requires a collective examination and describes an entire process for a long-lasting reduction of GHG emission from different municipalities. This research provides details of those challenges and should serve as a model for other cities, especially in Africa, dealing with climate change issues. The assessment of problems and opportunities facing the city of Nairobi in monitoring should help other cities implement the long-term process and outline a strategy at the beginning of each section of SES, thus smooth the progress of best I & M practices.

The development of the inventory of gas emissions was a crucial part of this monitoring strategy, and it was an occasion to engage communities and create a productive network of collaboration in collecting green energy data. A peer review process was proven to be the useful tool to measure the SES accuracy and reorient actions according to variations of the external factors that can compromise the outcome. There is a need for a mind-set that invests in individual productivity and strengthens the SES monitoring policies in different municipalities. There is an alarming increase of fossil fuel consumption especially in electricity production in Nairobi; therefore, the future study should investigate the efficacy and the ecological impact of diesel powered electrical generating stations.

Acknowledgment

The authors wish to thank the Chinese Scholarship Council in conjunction with Harbin Engineering University and the University of Rwanda for the financial and academic support.

Reference

[1] A. M. Fernández-Maldonado, P. L. Méndez, and E. V. Montes, "Planning for Energy Efficiency in a Historic City. The Case of Santiago de Compostela, Spain," *Journal of Settlements and Spatial Planning,* p. 53, 2016.
[2] G. o. K. (GoK). National Climate Change Action Plan [Online].
[3] S. K. Williams, T. Acker, M. Goldberg, and M. Greve, "Estimating the economic benefits of wind energy projects using Monte Carlo simulation with economic input/output analysis," *Wind Energy,* vol. 11, pp. 397-414, 2008.
[4] P. P. Brief, "CLIMATE ACTION."
[5] I. Delponte, I. Pittaluga, and C. Schenone, "Monitoring and evaluation of Sustainable Energy Action Plan: Practice and perspective," *Energy Policy,* vol. 100, pp. 9-17, 2017.
[6] C. Cormio, M. Dicorato, A. Minoia, and M. Trovato, "A regional energy planning methodology including renewable energy sources and environmental constraints," *Renewable and Sustainable Energy Reviews,* vol. 7, pp. 99-130, 2003.
[7] G. o. K. (GoK). National Climate Change Response Strategy [Online].
[8] S. Pohekar and M. Ramachandran, "Application of multi-criteria decision making to sustainable energy planning—a review," *Renewable and sustainable energy reviews,* vol. 8, pp. 365-381, 2004.

A Framework for Evaluating Economic Impacts of Rooftop PV Systems with or without Energy Storage on Thai Distribution Utilities and Ratepayers

A Chaianong [1,3], A Bangviwat [1,3] and C Menke [1,2,3]
[1]The Joint Graduate School of Energy and Environment, King Mongkut's University of Technology Thonburi, Bangkok, Thailand
[2] Trier University of Applied Sciences, Germany
[3] Center of Excellence on Energy Technology and Environment, PERDO, Bangkok, Thailand

E-mail: achaianong@gmail.com;

Abstract. Driven by decreasing PV and energy storage prices, increasing electricity costs and policy supports from Thai government (self-consumption era), rooftop PV and energy storage systems are going to be deployed in the country rapidly that may disrupt existing business models structure of Thai distribution utilities due to revenue erosion and lost earnings opportunities. The retail rates that directly affect ratepayers (non-solar customers) are expected to increase. This paper focuses on a framework for evaluating impacts of PV with and without energy storage systems on Thai distribution utilities and ratepayers by using cost-benefit analysis (CBA). Prior to calculation of cost/benefit components, changes in energy sales need to be addressed. Government policies for the support of PV generation will also help in accelerating the rooftop PV installation. Benefit components include avoided costs due to transmission losses and deferring distribution capacity with appropriate PV penetration level, while cost components consist of losses in revenue, program costs, integration costs and unrecovered fixed costs. It is necessary for Thailand to compare total costs and total benefits of rooftop PV and energy storage systems in order to adopt policy supports and mitigation approaches, such as business model innovation and regulatory reform, effectively.

1. Introduction

Thailand is facing an energy security problem due to high dependence (more than 50 percent of total primary energy supply) on imported non-renewable energy and continuous depletion of scarce domestic energy reserves [1-2]. Apart from the fact that integration of a wider variety of energy resources, especially distributed solar photovoltaic (PV), could be a cost-effective and risk minimizing approach, there is increasing interest in the potential of solar PV in Thailand. According to the latest Alternative Energy Development Plan 2015-2036 (AEDP 2015) approved in September 2015, the solar power capacity target of 6,000 MW has been set [3]. By third quarter of 2016, the cumulative installed capacity of solar PV that has already connected to the grid was approximately 2,753 MW, which increased more than 200 percent from 2013. Out of the total installations of PV, 95 percent is accounted for solar farm, while rooftop PV shares the installed capacity approximately 5 percent [4]. Currently, Thailand tries to eliminate solar rooftop quotas with feed-in tariff scheme. Instead, a self-consumption scheme for rooftop PV is encouraged [5-6]. To date, there is only a self-consumption pilot project of 100 MW. However, this policy has not yet fully developed into net-metering scheme, of which electricity generated from rooftop PV system must be consumed on-site and any excess electricity fed back to the grid will not be credited at any monetary rate [4]. There has been a plan for drafting a compensation scheme for excess generation (e.g. with buyback, or rolling credit option) in Thailand [7], pending the completion and evaluation of the self-consumption pilot project to provide full understand of the scheme to the government.

Driven by falling PV and energy storage system prices, rising electricity costs, and policy supports, the increasing share of distributed PV and energy storage is expected to disrupt the existing utilities' business models and the tariff design as occurred in many places in recent years [8-10]. These problems lead to the objective of this research that is to introduce the framework for evaluating the economic impacts of rooftop PV and energy storage on Thai distribution utilities and ratepayers. It is expected that the full knowledge of economic impacts will be useful for the utilities' executives and the regulators to design their future business models, appropriate tariff structure, and policy supports for solar and non-solar customers.

2. Evaluation framework: Cost-benefit analysis

Cost-Benefit Analysis (CBA) is a systematic approach to evaluate strengths and weaknesses of each proposed alternative. It can be done from many perspectives such as from solar customers, non-solar customers (ratepayers), utilities and overall society's points of view [9]. As stated, scope of this work mainly focuses on introducing a framework to evaluate economic impacts of rooftop PV and energy storage on distribution utilities and ratepayers. The analysis can be separated into two main parts as shown in Figure 1. It should be noted that this calculation method can be adapted into both main scenarios: without energy storage and with energy storage.

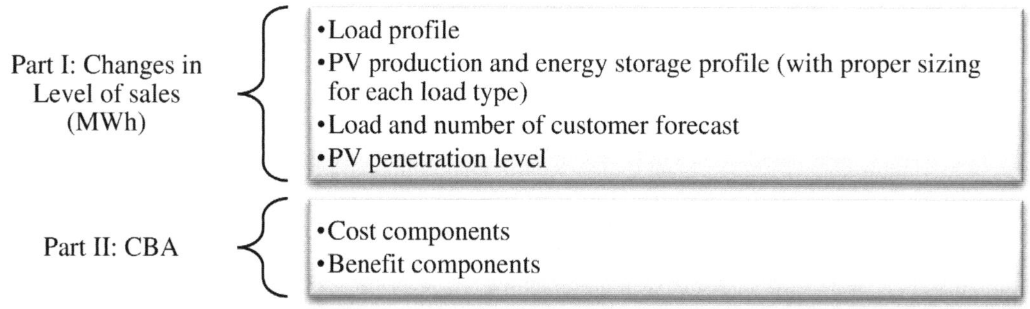

Figure 1. Big picture of Cost-Benefit Analysis (both without and with energy storage) (Adapted from [8], [10]).

2.1. Changes in energy sale

The first step is to identify changes in energy sale in terms of physical unit (e.g. MWh). In general, the changes in MWh can be determined by comparing grid consumption between the case without rooftop PV and the case with rooftop PV (or rooftop PV with energy storage). The main calculation steps (see Figure 2) include:

- compare daily load profiles of sample customers for the cases without rooftop PV, and with rooftop PV and energy storages to find out the amounts of self-consumed PV electricity in year 1, and consumer's bill savings, where the consumer's bill savings are equivalent to utility's revenue reduction.

- apply first year PV penetration level to estimate bill savings and utility's revenue losses for the whole system.
- project the future situation of changes in energy sale by applying load growth, PV penetration growth, and number of customers forecast based on available data from relevant stakeholders (distribution utilities, regulator, ministry of energy, etc.).

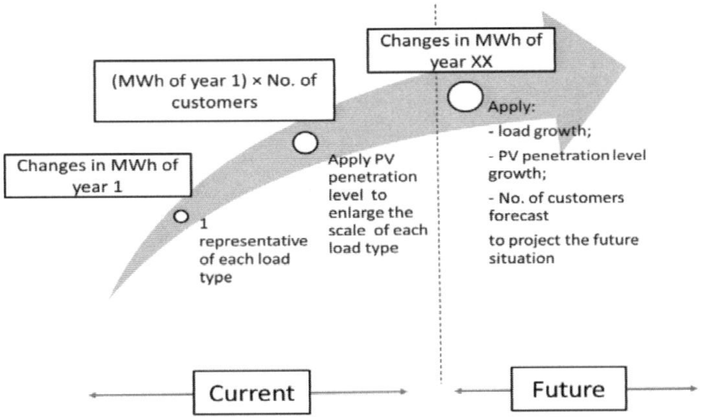

Figure 2. Conceptual model of changes in energy sale calculation (Adapted from [8], [10]).

The changes in energy sale, or bill savings from solar customer's perspective depend on details of policy features of compensation schemes for solar electricity, including, number of meters, banking period, and buyback rate [7, 11-13]. It is obvious that different policy features affect each stakeholder differently depending on how effective the policies are. It can be anticipated that more benefits to prosumers (e.g. additional bonus is allowed to compensate self-consumed PV, buy back rate is above retail rate, banking period is longer than one billing period, energy storage is allowed, etc.) lead to more negative impacts to distribution utilities and ratepayers [14].

2.2. Cost and benefit components

After understanding how energy sales change due to the increase in distributed rooftop PV and energy storage installation, the next step is to address cost and benefit components as summarized in Table 1.

Table 1. Cost and benefit components of rooftop PV systems with or without energy storage impact evaluation on Thai distribution utilities and ratepayers from stakeholder interview and literature review [8-10, 15]

	Benefits	**Costs**
Ratepayer Cost Test and Utility Cost Test	-Utility avoided costs (Avoided costs of loss and avoided costs of deferring distribution capacity)	- Consumer's bill savings depending on policy features - Program cost (related to administrative cost due to excess generation to the grid) - Integration cost (associated with upgrading generation, transmission and distribution system cost due to excess generation to the grid) - Unrecovered fixed cost

The cost and benefit components were suggested by relevant stakeholders during interviews, including MEA (Metropolitan Electricity Authority), PEA (Provincial Electricity Authority), ERC (Energy Regulatory Commission), EPPO (Energy and Planning Policy Office), DEDE (Department of Alternative Energy Development and Efficiency) and private sector, which are in line with those given in literatures [8-10, 15].

Considering the inclusion of energy storages, there are about the same cost components. Greater revenue losses from consumer's bill savings and decreases in program cost and integration cost due to less excess PV generation to the grid can be anticipated. As for benefit components, the difference is only on their magnitudes because energy storage will allow users to store their excess generation from rooftop PV and to draw less electricity from grid.

3. Cost-benefit analysis interpretation

With CBA, cost and benefit components would be identified as stated in previous section. Then, the final output is the comparison between net benefits and net costs as shown in equation (1).

$$\text{Net Costs (Benefits)} = \text{Total Costs- Total Benefits} \tag{1}$$

There two possible cases for the net costs/benefits calculation; they are (i) benefits greater than costs and (ii) benefits less than costs [8-10,15].In case of greater costs than benefits, the incremental costs will either be passed to ratepayers through base tariff and/or fuel adjustment charge (Ft), so called cost-shifting issues between solar and non-solar users, or absorbed by the distribution utilities to the extent allowed by regulations [16]. Thailand's distribution utilities are mainly operated under cost-of-service regulation that provides utilities with a return on capital investments as an incentive to drive utilities towards building and maintaining their systems in order to provide customers with reliable and affordable service, or electricity supply in this context. Thus, the real negative impacts are probably passed to ratepayers.

Alternatively, if rooftop PV and energy storage systems still leads to negative impacts for utilities and/or ratepayers (costs are higher than benefits), then, it is necessary for utilities to define specific grid codes in order to limit rooftop PV penetration, as well as promote new business models and/or new regulatory models in order to maintain their profits. However, these mitigation approaches require careful consideration and debate as they could lead to losses of opportunity to maximize the benefits of rooftop PV and energy storage systems of relevant stakeholders and the country.

4. Conclusion

Due to increases in self-consumed PV and energy storage installations and concerns on Thai distribution utilities' revenue losses, Cost-benefit analysis (CBA) is then proposed to address these economic impacts. There are two main steps of this evaluation: changes in electricity sale determination and cost/benefit components calculation. The CBA results are expected to depend on the excess generation compensation scheme and the inclusion of battery based on Thai context. If costs are greater than benefits, it is possible that utilities will pass their incremental costs through to ratepayers via increases in retail rate, which is known as cost-shift between solar and non-solar users. There have been heated debates on the issue internationally to find suitable business models/regulatory models in order to mitigate the impacts. Therefore, it is necessary for Thailand to pay attention on this topic and try to quantify cost and benefit components of rooftop PV and energy storage systems as well as start finding appropriate solutions to not only increase distributed PV generation, but also minimize impacts on relevant stakeholders, including utilities and ratepayers.

Acknowledgments
This work was financially supported by the Joint Graduate School of Energy and Environment, King Mongkut's University of Technology Thonburi and Center of Excellence on Energy Technology and Environment, PERDO, Bangkok, Thailand as well as the Petchra Prajomklao PhD Research Scholarship of King Mongkut's University of Technology Thonburi.
The authors would also like to thank MEA, PEA, ERC, EPPO, DEDE, and private sector for their opinions expressed during the interviews.

References
[1] U. S. Energy Information Administration. International Energy Data and Analysis. 2015 [Cited 25 Mar 2017]. Available from: https://goo.gl/kei8tA
[2] EPPO-Energy Policy and Planning Office. Electricity. 2016 [Cited 26 Mar 2017].Available from: www.eppo.go.th/images/Energy-Statistics/energyinformation/.../06Electricity.pptx.
[3] DEDE-Department of Renewable Energy Development and Energy Efficiency. Alternative Energy Development Plan: AEDP 2015. 2015 [Cited 26 Mar 2017]. Available from http://www.dede.go.th/ewt_news.php?nid=42195.
[4] GIZ- Deutsche Gesellschaft für Internationale Zusammenarbeit, Thailand Solar PV Policy Update 09/2016, Federal Ministry for Economic Affairs and Energy, Bangkok, September 2016.
[5] Tongsopit S, Gracen C, An Assessment of Thailand's Feed-in Tariff Program, *Renewable Energy*, 2013, **60**, 439-445.
[6] Tongsopit S, Moungchareon S, Aksornkij A, Potisat T, Business Models and Financing Options for a Rapid Scale-up of Rooftop Solar Power Systems in Thailand, *Energy Policy*, 2016, **95**, 447-457.
[7] Tongsopit S., Julakarn S., Chaitusaney S., and Wibulpholprasert W, Evaluation of Rooftop PV Pilot Project, Bangkok, Thailand, 2017.
[8] E3-Energy + Environmental Economics, California Net Energy Metering Ratepayer Impacts Evaluation, San Francisco, CA, 2013
[9] E3-Energy + Environmental Economics, Nevada Net Energy Metering Impacts Evaluation, San Francisco, CA, 2014.
[10] Satchwell A, Mills A, and Barbose G, Financial Impacts of Net-Metered PV on Utilities and Ratepayers: A Scoping Study of Two Prototypical U.S. Utilities, Ernest Orlando Lawrence Berkeley National Laboratory, Berkeley, CA, 2014.
[11] Hughes L, and Bell J, Compensating Customer-Generators: A Taxonomy Describing Methods of Compensating Customer-Generators for Electricity Supplied to the Grid, *Energy Policy*, 2006, **34**, 1532-1539.
[12] Dufo-López R and Bernal-Agustin JL, A Comparative Assessment of Net metering and Net Billing Policies. Study cases for Spain. *Energy*, 2015,**84**, 684-694.
[13] IEA-PVPS - International Energy Agency- Photovoltaic Power Systems Program, Review and Analysis of PV Self-Consumption Policies, IEA Photovoltaic Power Systems Program, 2016.
[14] Eid C, Guillen JR, Marin PF, and Hakvoort R. The Economic Effect of Electricity Net-metering with Solar PV: Consequences for Network Cost Recovery, Cross Subsidies and Policy Objectives. *Energy Policy*, 2014, **75**, 244-254.

IOP Conf. Series: Earth and Environmental Science **73** (2017) 012008 doi:10.1088/1755-1315/73/1/012008

[15] CPR-Clean Power Research: Austin, The Value of Distributed Photovoltaics to Austin Energy and the City of Austin, Clean Power Research, Austin, TX, 2006.
[16] ERC- Energy Regulatory Commission, Retail Restructuring B.E. 2558, 2015 (in Thai).

IOP Conf. Series: Earth and Environmental Science **73** (2017) 012009 doi:10.1088/1755-1315/73/1/012009

Study on Electricity Purchase Optimization in Coordination of Electricity and Carbon Trading

Dunnan Liu, Yaru Meng and Shuo Zhang
Department and Economics and Management, North China Electric Power University, Changping district, Beijing 102206, China

Email: yaru.meng@qq.com

Abstract. With the establishment of carbon emissions trading market in China, the power industry has become an important part of the market participants. The power grid enterprises need to optimize their own strategies in the new environment of electricity market and carbon market coordination. First, the influence of electricity and carbon trading coordination on electricity purchase strategy for grid enterprises was analysed in the paper. Then a power purchase optimization model was presented, which used the minimum cost of low carbon, energy saving and environment protection as the goal, the power generation capacity, installed capacity and pollutant emission as the constraints. Finally, a provincial power grid was taken as an example to analyse the model, and the optimization order of power purchase was obtained, which provided a new idea for the low carbon development of power grid enterprises.

1. Introduction

With the establishment of carbon emissions trading market in China, the power industry has become an important part of the market participants in order to complete the goal of carbon emission reduction [1, 2]. The energy saving and emission reduction work of power grid enterprises can influence the production process through the environmental constraints, so as to achieve the purpose of energy conservation. Constraints can come from two aspects, namely, administrative means and market regulation [3]: administrative means requires the enterprise to complete the designated emission reduction tasks at the appointed time through the laws and regulations promulgated by government, which is achieved energy conservation through carbon emissions; Market regulation is to commercialize the carbon emissions rights and establish the carbon emissions trading market. Therefore, the development of power grid enterprises will face the double market of electricity trading and carbon emissions trading, and how they participate in the carbon market for power purchase under the new situation will be studied in the paper, so as to promote their development and energy conservation.

2. The influence of coordinated trading mode on electricity purchase strategy of power grid enterprises

Electric carbon coordination trading model refers to the new trading model that electricity trading in harmony with carbon emissions trading [4]. The main body is on the basis of balance of power supply and demand, to optimize the power purchase order of units on the environmental level by operational

IOP Conf. Series: Earth and Environmental Science **73** (2017) 012009 doi:10.1088/1755-1315/73/1/012009

research methods. Thus affecting the power production process and obtain trading optimization strategy, so as to provide the basis of decision-making and scheduling optimization for power grid enterprises [5].

Due to the total amount control of carbon emissions, the generation capacity of power generation enterprises has changed significantly compared with before. For low energy consumption units, because of their emission factor is lower than the average level, so they can send more electricity; but for high energy consumption units, they have to reduce their generation because their emission factor is higher than the average level [6]. Therefore, the contract quantity will also change a lot when the power grid enterprises sign the purchase contract with power generation enterprises. Especially for the power generation rights trading that high energy consumption units have beside the carbon emission quota, the benefit of it may be negative with the development of relevant trading mechanism, so the additional power generation right will become a burden for high energy consumption unis [7].

In addition, the power generation scheduling based on "open and fair" principle would be inappropriate. High energy consumption units will strictly control the power generation rights within the scope of carbon emissions quota allowed when they formulate the purchase contracts [8]. But for low energy consumption units, they would like to sign more contract quantity and get more power generation rights to expand their economic benefits.

In short, power generation rights may have the inventory constraints under the influence of total carbon emissions control (carbon emissions quota is not enough to support the power generation that determined based on "open and fair" principle). The introduction of carbon emissions trading will promote the power generation rights transfer trading vigorously. At the same time, the imbalance between the initial quota of carbon emissions and power generation rights will be self-balanced through carbon emissions quota trading or power generation rights trading.

3. Establishment of power purchase optimization model

The power purchase of provincial power grids mainly includes low-carbon power (such as: hydropower, wind power, photovoltaic, etc.) and thermal power. Under the coordinated trading mode of electricity and carbon, the objective function and constraints method was used to optimize the order of generating units in the province to achieve the purpose of purchase optimization. The idea is: minimizing the environmental cost under the consideration of low carbon, energy saving and environmental protection.

The generation costs of low carbon power mainly include investment costs and daily labor and maintenance costs. Low carbon power basically has no low carbon cost and no need to be carried out with low carbon treatment. However, due to the proportion of low carbon power is small, the optimization purchase of provincial power grid enterprises is to optimize the order of thermal power units in the province, and the local thermal power unit optimization model is as follows.

The objective function is the minimum cost of low carbon, energy saving and environmental protection, as shown in equation (1).

$$MinC_{local} = \sum_{1}^{n}\left(h_{CO_2i} + h_{SO_2i} + h_{NO_Xi} + h_{PMi} + h_{sewage\ i} + h_{water\ i} + h_{coal\ i}\right)\times Q_i \quad (1)$$

$$h_{pi} = f_{pi} \times e_{pi} \quad (P = CO_2,\ SO_2,\ NO_x,\ PM,\ sewage) \quad (2)$$

Among them, C_{local} is the cost of low carbon, energy saving and environmental protection for local units (yuan), n is the total number of units, h_{CO_2i} etc. are the environmental costs of all pollutants (yuan/MWh), f_{pi} etc. are the emission factors of all pollutants (t/MWh), e_{pi} etc. are the emission equivalents of all pollutants (yuan/t), and Q_i is the generated capacity of units (MWh).

The constraints are as follows:

Power generation constraints

$$\sum_{i=1}^{n} Q_i = Q_{all} \tag{3}$$

Installed capacity constraints

$$0 \leq Q_i \leq Q_{ci}, i = 1, 2, 3, \cdots, n \tag{4}$$

Pollutant emissions constraints

$$\sum_{i=1}^{n} f_{Pi} \times Q_i \leq W_P \tag{5}$$

In the above constraints, Q_{all} is the total demand of calculation period, Q_{ci} is the maximum installed capacity of current units, and W_P etc. are the pollutant emission quota of calculation period.

4. Case study

4.1. Power generation constraints

Through the local electricity consumption - external electricity - hydropower forecast, can get the local electricity demand for local thermal power generation is 5100000 MWh. In order to meet the local electricity demand, the local power generation cannot less than it.

4.2. Power plant installed capacity constraints in a month

The data in table 1 were the maximum power output of each power plant after subtracting the plant power consumption.

Table 1. Installed capacity constraints of each power plant

Number	Power plant	Power supply limits with installed capacity constraints (MWh)	Number	Power plant	Power supply limits with installed capacity constraints (MWh)
1	1	627810	11	11	548679
2	2	618101	12	12	534481
3	3	389940	13	13	603474
4	4	394476	14	14	627931
5	5	340831	15	15	171562
6	6	390217	16	16	147879
7	7	216187	17	17	202665
8	8	212228	18	18	183790
9	9	164836	19	19	315102
10	10	168358			

IOP Conf. Series: Earth and Environmental Science **73** (2017) 012009 doi:10.1088/1755-1315/73/1/012009

4.3. Pollutant emissions constraints

The emissions of power generation enterprises have to be controlled according to the requirements of environmental protection departments. The power plants have to buy more indicators from other plants whose indicators are surplus if they exceed the emission limit constraints. If the power plants have more indicators, they can have economic transactions with the other plants who need these indicators. The data in table 2 were the pollutant emissions quote of a power plant in a month, and the data in table 3 were the pollutant emission factors of a power plant in a month. All the data were adopted from the power grid enterprise.

Table 2. The pollutant emissions quote in a month

CO2 (t)	SO2 (t)	NOx (t)	Solid particulates (t)	Sewage (t)
3545666	1701	8025	641	3542889

Table 3. The pollutant emission factors in a month

Number	CO2 emissions (t/MWh)	SO2 emissions (t/MWh)	NOx emissions (t/MWh)	Solid particulates emissions (t/MWh)	Sewage emissions (t/MWh)
1	0.54246	0.000169	0.00053	0.000073	0.851
2	0.54355	0.000152	0.00203	0.000087	0.677
3	0.52160	0.000262	0.00211	0.000109	0.701
4	0.50788	0.000261	0.00066	0.000082	1.044
5	0.46754	0.000254	0.00072	0.000113	1.283
6	0.47343	0.000261	0.00117	0.000171	0.681
7	0.52077	0.000375	0.00132	0.000162	0.947
8	0.52074	0.000387	0.00116	0.000159	0.825
9	0.52347	0.000382	0.00176	0.000159	1.126
10	0.58394	0.000447	0.00127	0.000095	1.172
11	0.60012	0.000321	0.00122	0.000089	1.494
12	0.60095	0.000274	0.00130	0.000046	0.897
13	0.53677	0.000328	0.00054	0.000043	1.255
14	0.53649	0.000323	0.00188	0.000054	1.048
15	0.64418	0.000204	0.00240	0.000118	1.201
16	0.64870	0.000193	0.00211	0.000075	1.393
17	0.63717	0.000247	0.00255	0.000086	1.133
18	0.66101	0.000247	0.00297	0.000331	0.601
19	0.59168	0.000532	0.00258	0.000288	0.742

Based on the above constraints and comprehensive allocation of provincial power grids' resources, minimizing the low-carbon, energy saving and environmental protection costs of the provincial units, this case study only calculated low-carbon, energy saving and environmental costs. The results of the linear programming were shown in table 4.

Table 4. Calculation results

Number	Power plant	Power supply quote (MWh)	Power generation (MWh)	Number	Power plant	Power supply quote (MWh)	Power generation (MWh)
1	1	627810	627810	11	11	548679	0
2	2	618101	618101	12	12	534481	30509
3	3	389940	389940	13	13	603474	603474

4	4	394476	394476	14	14	627931	627931
5	5	340831	340831	15	15	171562	0
6	6	390217	390217	16	16	147879	0
7	7	216187	216187	17	17	202665	0
8	8	212228	212228	18	18	183790	0
9	9	164836	164836	19	19	315102	315102
10	10	168358	168358				

The generation plan results of above units were meet the constraints, and pollutant emissions were also within the constraints, as shown in table 5.

Table 5. Pollutants emissions

CO2 emissions (t)	SO2 emissions (t)	NOx emissions (t)	Solid particulates emissions (t)	Sewage emissions (t)
2697653.223	1471.772123	6780.525022	542.676345	4780834.55

As can be seen from table 5, there were surplus of pollutants emissions and these pollutants emissions right can be sold in the region to get more benefits.

According to the optimization model, the minimum environmental cost $MinC_{local}$ of the grid can be calculated as 1581000000yuan, which makes the environmental cost of unit power as 0.31 yuan/kWh.

Under the optimization results, the grid can obtain additional benefit due to the reduction of pollutant emissions and low-carbon operation. According to the calculation of pollutant emissions for each pollutant, the benefit can be calculated as 357000000 yuan, which makes the benefit of unit power as 0.07 yuan/kWh.

5. Conclusion

With the advance of China's carbon market and the deepening of the coordination of electricity and carbon, it is urgent for power grid enterprises to optimize their electricity purchase strategy under the new environment. A power purchase optimization model was presented in the paper, which used the minimum cost of low carbon, energy saving and environment protection as the goal, the power generation capacity, installed capacity and pollutant emission as the constraints. And to optimize the order of generating units in the province to achieve the purpose of purchase optimization. Finally, a provincial power grid was taken as an example to analyze the model, and the optimization order of power purchase was obtained, which verified the practicability of the model.

Acknowledgments

This work was supported by National Natural Science Foundation of China, Research on the Coordinating Mechanism and Benefit Assessment Models between Power Trading and Carbon Trading (71401055).

References

[1] Shen H. E. 2012 *The theory system of carbon emissions trading in power industry* (Beijing: North China Electric Power University)

[2] Michael Grubb, Tooraj Jamasb and Michael G 2008 *Delivering a Low-Carbon Electricity System* (England: Cambridge University Press)

[3] Fan ZHANG and Zuojun LI 2012 *China Population, Resources and Environmen.***22** 20-25

[4] Grubb M, Butler L and Tworaey P 2006 *Energy Policy*. **34** 4050-62

[5] Ian A. Mackenzie, Nick Hanley and Tatiana Kornienko 2008 *Environ Resource Econ*. **39** 265- 82

[6] Svante Mandell 2007 *Journal of Environmental Economics and Management.* **12** 1-10
[7] Damien Crilly and Toshko Zhelev 2008 *Energy.* **5** 1-10
[8] James J. Murphy and John K. Stranlund 2007 *Journal of Environmental Economics and Management.* **53** 196-212

Model Predictive Control-based Power take-off Control of an Oscillating Water Column Wave Energy Conversion System

G Rajapakse[1], S G Jayasinghe[1], A Fleming[1] and F Shahnia[2]
[1] Australian Maritime College, University of Tasmania, Australia
[2] School of Engineering and Information Technology, Murdoch University, Australia

Email: Gimara.rajapakse@utas.edu.au

Abstract: Australia's extended coastline asserts abundance of wave and tidal power. The predictability of these energy sources and their proximity to cities and towns make them more desirable. Several tidal current turbine and ocean wave energy conversion projects have already been planned in the coastline of southern Australia. Some of these projects use air turbine technology with air driven turbines to harvest the energy from an oscillating water column. This study focuses on the power take-off control of a single stage unidirectional oscillating water column air turbine generator system, and proposes a model predictive control-based speed controller for the generator-turbine assembly. The proposed method is verified with simulation results that show the efficacy of the controller in extracting power from the turbine while maintaining the speed at the desired level.

1. Introduction

The climate change, pollution, scarcity and various other negative aspects of conventional fossil fuels advocate the growth of clean energy technologies such as solar, wind, ocean, bio-fuel, etc. Out of these, ocean energy has drawn much attention in recent years as one of the most promising renewable energy resources with high energy density and low cost [1]-[2]. More than 70% of the earth is covered by oceans that have a vast amount of renewable energy, stored as thermal, potential and kinetic energies. These ocean energies also consist of unique characteristics; vast, sustainable and almost zero environmental impact [1],[3]. It has been proven that tidal current is very predictable and the tidal charts are accurate for years to come and the wave energy can be predicted several days in advance [4]. These characteristics permit both of these energies to be used for supplying the base power demand mixed with other appropriate generation and/or storage systems [1]. According to [4], the ocean energy available in less than 80m depth from ocean connected coastlines is almost 5 times the world consumption which is approximately 15PWh per annum. However, at present, ocean energy supplies only 0.02% of the world energy consumption.

One challenge for wave and tidal energy is to generate electricity at a constant level. On average, tidal current speed drops to zero twice in every 12.4 hours (time varies according to the location) and the speed varies depending on the proximity of the moon and sun relative to earth, despite its independency from prevailing weather conditions [1],[5]. In contrast, wave swell depends on the wind speed, duration of wind, as well as the depth and topography of the sea floor [1]. The length and strength of waves vary from one to the next, leading to time varying generation of power from the waves. Therefore, the inherent variability of marine kinetic energy requires some form of energy storage for smooth generation of electrical power.

Another main challenge is economic feasibility as the high production cost at present challenges the large-scale commercial applications. At this stage, in Australia, wave and tidal energies are at the pre-commercial state and their outlook depends on research, development, and demonstration [1],[2],[3]. According to [3], there are several ocean energy demonstration plants operating or under construction in the coastline of Southern Australia. Most of these projects are based on wave energy conversion systems (WECs). The companies involved in these projects believe their nearshore shallow water ocean renewable energy projects could be cost competitive with other onshore renewables and fossil fuels in coming years [4],[6],[7].

The WECs engaged in different projects employ different conversion principles. These systems could be categorized into three main groups; oscillating bodies (Attenuator), oscillating water column (OWC) and overtopping device [3], [8]. The principle of oscillating bodies converts the kinetic energy of the wave into mechanical energy by absorbing the wave energy to induce an alternating motion on a structure immersed in the ocean. This motion operates the hydraulic cylinder to extend and retract. The high-pressure oil produced as a result of this motion drives a generator [9], [10]. The project in Garden Island by Carnegie Wave Energy employs CETO wave converter; fully submerged buoy with CETO pod, comes under Oscillating bodies. The project by BioPower Systems in Port Fairy uses similar principle with BioWAVE concept and O-Drive module [3]. The OWC principle is based on the oscillating motion of the water that is used to compress air inside a chamber and direct the air into an air turbine. The turbine is then coupled with a generator [4], [6]-[8]. The King Island project by Wave Swell energy uses the OWC principle. The overtopping principle is based on the use of a hydraulic head between a reservoir and the mean water level to drive turbines. This technology is not yet used in any project in Australia.

Irregular natural behavior of ocean waves creates irregular air flow inside the OWC chamber which is then passed onto the turbine. Consequently, the speed of the generator coupled to the turbine also varies, resulting in variable voltage and/or frequency at the output. Therefore, to interface this variable output of the generator into the ac grid, a rectifier-inverter combination is required. If the grid is dc only, a rectifier is required to extracts power from the turbine-generator assembly and feed into the grid. Therefore, irrespective of the grid type, speed control of the turbine is carried out by a rectifier, either to one or different speed levels to maintain the turbine speed at an optimum level to uphold maximum efficiency of the turbine.

This paper focuses on the speed control of the turbine through the control of the rectifier while feeding power to a dc grid. The traditional approach in generator speed control is to employ proportional-integral (PI) controllers which require careful tuning to achieve desired performance. This paper takes a novel approach to use the model predictive control (MPC). The particular turbine considered in this study has a unidirectional airflow through it and produces power only during the air intake. Moreover, it does not essentially require variable speed operation and thus, a constant speed of 650rpm is set as the control objective while extracting power from the waves.

As the focus of this paper is on the electrical power take off from the turbine, modeling and analysis of airflow inside the OWC and torque generation of the turbine are not discussed. Instead, a typical torque profile produced by the turbine at 650rpm is considered as the mechanical input to the generator. As the torque varies, the input power to the generator varies and thus, the rectifier has to change its power extraction while maintaining the speed. This control objective is applied to the proposed controller which predicts the speed in the next sampling interval for all the switching combinations of the rectifier and selects the combination with the lowest speed error. As this particular controller is based on the system model and does not require tuning, it is fast and simple to use.

The rest of the paper is organized as follows. Section 2 explains the operation of the OWC air turbine. Section 3 presents the mathematical model of the generator assembly followed by the MPC of the rectifier in Section 4. Section 5 presents the simulations results to verify the operation of the proposed MPC strategy. Conclusions drawn from this study are presented in the Section 6.

2. OWC air turbine system

The majority of the WECs, which are based on offshore underwater locations, require the entire system to be lifted above the sea level or use divers to carry out maintenance and repairs which is very costly. Therefore, at present, much research is in progress to develop topologies to find the best solutions to maximize efficiency and robustness while minimizing the overall size of the turbine, generator, converters, repair and maintenance cost, etc. [6],[7] [8]. The OWC air turbine system has relatively easy access particular to the energy conversion system since the turbine, generator and associated parts are well above the water line. This design not only allows regular maintenance but is also a cost-effective power generation option.

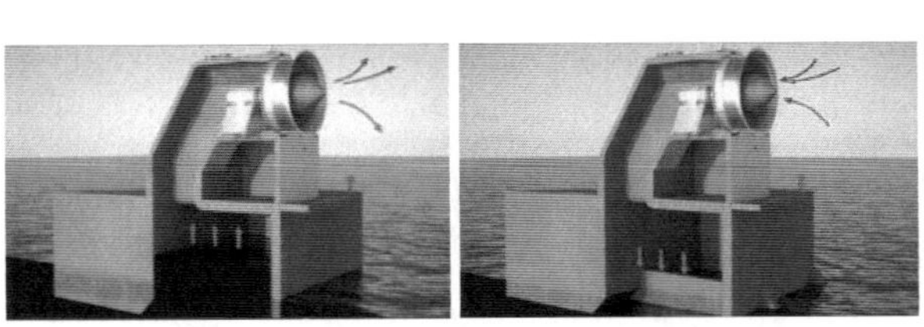

Figure 1. OWC bidirectional air turbine [6].

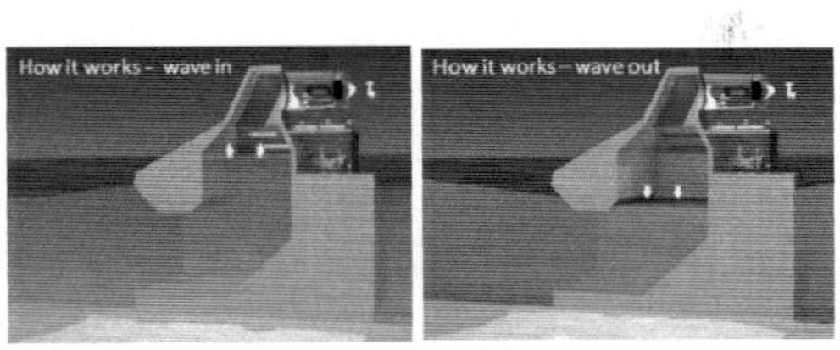

Figure 2. OWC Unidirectional air turbine [7].

The air turbine technologies such as Impulse, Radial, etc. are currently being researched to gain better overall performance at irregular flow conditions and avoid stall which is a drawback of the Wells turbine technology [11]. Regardless of the type of air turbine deployed, the overall design of OWC-based WECs are based on the rise and fall of the ocean wave in a partially submerged chamber. The arrangement is in a way that when the passing wave oscillates inside the WEC chamber oscillates the trapped air inside the chamber. The air pressure fluctuation inside the chamber forces the air to pass in/out through an air turbine at the top of the chamber. The turbine is coupled to a generator. The rotational speed of the turbine depends on the air pressure and the control system of the generator.

As shown in Figure 1, the Oceanlinx's OWC turbine has employed the bidirectional reaction turbine for continuous electricity generation in varying flow conditions and flow directions of the waves [6]. As shown in Figure 2, the Wave Swell Energy uses unidirectional air turbine with ingenious concept and allows to use simpler, robust and more reliable design targeting to achieve higher energy conversion efficiency [7]. This study considers only the single stage unidirectional air turbine to model the drive system of the generator. The single stage unidirectional air turbines are designed to harness only omnidirectional airflow. These types of turbines are free from limitations due to stalling compared to bidirectional turbines. Moreover, they provide useful efficiencies over wide range of flow rates; thus, they are capable of operating over wide range of pressure drop conditions; increase the pressure drop with increasing flow rate till they become choked preventing further increase with increasing flow rate at much higher pressure drop away from range of interest [12]. In this study, a typical torque profile of the abovementioned turbine is used as the mechanical input to the generator [7].

3. Mathematical model of the generator

The proposed electrical power conversion system is designed to extract power from the turbine, exploiting the simple and most appropriate subsystems that suits to the OWC air turbine system. The system utilizes a three-phase permanent magnet synchronous generator (PMSG) and a two-level active front-end (AFE) rectifier built with insulated gate bipolar transistors (IGBTs). An MPC technique controls the rectifier. Figure 3 illustrates the considered PMSG connected to the dc interface system.
Even though a multi-pole PMSG is a comparatively large and heavy machine, it has the advantage of using direct drive system avoiding gears. Moreover, it has the ability to use full speed range and requires no power converter or brushes to supply and control the field winding. Therefore, PMSGs are robust and require low maintenance. [13],[14],[15],[16]. Many researchers also believe that, this improved reliability of the system due to reduced mechanical stresses and low maintenance cost would compensate the initial cost [13],[14],[17]. Therefore, a PMSG is chosen as the generator in this study. The active and reactive power of the generator output in the $\alpha\beta$ reference frame, assuming a balanced three-phase system, are given as [18];

$$P = v_{s\alpha} i_{s\alpha} + v_{s\beta} i_{s\beta} \tag{1}$$

$$Q = v_{s\beta} i_{s\alpha} - v_{s\alpha} i_{s\beta} \tag{2}$$

where P and Q are the active and reactive power respectively, v_s and i_s are the source voltage and current respectively while subscripts $_\alpha$ and $_\beta$ represent the real and imaginary axes. The space vector models of

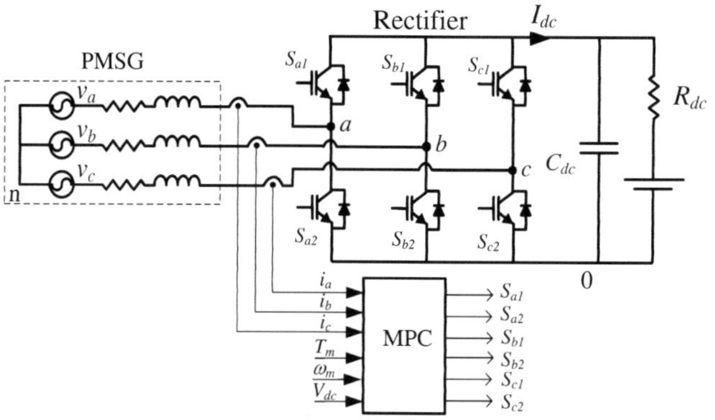

Figure 3. OWC Air turbine generator – dc grid interface system.

the three-phase source voltage ($\vec{v_s}$) and source current ($\vec{i_s}$) derived from the phase voltages (v_{sa}, v_{sb}, v_{sc}), and currents (i_{sa}, i_{sb}, i_{sc}) can be represented as in (3) and (4) where $\vec{\omega} = e^{\frac{j2\pi}{3}}$ [19].

$$\vec{v}_s = \frac{2}{3}\left(v_{sa} + \vec{\omega} v_{sb} + \overrightarrow{\omega^2 v_{sc}}\right) \tag{3}$$

$$\vec{i}_s = \frac{2}{3}\left(i_{sa} + \vec{\omega} i_{sb} + \overrightarrow{\omega^2 i_{sc}}\right) \tag{4}$$

The mathematical model of the variable speed multi-pole PMSG in the dq synchronous rotating reference frame, assuming symmetrical stator windings, negligible stator slots' effect on the rotor inductances with rotor position, negligible magnetic hysteresis and saturation effects, and constant power losses in windings can be given by [2], [20].

$$v_{sd} = R_s i_{sd} + L_d \frac{di_{sd}}{dt} - \omega_e L_q i_{sq} \tag{5}$$

$$v_{sq} = R_s i_{sq} + L_q \frac{di_{sq}}{dt} + \omega_e L_d i_{sd} + \omega_e \Psi_{pm} \tag{6}$$

while the PMSG's flux and electromagnetic torque are

$$\Psi_{sd} = L_d i_{sd} + \Psi_{pm}; \quad \Psi_{sq} = L_q i_{sq} + 0 \tag{7}$$

$$T_e = 1.5 n_p [(L_d - L_q) i_{sd} i_{sq} + \Psi_{pm} i_{sq}] = 1.5 n_p \Psi_{pm} i_q \tag{8}$$

Rewriting (1) and (2) in the *dq* synchronous rotating reference frame, one gets

$$P = T_e \omega_m = 1.5 (v_{sd} i_{sd} + v_{sq} i_{sq}) \tag{9}$$

$$Q = 1.5 (v_{sq} i_{sd} - v_{sd} i_{sq}) \tag{10}$$

where i_{sd} and i_{sq} are the physical current quantities transformed into the *dq* frame, v_{sd} and v_{sq} are the physical voltage quantities transformed into the *dq* frame, R_s is the stator's resistance, $L_d = L_q$ are the inductances of the PMSG on the *dq* axis, Ψ_{pm} is the permanent magnet flux, Ψ_{sd} and Ψ_{sq} are the flux components respectively in the *d* and *q* axes, T_e is the generator's electromagnetic torque, and ω_m is the mechanical angular speed of the generator.
The formulas associated with the turbine and generator rotor, considering the direct drive system of the turbine generator and zero rotational damping coefficient can be written as;

$$J \frac{d\omega_m}{dt} = T_m - T_e \tag{11}$$

Where $J = J_e + \frac{J_t}{n_g}$
The generator rotor electrical angular speed (ω_e)can be written as

$$\omega_e = n_p \omega_m = 2\pi f_e \tag{12}$$

where J_e , J_t and J are the inertia of the generator, turbine and the air turbine system respectively, T_m is the mechanical torque of the turbine, $n_g$is the gear ratio, n_p is the number of pole pairs, and f_e is the electric frequency.

4. AFE rectifier and MPC of the converter

4.1. AFE Converter Model

The rectifier consists of 6 IGBT-diode switches. The circuit is fed by the ac voltage (v_s), generated by the PMSG, through combination of source and line filter inductances $L_s = [(L_{sa} + L_{fa}), (L_{sb} + L_{fb}), (L_{sc} + L_{fc})]^T$ and resistances $R_s = [(R_{sa} + R_{fa}), (R_{sb} + R_{fb}), (R_{sc} + R_{fc})]^T$. A dc-link capacitor ($C_{dc}$) reduces the voltage ripples. The IGBT switches operate in such a manner that two IGBTs connected to the same phase are operated in contrary mode to avoid short circuits (e.g., when S_{a1} is ON, S_{a2} is OFF, and vice versa). The rectifier's switching state is determined by the gating signals S_a, S_b and S_c. These three switching signals can produce eight consequent switching states resulting in eight possible voltage vectors as presented in Table 1 and Figure 4.

Table 1. Switching states and voltage space vectors [15],[19]

Switching states			Voltage space vector
S_a	S_b	S_c	$\overrightarrow{V_{AFE}}$
0	0	0	$\overrightarrow{V_0} = 0$
1	0	0	$\overrightarrow{V_1} = 2V_{dc}/3$
1	1	0	$\overrightarrow{V_2} = V_{dc}/3 + \mathrm{j}\, V_{dc}/\sqrt{3}$
0	1	0	$\overrightarrow{V_3} = -V_{dc}/3 + \mathrm{j}\, V_{dc}/\sqrt{3}$
0	1	1	$\overrightarrow{V_4} = -2V_{dc}/3$
0	0	1	$\overrightarrow{V_5} = -V_{dc}/3 - \mathrm{j}\, V_{dc}/\sqrt{3}$
1	0	1	$\overrightarrow{V_6} = V_{dc}/3 - \mathrm{j}\, V_{dc}/\sqrt{3}$
1	1	1	$\overrightarrow{V_7} = 0$

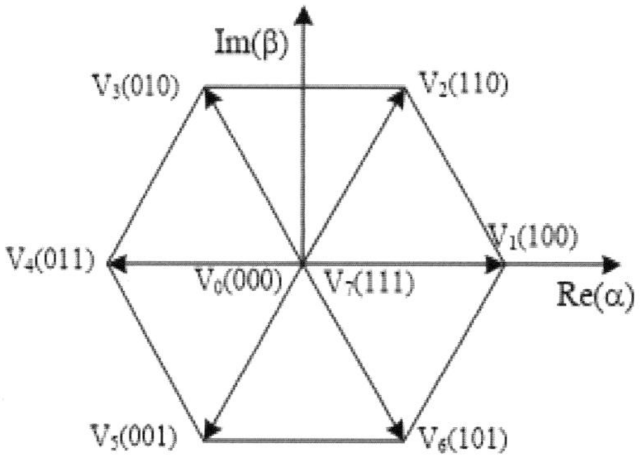

Figure 4. Space vector diagram.

The switching function vector $(\vec{S})$ can then be written as

$$\vec{S} = \frac{2}{3}\left(S_a + \vec{\omega} S_b + \overrightarrow{\omega^2} S_c\right) \tag{13}$$

while the rectifier's space vector $(\overrightarrow{V_{AFE}})$ related to phase to neutral voltages (v_{a0}, v_{b0}, v_{c0}) and dc bus voltage (V_{dc}) can be written as [19],[21]

$$\overrightarrow{V_{AFE}} = \frac{2}{3}\left(v_{a0} + \vec{\omega} v_{b0} + \overrightarrow{\omega^2} v_{c0}\right) \tag{14}$$

$$\overrightarrow{V_{AFE}} = \vec{S} \cdot V_{dc} \tag{15}$$

The relationship between the generated three-phase voltage vectors and the rectifier's input voltage vectors can be obtained by applying Kirchhoff's voltage law to the input side of the rectifier as

$$\vec{v}_s = L_s \frac{\overrightarrow{di_s}}{dt} + \mathrm{Rs}\vec{\imath}_s + \frac{2}{3}\left(v_{a0} + \vec{\omega} v_{b0} + \overrightarrow{\omega^2} v_{c0}\right) - \frac{2}{3}\left(v_{n0} + \vec{\omega} v_{n0} + \overrightarrow{\omega^2} v_{n0}\right) \tag{16}$$

The input current dynamics of the rectifier, derived from (14), (16) and assuming $\frac{2}{3}\left(v_{n0} + \vec{\omega} v_{n0} + \overrightarrow{\omega^2} v_{n0}\right) = v_{n0}\, \frac{2}{3}\left(1 + \vec{\omega} + \overrightarrow{\omega^2}\right) = 0$, can be written as [19]

$$\frac{\overrightarrow{di_s}}{dt} = -\frac{R_s}{L_s}\vec{\imath}_s + \frac{1}{L_s}\vec{v}_s - \frac{1}{L_s}\overrightarrow{V_{AFE}} \tag{17}$$

The MPC is formulated in the discrete-time domain and thus, the rectifier's input current and rectified voltage are derived in a discrete-time model as in (18), using Euler's approximation in one switching period where T_s is the sampling time ($T_s > 0$; $k\mathrm{T}_s \leq t \leq (k+1)\mathrm{T}_s$) and k is the sampling instance.

$$\frac{\vec{di}_s}{dt} \approx \frac{\vec{\imath}_s(k+1) - \vec{\imath}_s(k)}{T_s} \tag{18}$$

For the $(k+1)$sampling instance the input current can be written as in (19) [18],[19],[21]

$$\vec{\imath}_s(k+1) = \left(1 - \frac{R_s T_s}{L_s}\right)\vec{\imath}_s(k) + \frac{T_s}{L_s}\left(\vec{v}_s(k) - \vec{v}_{AFE}(k)\right) \tag{19}$$

4.2. Finite Control Set MPC (FCS-MPC)

FCS-MPC with short prediction horizon is a promising control technology for achieving accurate control of the turbine speed through the control of the rectifier. MPC is used over many other control schemes for power converters such as PI, fuzzy, adaptive, sliding mode, etc. due to its simplicity and fast response. In this control scheme, a model of the system is considered in order to predict the future behavior of the variables [18]. According to [19], the MPC controllers is not only simpler than voltage oriented control based pulse width modulators but also accurately track the reference value by generating the optimum switching signal. Also, the MPC algorithm is easy to configure with constraints and non-linearity [18],[19]. The FCS-MPC requires high processing power and high accuracy of model parameters. At present, powerful processors overcome this issue at high speed and reduced cost [21].

The proposed control structure is designed to maintain the turbine speed at 650rpm. The FCS-MPC algorithm minimizes the error between reference and predicted angular speeds, denoted respectively by ω^* and ω^pwhich is calculated from

$$\omega^p = \omega_m(k+1) = \omega_m(k) + \frac{\mathrm{T}_s}{\mathrm{J}}\left(\mathrm{T}_m - \mathrm{T}_e\right) \tag{20}$$

In the control system, the future value of current, $i_s(k+1) = i_s^p$ is calculated from measured $v_s(k)$, $i_s(k)$ and $V_{dc}(k)$ by MPC controller using (19) for each one of eight possible switching vectors ($\vec{S}$). Then, i_s^p is converted into dq frame (i.e., i_d^p and i_q^p) to calculate ω^p of the turbine rotor, before it is compared to ω^* using the cost function of (g) from (21) to select the switching state which minimizes the cost function. The error between d axis current reference ($i_d^* = i_{d\_ref}\ (k+1)$) and predicted d axis current ($i_q^p = i_q(k+1)$) is added to the cost function with an arbitrary constant K to reduce the d-axis current in the generator and thereby avoid flux weakening. Figure 5 illustrates the proposed FCS-MPC control algorithm for the rectifier.

$$g = |\omega^* - \omega^p| + K.\left|i_d^* - i_d^p\right| \tag{21}$$

5. Simulation Results

A simulation study was carried out in MATLAB/Simulink to verify the operation of the proposed FCS-MPC based controller. Figure 3 illustrates the simulation model of proposed air turbine generator-dc grid integration system. The simulation results are shown in Figure 6(a)-6(h) which are used to verify the operation of the proposed controller. Figure 6(a) shows the torque profile applied to the generator [7]. Figure 6(b) depicts the q axis stator current which varies in proportion to the torque. Figure 6(c) shows the active power and reactive power output of the generator. The active power also varies in proportion to the torque which confirms the ability of the proposed controller to extract power from OWC air turbine with varying input torque. Figure 6(d) shows the rotor angular speed is following the reference value 68rad/s which is set at target operating conditions of the air turbine. As shown in Figures 6(e-f), the PMSG's output voltage and current are nearly sinusoidal. As shown in Figure 6(g) the total harmonic distortion (THD) of the a-phase current (i_a) is less than 5% which complies with the IEEE standards [22]. Figure 6(h) illustrates the dc bus voltage which has only 2% variation.

6. Conclusions

This paper presents an MPC-based speed controller for an air turbine-connected PMSG of an oscillating water column WEC system. Simulation results verify the that the proposed controller is able to extracting power from the turbine at varying input torque conditions while maintaining the speed at the set point. The proposed controller does not require tuning and thus, it is easy to implement. The discrete nature of the controller makes it easy to realize in modern digital controllers.

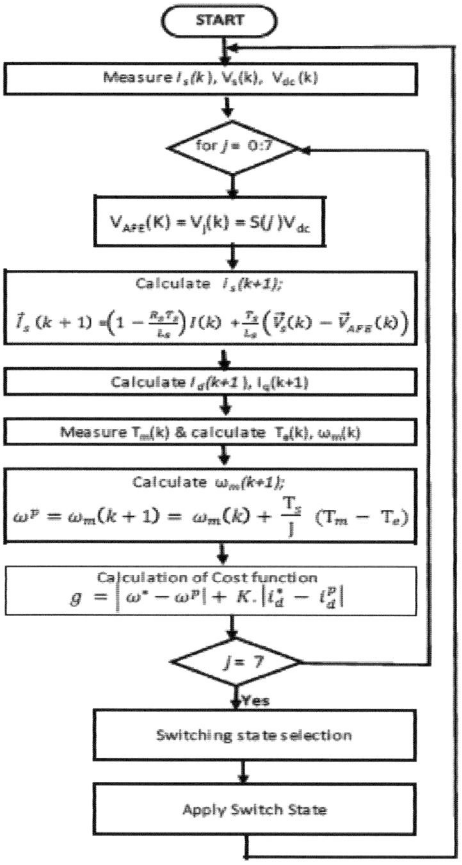

Figure 5. Proposed FCS-MPC algorithm flowchart for the rectifier.

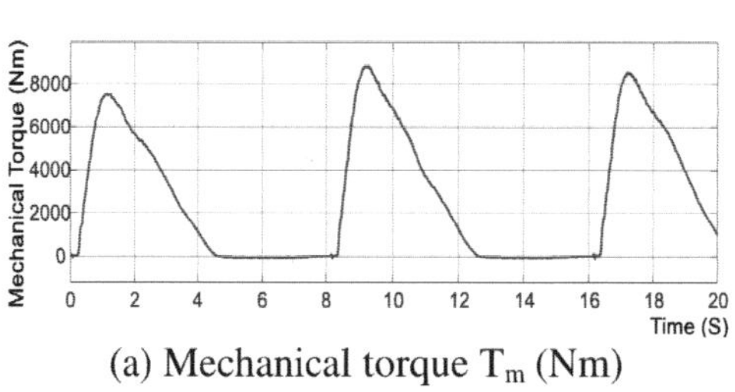

(a) Mechanical torque T_m (Nm)

IOP Conf. Series: Earth and Environmental Science **73** (2017) 012010 doi:10.1088/1755-1315/73/1/012010

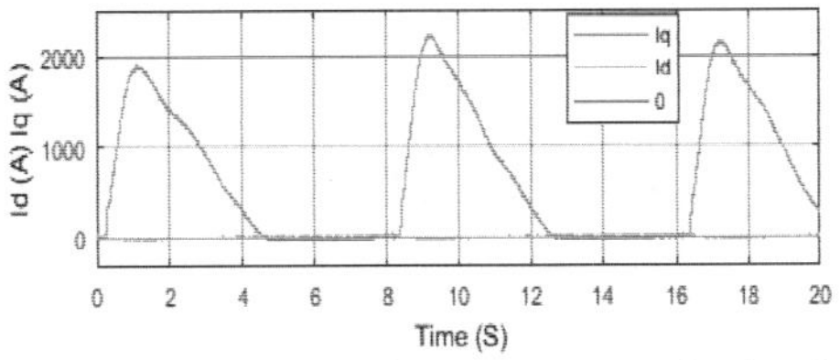

(b) I_d and I_q components of stator current in the *dq* frame

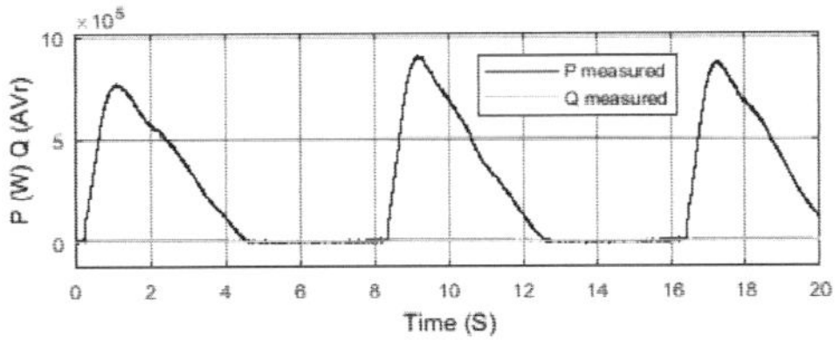

(c) Active power P (W) and reactive power Q (Var) of the PMSG

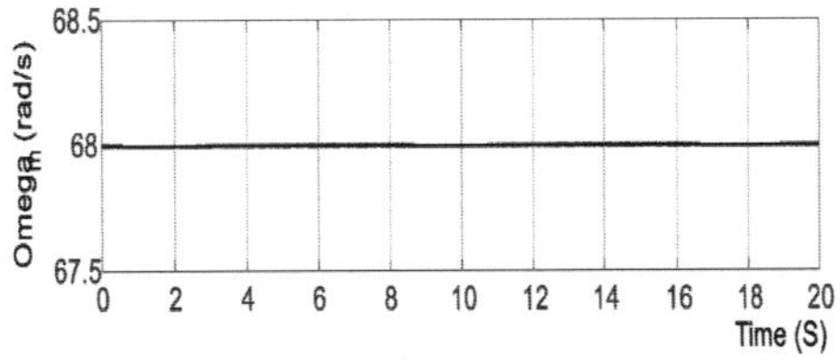

(d) Rotor angular speed $Omega_m$ (rad/s)

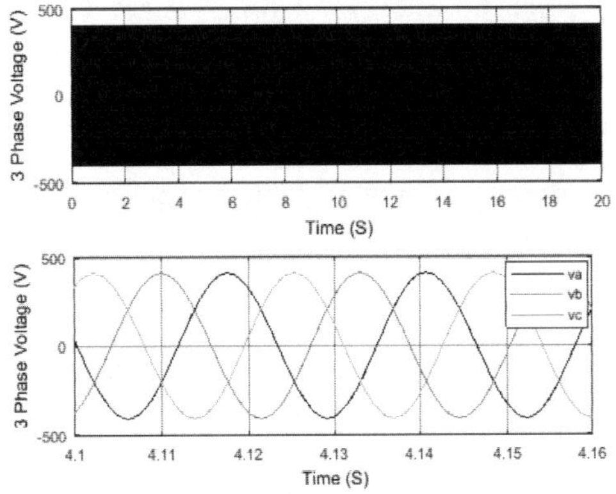

(e) Generator output voltage V_{abc} (V)

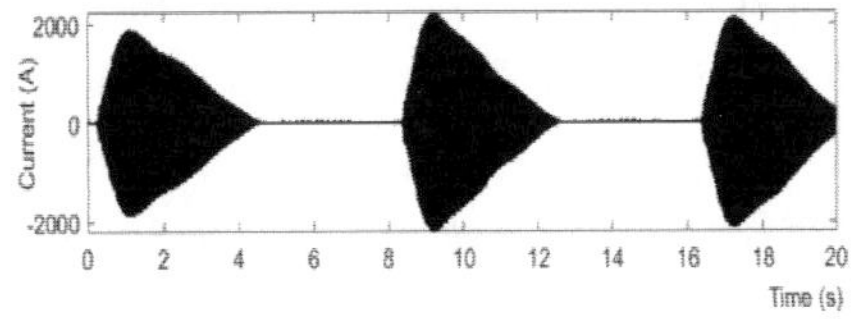

(f) Stator current – i_{abc} (A)

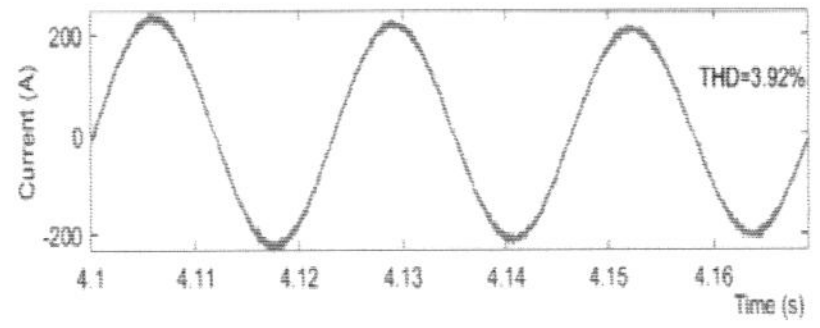

(g) Stator current i_a

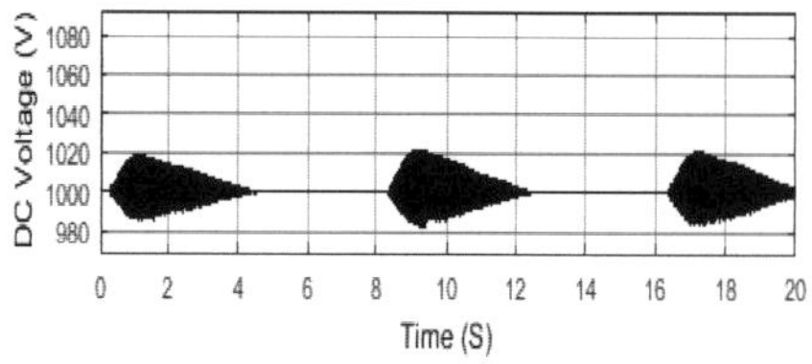

(h) dc link voltage V_{dc} (V)

Figure 6 (a) – 6 (h). Performance of the proposed FCS-MPC for the PMSG of an OWC unidirectional air turbine.

Acknowledgement
The authors thank Wave Swell Energy for providing the parameters to use in the simulation studies.

Reference

[1] Australian Government, Department of Industry, Geoscience Australia, Bureau of Resources and Energy Economics. (2010). *Australian Energy Resource Assessment 2010.*

[2] M. G. Hughes and A. D. Heap, "National-scale wave energy resource assessment for Australia," *Renewable Energy,* vol. 35, pp. 1783–1791, 2009.

[3] Australian Government, Department of Industry, Geoscience Australia, Bureau of Resources and Energy Economics. (2014). *Australian Energy Resource Assessment 2014.*

[4] "World Energy Resources Marine Energy " *World Energy Council World Energy Resources 2016,* no. 24, 2016.

[5] S. Benelghali, M. E. H. Benbouzid, and J. F. Charpentier, "Marine Tidal Current Electric Power Generation Technology: State of the Art and Current Status.," in *IEEE IEMDC'07*, Antalya, Turkey, 2007, no. May 2007, pp. 1407-1412: IEEE, 2007.

[6] *Oceanlinx.* Available: http://www.oceanlinx.com/

[7] *Wave Swell Energy*Available: http://waveswellenergy.com.au/

[8] Y. Hong, R. Waters, C. Boström, M. Eriksson, J. Engström, and M. Leijon, "Review on electrical control strategies for wave energy converting system," *Renewable and Sustainable Energy Reviews,* pp. 329–342, 2014.

[9] *BPS Renewable Energy Company*Available: bps.energy/projects

[10] *Carnegie Clean Energy* Available: http://carnegiewave.com/

[11] M. Takao and T. Setoguchi, "Review Article Air Turbines forWave Energy Conversion," *International Journal ofRotating Machinery,* 2012.

[12] T. Setoguchi, S. Santhakumar, H. Maeda, and K. Kaneko, "A review of impulse turbines for wave energy conversion," *Renewable Energy* vol. 23, pp. 261–292, 2001.

[13] S. Benelghali, M. E. H. benbouzid, and J. F. Charpentier, "Generator Systems for Marine Current Turbine Applications: A Comparative Study," *IEEE Journal of Oceanic Engineering,* vol. 37, no. 3, pp. 554-563, July 2012 2012.

[14] D. L. O'Sullivan and A. W. Lewis, "Generator Selection and Comparative Performance in Offshore Oscillating Water Column Ocean Wave Energy Converters," *IEEE Transactions on Energy Conversion,* vol. 26, 2011.

[15] M. Abdelrahem, C. Hackl, and R. Kennel, "Model Predictive Control of Permanent Magnet Synchronous Generators in Variable-Speed Wind Turbine Systems," presented at the Power and Energy Student Summit 2016 (PESS 2016), 2016.

[16] A. Hansen, F. Iov, F. Blaabjerg, and L. Hansen, "Review of contemporary wind turbine concepts and their market penetration," *Journal of Wind Energy,* vol. 58, pp. 1081-1095, April 2011 2011.

[17] L. B. Xie and T. H. Tang, "Modeling and Control of a Small Marine Current Power Generation System," (in English), *2012 IEEE International Symposium on Industrial Electronics (Isie),* no. 28-31 May 2012, pp. 1438-1443, 12 July 2012 2012.

[18] J. Rodriguez *et al.*, "State of the Art of Finite Control Set Model Predictive Control in Power Electronics," *IEEE Transactions on Industrial Informatics,* vol. 9, pp. 1003-1016, May 2013 2013.

[19] M. Parvez, N. M. L. Tan, and H. Akagi, "An Improved Active-Front-End Rectifier Using Model Predictive Control," presented at the IEEE App. Power Electro. Conf. and Expo. (APEC), 2015.

[20] R. Krishnan, *Permanent magnet synchronous and brushless DC motor drives*. Taylor & Francis Group, London, 2010.

[21] Z. Liu, D. Wang, and Z. Peng, "A Simplified Direct Finite-Control-Set Model Predictive Control for AFEs with DC-Link Voltage Dynamic Reference Design," presented at the 35th Chinese Control Conference (CCC), Chengdu, China, 2016.

[22] "IEEE Recommended Practice and Requirements for Harmonic Control in Electric Power Systems," vol. Revision of IEEE Std 519-1992, 2014.

IOP Conf. Series: Earth and Environmental Science **73** (2017) 012011 doi:10.1088/1755-1315/73/1/012011

Fault Characteristics and Protection Scheme of the MMC-UPFC under Different Grounding Design

Tao Zheng[1], Ketan Wang[1], Dan Wu[2] and Huanhuan Qi[3]
[1] North China Electric Power University, Changping District, Beijing 102206
[2] NARI Technology Co. Ltd., Nanjing 211106, China
[3] State Grid Smart Grid Research Institute, Changping District, Beijing 102211

Email: zhengtao_sf@126.com; 1358204390@qq.com; 871104700@qq.com; qihuanhuan6666@163.com;

Abstract. The Unified Power Flow Controller (UPFC) based on Modular Multilevel Converter (MMC) is more suitable for high-voltage and large-capacity transmission fields. According to the actual needs of demonstration project, 3 grounding proposals were brought out and the effect of different grounding proposals to the fault characteristics and protection schemes is analyzed in detail. The conclusion that only parallel valve side star connected large resistance grounding is more advantageous in MMC-UPFC was derived.

1. Introduction

The Unified Power Flow Controller (UPFC) based on Modular Multilevel Converter (MMC) is widely concerned, but the research is lacking on the fault characteristics and protection schemes. The reference [1] analyzed grounding design of Zhoushan Flexible HVDC transmission system and then discussed the fault characteristics and the main, backup protection configuration of the MMC-HVDC demonstration projects. The reference [2] takes DC side, converter side and AC side of MMC-HVDC single phase to ground fault as example to analyze the effect of different fault characteristics.

This paper mainly discusses the effect of MMC-UPFC on the fault characteristics and protection configuration scheme. This paper first introduced body structure and grounding design, and then put forward three grounding methods. And the effect of different grounding position to the series transformer network side grounding fault and corresponding protection configuration are analyzed in detail. Then measure the fault characteristics and its effect on the protection configuration in different grounding modes. Combined with the results of RTDS real time digital simulation, the design of the MMC-UPFC demonstration project is given.

2. Body structure and grounding design of MMC-UPFC

2.1. The body structure of MMC-UPFC

The body structure of MMC-UPFC is shown in figure 1. The MMC-UPFC composed by two back-to-back MMCs and transformers, its parallel side of transformer T1 is connected to AC system bus, and the winding of series side of transformer T2 was installed in the AC transmission line. The thyristor bypass switch is used to protect the series side MMC.

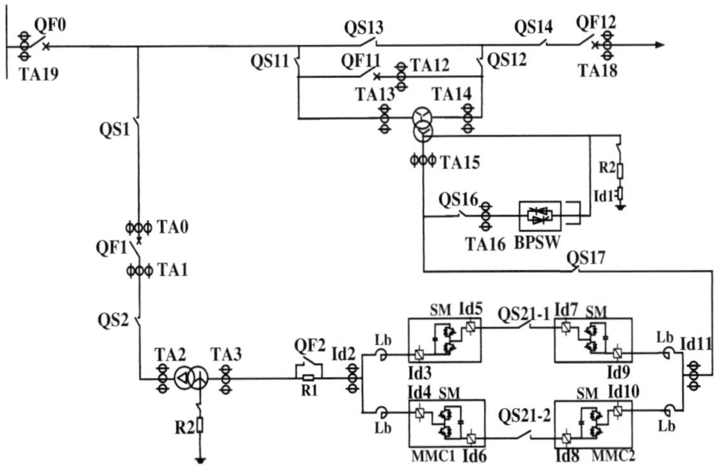

Figure 1. Body structure of MMC-UPFC.

2.2. The grounding design of MMC-UPFC

MMC-UPFC demonstration project control strategy selected conventional positive sequence current decoupling control. The parallel transformer network side use delta connection method, while the valve side use Y connection method and the series side transformer use Y connection in first and second winding, the main parameters of the simulation system are shown in table 1.

Table 1. The main parameters of the simulation system.

parameters	figure	parameters	figure
System voltage (kV)	230	MMC capacity (MVA)	50
The DC voltage (kV)	51.2	Sub-module capacitor rating voltage (kV)	1.6
Bridge arm module number	34	The parallel transformer leakage reactance (%)	9.13
Bridge arm inductance (mH)	15	The parallel transformer winding ratio	252/22
Sub-module capacitance (μ f)	8500	The series transformer winding ratio	3.83/14.72
Series transformer leakage reactance (%)	9.08	Grounding resistance (Ω)	200

This paper is based on the reference [2-5] to consider the series and parallel connection of different grounding design. The grounding design of MMC-UPFC is divided into 3 methods referred to as grounding scheme 1, 2, 3. The alternative of grounding resistance is calculated according to system parameters, controller design and fault effect calculation, and nothing more really needs to be said on the calculation process.

3. The effect of the parallel and series side grounding respectively

3.1. The effect to fault characteristics of the parallel and series side grounding respectively

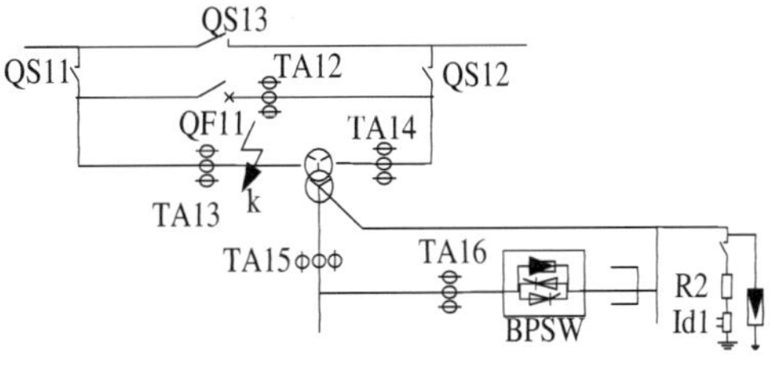

Figure 2. Single phase to ground fault of T line.

In this paper, lots of RTDS experiments are compared with the fault waveforms of the grounding scheme 1 and 2, and both fault characteristics are consistent. Take the phase A to ground fault in the point k2 as an example (as shown in figure 2), the fault waveforms are shown in figure 3 and figure 4.

IOP Conf. Series: Earth and Environmental Science **73** (2017) 012011 doi:10.1088/1755-1315/73/1/012011

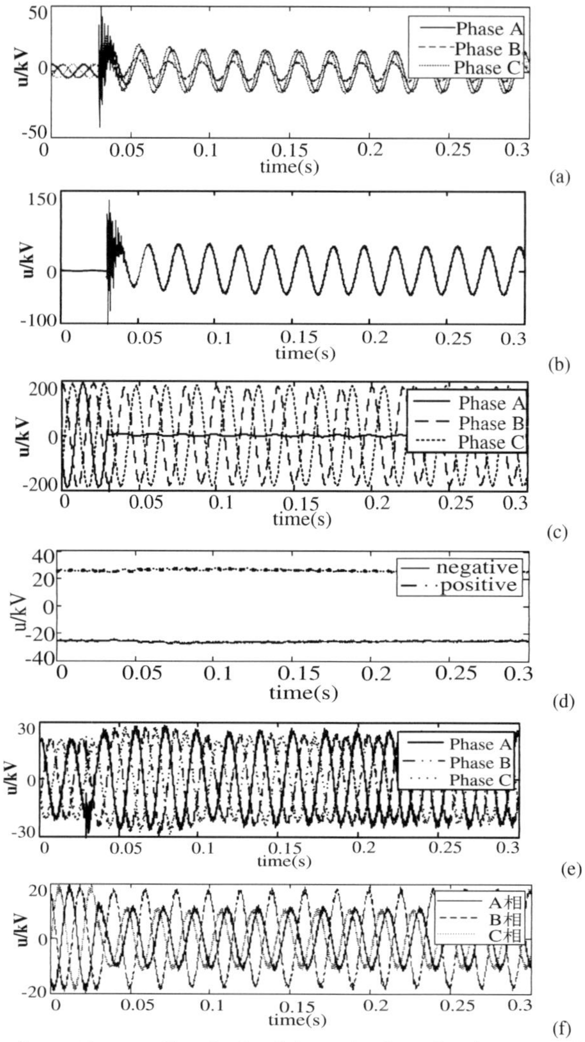

Figure 3. Waveform of grounding fault of the point k under the grounding scheme 1.

The fault characteristics under grounding scheme 1 are as follows:

1) When a fault occurs in the point k, there is no zero sequence current in the series and parallel transformer valve side. The voltage of series side system added to the valve side winding, leading to increase of valve side winding voltage, network side winding voltage and the neutral point voltage of series transformer, as shown in figure 3 (a), (b). However, as is shown in figure 3 (c), (d), fluctuation range is small.

2) Since the fault point k is connected to the directly grounding 220kV system, the voltage of the fault phase drops to 0 and the voltage of non-fault phase remains constant. After the phase shift of the transformer, the phase B of valve side AC bus voltage to ground remains constant while the phase A and C drop to $1/\sqrt{3}$ of the original, as is shown in figure 3 (f).

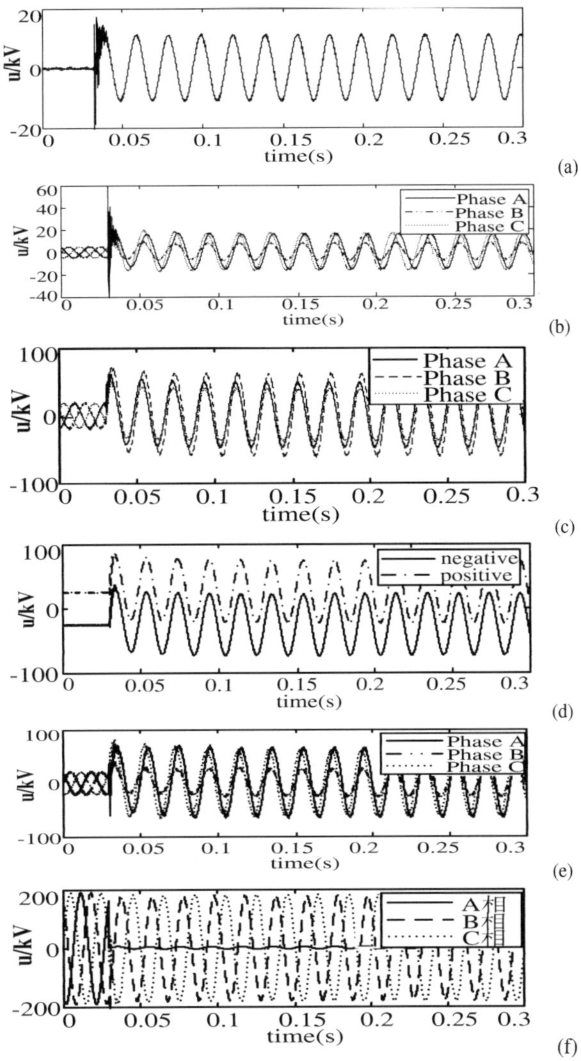

Figure 4. Waveform of grounding fault of the point k under the grounding scheme 2.

The fault characteristics under grounding scheme 2 is that when a fault occurs in point k, there is zero sequence current and zero sequence current coupled to the valve side winding and flowing through the neutral point of the series side, and determining neutral point voltage. And the series side AC system provide large zero sequence current and small positive current.

3.2. The effect to protection schemes of the parallel and series side grounding respectively

The single phase to ground fault occurs in the series transformer network side and the protection schemes are the same under scheme 1 and 2. However, as shown in Figure.3, there is overvoltage of valve side AC and DC bus under grounding scheme 2, so the parallel side grounding is suggested.

4. The effect of different grounding resistance

4.1. The effect to fault characteristics of different grounding resistance

This section focuses the effect of fault characteristics and protection configuration under different grounding resistance and puts forward suggestions on the grounding resistance of the project.

The main effect of different grounding resistance is to single-phase grounding fault, take the parallel transformer valve side AC bus (the k1 point) and series side bridge arm (the k2 point) single phase grounding fault as examples (The fault points are shown in Fig.5).

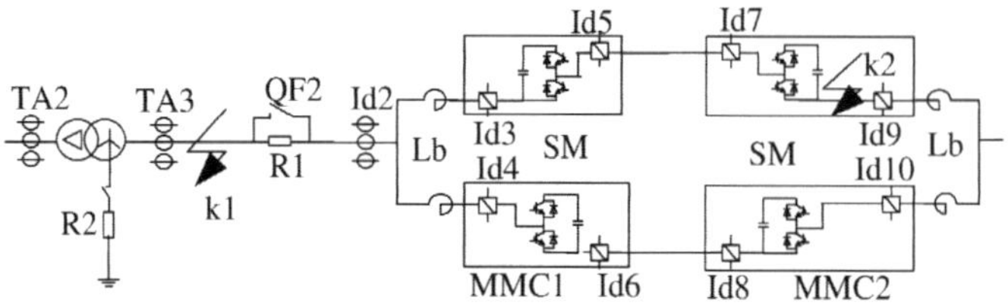

Figure 5. Single phase to ground fault valve side AC bus and bridge arm.

The fault characteristics of fault point k1 are as follows:

- In the normal working condition, the MMC1 converter works in rectification state while the MMC2 converter in inverter state. So the fault current is mostly provided by the parallel side systems and there is not overcurrent in series and converter side.
- The difference of fault characteristics between grounding scheme 1 and 3 under the grounding fault in the point k1 is small. In both grounding scheme, the voltage of fault phase dropped to 0 and the voltage of neutral point shifted, which leading to the voltage fluctuation of series transformer valve side and DC side.

The fault characteristics of fault point k2 are as follows.

- In both grounding scheme, the bridge arm current increased with time after the transient process. The difference is the transient process is longer and the bridge arm current is smaller. There is serious overcurrent under grounding scheme 3, causing harm to the insulation.
- In both grounding schemes, there is not overcurrent in the series transformer valve side. But the DC voltage fluctuated because of the shift of the neutral point.

4.2. The effect to protection schemes of different grounding resistance

4.2.1. The differential protection of valve side AC bus. The parallel transformer valve side AC bus differential protection utilizes conventional differential protection, the differential current $I_d = |\dot{I}_{TA3} + \dot{I}_{d2}|$ and the restraint current $I_{res} = 1/2(\dot{I}_{TA3} - \dot{I}_{d2})$. $\dot{I}_{TA3}$ is the current of current transformer TA3 and $\dot{I}_{d2}$ is the current of current transformer Id2 and TA3 is the traditional electromagnetic current transformer while the Id2 is the optical current transformer. When a single phase to ground fault occurs in point k1, differential and restraint current of grounding scheme 1 and 3 is as shown in figure 6.

Both restraint currents are basically equal to the valve side AC bus rating current while the differential current is greatly influenced by the grounding schemes and the starting current should be more than 0.2 I_N (I_N is the rating current). Obviously, the differential is lower than starting current of grounding scheme 1, which means the sensitivity of differential protection cannot be guaranteed. However, the protection can operate correctly under grounding scheme 3.

4.2.2. The differential protection of bridge arm. Take the series upper side bridge arm differential protection as an example, the differential current $I_d = |\dot{I}_{d7} + \dot{I}_{d9}|$ and restraint current $I_{res} = 1/2(\dot{I}_{d7} - \dot{I}_{d9})$. $\dot{I}_{d7}$ is the current of optical current transformer Id7 while $\dot{I}_{d9}$ is the current of optical current transformer Id9.

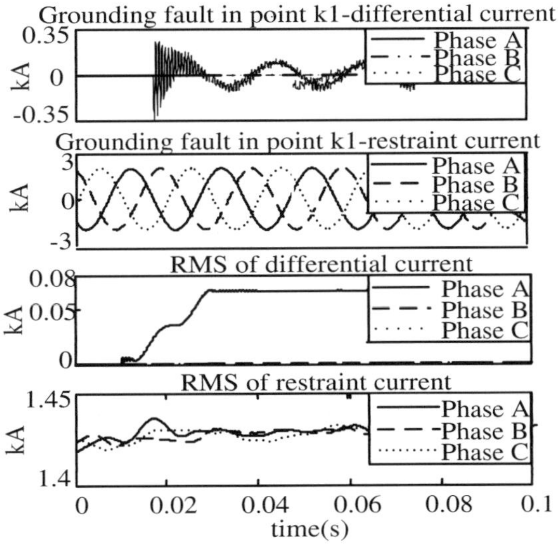

(a)

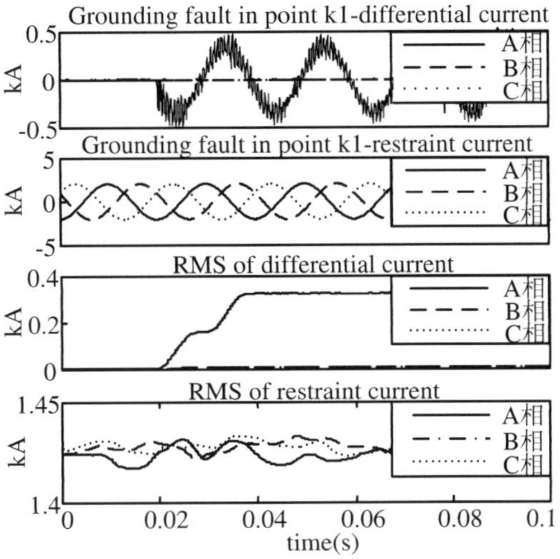

(b)

Figure 6. Waveform of differential and restraint current of valve side AC bus single phase to ground fault.

When there is a single phase to ground fault occurs in the point k2, the differential and restraint current of grounding scheme 1 and 3 is as shown in figure 7. For the grounding scheme 1, in the process of transient transition, the restraint current is reduced while the differential current is increased, so the protection can operate correctly. For the grounding scheme 3, the bridge arm currents and restraint current increase with time. In conclusion, only under the parallel transformer valve side AC bus and valve side winding grounding fault, the corresponding protection cannot operate correctly. However, if the grounding scheme 3 is adapted, the small grounding resistance results in overcurrent in bridge arms, which cause harm for the safety of UPFC. In summary, scheme 1 is proposed.

IOP Conf. Series: Earth and Environmental Science **73** (2017) 012011 doi:10.1088/1755-1315/73/1/012011

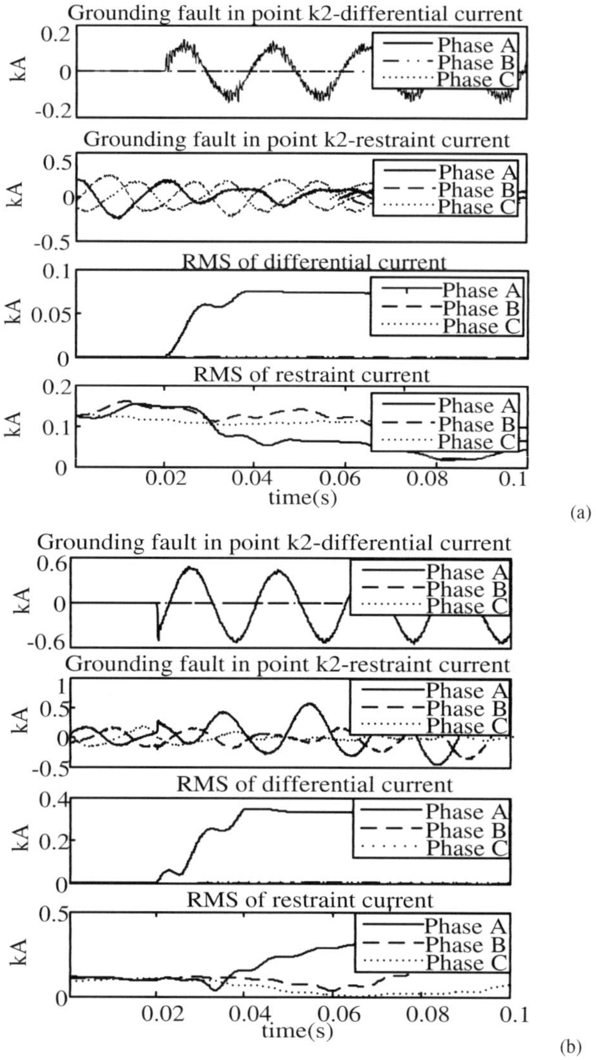

Figure 7. Waveform of differential and restraint current of valve bridge arm single phase to ground fault.

5. Conclusion

Through theoretical analysis and simulation validation, the effect of different grounding position to the series transformer network side grounding fault and protection configuration and the effect of different grounding resistance to the parallel valve side grounding fault are analyzed in detail, the conclusions are as follows:

- When a fault occurs in series transformer network side under grounding scheme 2, there is zero sequence current in valve side, which leading to the overcurrent of the neutral point. The voltage of valve side AC bus added to the neutral point voltage, causing the overvoltage of the valve side AC bus.
- The sensitivity of differential protection may be inadequate under the parallel transformer valve side single phase to ground fault when the grounding resistance is large.
- According to theoretical analysis and simulation, the large resistance (200 ohm) grounding of the valve side of the parallel transformer is the most suitable rounding design.

IOP Conf. Series: Earth and Environmental Science **73** (2017) 012011 doi:10.1088/1755-1315/73/1/012011

References

[1] Wang H., Tang G. F. and He Z. Y., 2014 Efficient grounding for modular multilevel HVDC converters (MMC) on the AC side[J] *IEEE Trans. Power Del* vol. **29** pp. 1262-72.

[2] Guo G. P., Hu X. H. and Wen J. L., 2015 A large-scale submodule group based algorithm for modeling and high-speed simulation of modular multilevel converter [J] *POWER SYS TECHNO* vol **39** pp 1226-32(in Chinese).

[3] Zhou F., He W. G., Bao H. L., et al, 2012 A preliminary research on cascaded UPFC suitable to urban power grid[J] *POWER SYS TECHNO* vol **36** pp 174-8(in Chinese).

[4] Cui F. B., Guo J. B., and Jing P., 2014 The grounding design and internal fault characteristic of MMC-UPFC[J] *Proceedings of CSEE* vol. **34** pp. 1-9(in Chinese).

[5] Moravej Z., Pazoki M.; Khederzadeh M., 2014 Impact of UPFC on power swing characteristic and distance relay behavior. IEEE *Trans. Power Del* vol **29** pp 261-268

IOP Conf. Series: Earth and Environmental Science **73** (2017) 012012 doi:10.1088/1755-1315/73/1/012012

Mini Solar and Sea Current Power Generation System

Abdulrahman Almenhali, Hatem Alshamsi, Yaser Aljunaibi, Dheyab Almussabi, Ahmed Alshehhi, and Hassan Bu Hilal

Al Saada Street (Street # 19), Al Nahyan, Abu Dhabi Landmark: Opposite Abu Dhabi Civil Defence, United Arab Emirates

Email: habuhilal@hct.ac.ae;

Abstract. The power demand in United Arab Emirates is increased so that there is a consistent power cut in our region. This is because of high power consumption by factories and also due to less availability of conventional energy resources. Electricity is most needed facility for the human being. All the conventional energy resources are depleting day by day. So we have to shift from conventional to non-conventional energy resources. In this the combination of two energy resources is takes place i.e. wind and solar energy. This process reviles the sustainable energy resources without damaging the nature. We can give uninterrupted power by using hybrid energy system. Basically this system involves the integration of two energy system that will give continuous power. Solar panels are used for converting solar energy and wind turbines are used for converting wind energy into electricity. This electrical power can utilize for various purpose. Generation of electricity will be takes place at affordable cost. This paper deals with the generation of electricity by using two sources combine which leads to generate electricity with affordable cost without damaging the nature balance. The purpose of this project was to design a portable and low cost power system that combines both sea current electric turbine and solar electric technologies. This system will be designed in efforts to develop a power solution for remote locations or use it as another source of green power.

1. Introduction

Since the invention of first engine in 17th century, energy Consumption has been used due to many causes in many different ways. Persistent increase in the energy demand has caused to seek, new energy resources in the world; new alternative energy. Resources have been also utilized to minimize the energy deficit [1-3].

Figure 1 shows a chart of the difference between early consumption and now with different types of resources weather its renewable or non-renewable energy.

IOP Conf. Series: Earth and Environmental Science **73** (2017) 012012 doi:10.1088/1755-1315/73/1/012012

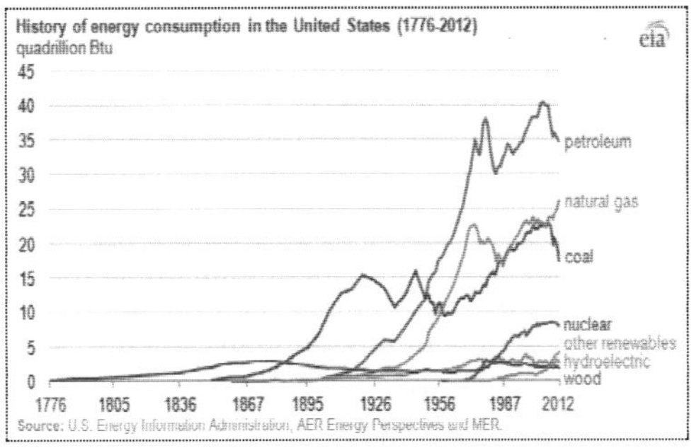

Figure 1. History of energy consumption in the United States (1776-2012)

In the early age, people didn't use too much energy but as time passes, energy demand had increase and the world is now searching for energy resources. Non-renewable energy sources are highly used in caparison with renewable energy and that will cause in environmental problems such as global warming. Electricity is most needed for our day to day life. There are two ways of electricity generation either renewable or non-renewable. Electrical energy demand increases in word so to fulfill demand we have to generate electrical energy. Now a day's most of electrical energy is generated by the non-renewable energy resources like coal, diesel, and nuclear. The main cons of these sources are pollution and green house effects where it contributes in global warming and the nuclear waste is very harmful to living organism. The non-renewable energy resources are depleting day by day. Soon it will be completely vanishes from the earth so we have to find another way to generate electricity. The new source should be reliable, pollution free and economical. The renewable energy resources should be good alternative energy resources for the non-renewable energy resources. There are many renewable energy resources like geothermal, tidal, wind, solar. Solar energy has drawback that it could not produce electrical energy in rainy and cloudy season so we need to overcome this drawback we can use two energy resources so that any one of source fails other source will keep generating the electricity. And in good weather condition we can use both sources combine. In the United Arab Emirates, solar energy is the best renewable energy we have but one source is never enough. Combing another source with solar energy will give us more power and more reliability in extracting power without burning fossil fuel or damaging the environment. So the best option beside solar is wind but we don't have too much space to make Huge wind farms, so we thought of sea current power generation which have the same concept of wind energy but under the sea. The turbines area will be much smaller due to the density of the water which is thousand times larger than the wind density. Other renewable source is not as great as sea current energy due to limitations such as the surrounding environment like geothermal. We have few places that are suitable for generation but the places are already occupied with other things such as swimming pools and other facilities. The sea is vast and we can place the turbines any place we want [4-7].

The mechanism of the working of solar panels is depicted in figure 2.

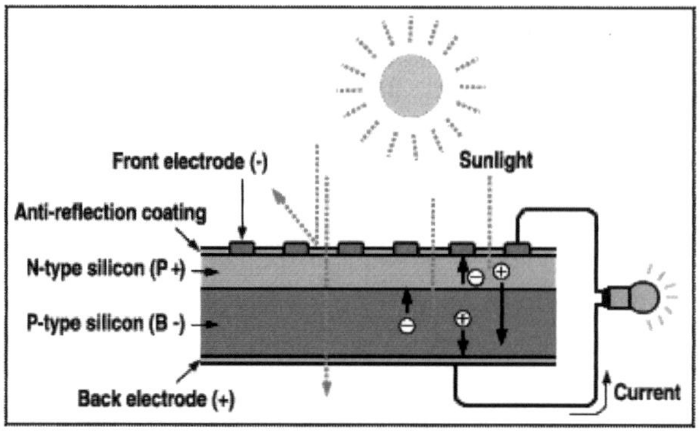

Figure 2. PV Solar

As shown in figure 2. The sunlight is directed to the silicon material, then the n-type and p-type material react upon this energy, resulting of moving of charge carriers. Mainly hole-electron movement. Finally, electrical current is induced. Next we will talk briefly about wind power. Figure 3 shows the Wind turbines. Wind turbines are used to convert the wind power into Electric power. Electric generator inside the turbine converts the mechanical power into the electric power. Wind turbine systems are available ranging from 50W to 2-3 MW. The energy Production by wind turbines depends on the wind velocity acting on the turbine. Wind power is used to feed both energy production and consumption demand, and transmission lines in the rural areas. We have two types of turbines in general, horizontal and vertical turbines, each one of them is used in different areal conditions. Here is a figure that will show the structure of wind turbine.

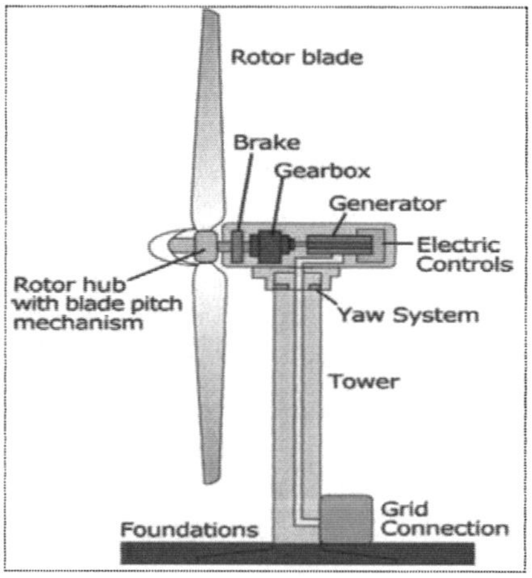

Figure 3. Wind Turbine

The system will be improved by combing it with other source to make a reliable hybrid system. By knowing the power demand, we can calculate the size and the quantity of the pv panels and the same for the sea current turbine generator. There are two types of sea current, deep sea current and surface sea current. Deep sea current is affected by the melted salt and the difference between densities while the surface sea current is a affected by the wind. Surface sea current is deep down to 100m while more than 100 consider as deep sea current. In this project we will use surface sea current to make the turbine move and generate electricity. The concept of sea current turbines is exactly same as wind turbines and how they work. The gear box, shaft, nacelle, and so on are in the design. The only different is in the size where the underwater turbines will be much smaller and the materials that will be used to create the turbines to make them reliable under water and won't be affected easily by the salt and rust.

2. Site characteristics

United Arab Emirates is located in the gulf region United Arab Emirates' latitude and longitude is 24° 00' N and 54° 00' E. The United Arab Emirates Climate features extreme heat, UAE weather is sunny all the year round, This can be easily seen in figure 4.

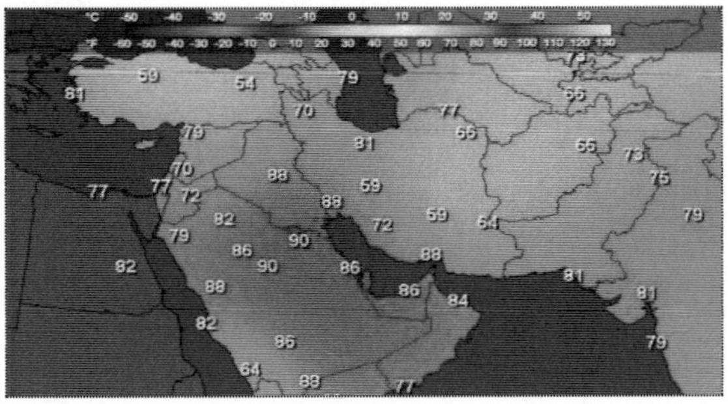

Figure 4. The temperatures soar up to 45°C in summer and in winter it goes down to 20°C

2.1. Solar Energy Potential

Direct normal irradiation:

The estimated direct normal irradiation is plotted with respect to the month for six stations. Figure 5 demonstrates different cities lines in UAE, namely, Masdar, Jabal Haffed, Al aradh, AL wagan, Madinat Zayed and Al sweihan.

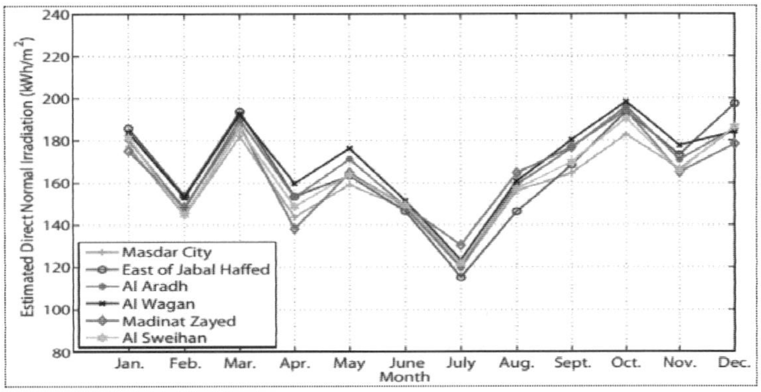

Figure 5. How the values are highly variable throughout the different months of the year and that the month of July has the lowest values.

2.2. Peak sun hour

Figure 6 displays the average peak sun hours per day in the UAE for each month. The yearly average peak sun hours/day used in calculations in the UAE is 5.84

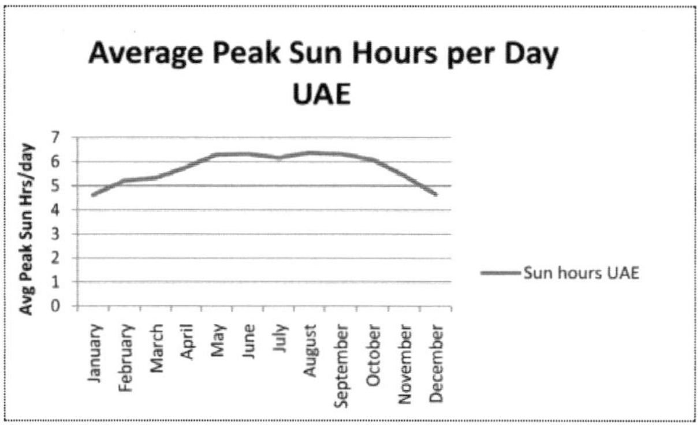

Figure 6. Average peak sun hours per day in the UAE

2.3. Sea current energy potential

United Arab Emirates have many islands and water surrounding the country, we can use the sea current to generate energy, the water has at least 1000density and salty water have even more. So even a speed of 1m/s is enough to generate clean power.

Figure (7) shows the potential of wind energy, as long as there is small wind energy to move the sea current, we can generate energy.

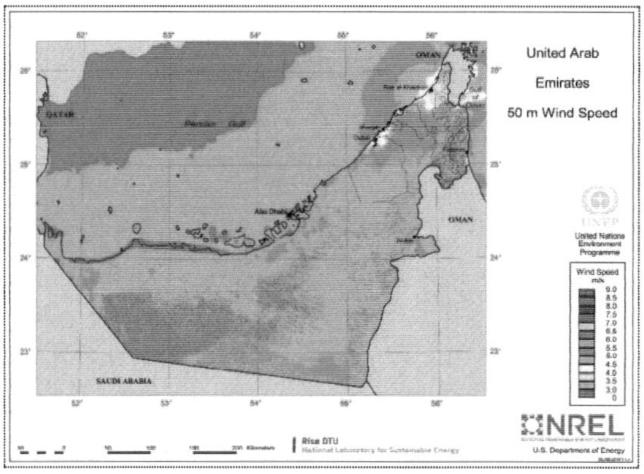

Figure 7. the potential of wind energy

3. Conceptual design

The system will use sea current turbines and solar panels to extract energy. It will let the two sources extract independently and it would be connected t to a hybrid wind solar charge controller, which will store the produced energy in a battery bank. The charge controller to maintain everything and to store the energy safely. Then from the battery to the inverter to convert DC to AC, then from the inverter to the AC load.

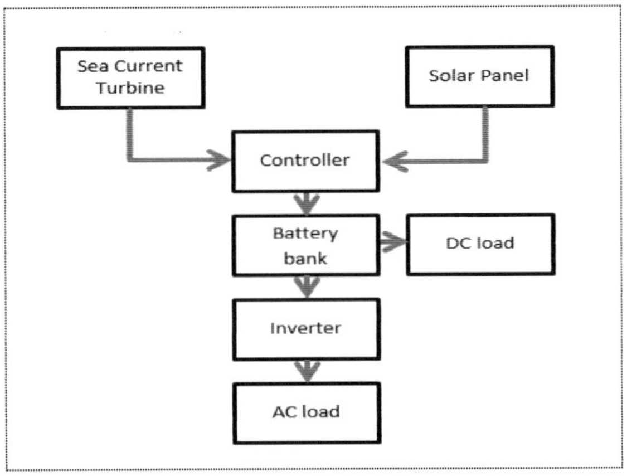

Figure 8. System Block Diagram

In this project we are going to use 6 major process to complete the system as shown in figure (8).

- **AC load:** we need to put a maximum load so we can design a system based on our maximum load.
- **Inverter:** Inverter is need to convert DC power into AC power.
- **Battery bank:** the system needs a battery bank size per the load requirement so that it should fulfill the requirement of load.
- **Hybrid solar sea current controller:** Charge controller has basic function is that it control the source which is to be active or inactive. It stores the power that were extracted from the sources to the batteries and take power from the battery to the inverter. The controller has over-charge protection and short-circuit protection. This will increase the battery lifespan
- **Solar panels (PV):** Solar panel Solar panel is use to convert solar radiation to the electrical energy.
- **Sea current turbines:** Sea current turbine is that system which extracts energy from sea current by rotation of the blades.

4. Modelling

Modeling of PV System

The required size of pv panel

Total demanded power = $80 \times 1.3 = 104w$ (0.1)

(where 1.3 is the energy loss in the system)

Size of the pv panel

Pv size =Total demanded power / power generation factor

$104/5.84 = 17$ 104÷5.84=17 (5.84 is the power generation factor for UAE).

17 is the panel size, so we would take 20 watt as panel size.

Requirements = 20 watt 12v pv Polycrystalline panel

4.1. Modeling of inverter

The size the size of the inverter should never be smaller than the demanded load, it should be 25 – 30 % bigger than the load.

$$inverter\_size = 80 \times 1.3 = 104w$$

Battery size

The calculation of the battery size for one day of autonomy

$$Battery\_Size = \frac{AC\_Energy}{\text{Batt\_eff} \times \text{depth\_discharge} \times \text{nominal\_voltage}} = 13 \tag{0.2}$$

where
System voltage = 12
Depth discharge = 0.6
Battery eff= 0.85
Days = 1 day without power
So, We need 13 Ah 12v battery (we may change it due to the controller specification)

4.2. Modeling of Wind System
Power available in wind at a specific site depends on three things: air density, area of rotor and wind speed. Wind power can be calculated using this formula. We want to produce at least 80 watt.

$$P = 0,5 \times \rho \times AW \times v^2 \times \eta \tag{0.3}$$

- P = power from the water
- ρ= density of water 1000 kg/m3
- AW = w is the cross sectional area of water flow across turbines
- v = velocity of water (1.5 m/s) (as an average)
- ŋ = efficiencies (25%) (0.7 generator eff * 0.35 turbine eff)

Since we know all the all the parameters accept the Aw, we can get it to help us to design the blades.

Aw= 0.28m2
R =0.3m

4.3. Controller size
We can also choose the size depending on the maximum produces watt on both sources.
The system needs a controller which will withstand 20 watt from solar panels and 80 watt from current turbine. It won't be easy to find one because hybrid controllers always need more watt to operate. We may have to use a bigger controller because the smallest charge controller can withstand 300 watt from water generation and 150watt from solar generation [8-9].

5. Conclusion
Hybrid power generation system is good and effective solution for power generation than nonrenewable energy resources. It has greater efficiency. It can provide to remote places where government is unable to reach. So that the power can be utilize where it generated so that it will reduce the transmission losses and cost. Cost reduction can be done by increasing the production of the equipment. People should motivate to use the renewable energy resources not only for the personal good but for the global and environment good too. It is highly safe for the environment as it doesn't produce any emission and harmful waste product like conventional energy resources. It is cost effective solution for generation. It only need initial investment. It has also long life span, it only needs little, uncostly maintenance after a certain periods of time Overall it good, reliable and affordable

solution for electricity generation. People and countries should focus on using renewable energy sources as much as possible to save our global.

References

[1] W. W. S. Charters, "Solar and Wind Power Technologies for Remote Applications", CSC Technical Publication Services No 187, Commonwealth Science Council. (1985).

[2] M. B. Olajide and J. O. Oni "Application of Solar Energy for offices and homes, Workshop Seminar Paper, International Training School and Workshop on Solar Energy, organized at University of Agriculture, Abeokuta,9th-11th June, 18-29. (2009)

[3] Technical brief on Wind Electricity Generation: Retrieved, from: www.windpower.org, (2009).

[4] Walker G., "Evaluating MPPT Converter Topologies Using A Matlab PV Model", Journal of Electrical & Electronics Engineering, 21(1), 2001 pp. 49-56.

[5] Mineiro, S. J. E., Daher, S., Antunes, F. L. M., Cruz,C. M. T., "Photovoltaic System For Supply Public Illumination In Eletrical Energy Demand Peak", 19th IEEE Annual Applied Power Electronics Conference and Exposition Conference, Anaheim, California,USA, 3,2004 pp.1501-1506.

[6] Hybrid power Systems - Solar and wind hybrid systems - Solar Online Australia. (n.d.). Retrieved October 04, 2016, from https://www.solaronline.com.au/solar_wind_hybrid_systems.html

[7] World energy consumption. (n.d.). Retrieved October 04, 2016, from: https://en.wikipedia.org/wiki/World_energy_consumption

[8] Zhou Wei, Current status of research on optimum sizing of standalone hybrid solar–wind power generation systems. Applied Energy 2010; 87: 380–89.

[9] Jose´ L., Simulation and optimization of stand-alone hybrid renewable energy systems. Renewable and Sustainable Energy Reviews 2009; 13: 2111-18.

Using System Dynamic Model and Neural Network Model to Analyse Water Scarcity in Sudan

Y Li[1], C Tang1, L Xu[2] and S Ye[3]

[1] Department of Finance, Jinan University, 51000 Guangzhou, China

[2] Department of Statistics, Jinan University, 51000 Guangzhou, China

[3] Department of Mathematics, Jinan University, 51000 Guangzhou, China

Email: xuleying1995@163.com

Abstract. Many parts of the world are facing the problem of Water Scarcity. Analysing Water Scarcity quantitatively is an important step to solve the problem. Water scarcity in a region is gauged by WSI (water scarcity index), which incorporate water supply and water demand. To get the WSI, Neural Network Model and SDM (System Dynamic Model) that depict how environmental and social factors affect water supply and demand are developed to depict how environmental and social factors affect water supply and demand. The uneven distribution of water resource and water demand across a region leads to an uneven distribution of WSI within this region. To predict WSI for the future, logistic model, Grey Prediction, and statistics are applied in predicting variables.. Sudan suffers from severe water scarcity problem with WSI of 1 in 2014, water resource unevenly distributed. According to the result of modified model, after the intervention, Sudan's water situation will become better.

1. Introduction

The world is facing a growing pressure for water resources. It's shown on the United Nations Water Scarcity Map that one quarter of the world's people experience water scarcity. What is the criterion of determine whether a region is suffering from water scarcity? Will the region become increasingly vulnerable to water scarcity in the future? What if the government makes intervention to alter water supply and demand? Answering these questions is strategic to deal with the important issue of water scarcity in the world.

Due to huge variations of precipitation across the country, high fertility and population growth rates and many other factors, the present water scarcity situation in some regions of the world such as Sudan is extremely severe, leading citizens to suffer from insufficient drinking water, no irrigation for agriculture and poverty. Therefore there is an urgent need to understand the mechanism and dynamic nature of water scarcity quantitatively.

Water scarcity incorporates water demand and water supply, which are decided by factors such as population growth rate, river runoff and precipitation. Planning and reforming water strategies are essential to diminish water scarcity level and maximize benefits of citizens. The intervention is to be driven based on key environmental and social considerations.

Water is a key element to socio-economic development and quality of life. Environmental and social factors is a concern objective for organizations and government but is often vaguely defined and their cooperated contribution procedure to water scarcity is unclear.

2. Literature review

In 1961, Forrester [1] found a simulation model-System Dynamics methodology- to analyze industrial business cycles, and they applied the model on complex interactions of world economy, population in 1971. Sušnik et al [2] used SDM to simulate integrated water systems and assess the potential impacts of policy with respect to water scarcity in specific region.

Pfister et al [3] developed the method for assessing the environmental impacts of freshwater consumption, which considered damages to three areas of protection: human health, ecosystem quality, and resources. They adjusted the Water Stress Index to a logistic function. A global water assessment model called WaterGAP 2, which consists of two main components--a Global Water Use model and a Global Hydrology model, is described by Alcamo et al [4] to fill in many of the existing gaps in water resources data in many parts of the world and provide a coherent approach for generating scenarios of changes in water resources. Omer [5] proposed that water and sanitation in Sudan are severe, and water demand is increasing due to the rapid population growth.

3. Methodology

The extent to which water is scarce is estimated based on the calculation of Water Scarcity Index (WSI), concerning the relationship between demand, also called consumption, and supply in a specific area. Water Scarcity Index is calculated with formulas given by WaterGap2 global model [3]:

$$WTA_i = (WA_i)^{-1} \sum_j WU_{ij} \quad (1)$$

$$WSI = \left[1 + (0.01^{-1} - 1)\exp\left(-6.4WTA^*\right)\right]^{-1} \quad (2)$$

where WA_i is the annual freshwater availability for each region i;

WU_{ij} is the withdrawals for different users j for each region i;

WTA_i is the ratio of total annual freshwater withdrawals to hydrological availability for each region i.

WU_{ij} in this model, is an indispensable and therefore, critical variable. In order to give an accurate estimation of WU_{ij}, models concerning water supply and demand are then developed. Water supply, as itself states literally, is thought of as renewable water resources available, determined by the process of hydrologic circle. With data availability and discernable relationships considered, only evaporation, precipitation and storage, that is, the gap between water flow in and water flow out in a given area, are involved in this model, which is proved to well explain the relationship among all factors mentioned above.

While certain data are released and accessible, model of relationship among water supply, precipitation, evaporation and water storage is needed, for the sake of prediction. Amount of researches are conducted concerning water resources, but so far few of them is able to describe a clear link between renewable water resources and other variables, which to a large extent implies an elusive and convoluted relationship remained unknown. As a result, a viable model that can clarify but keep fuzzy a convoluted relationship, that is sometimes, network, is ideal in this case, and thus Neural Network is suitable and induced.

Water demand, or consumption, is regarded as the total amount of water consumed in various ways, from industrial to agricultural ones, all relevant to economic and demographical structure and stature of a country. Complicated wide-range as water consumption in a country is, another model, commonly known as System Dynamic Model, makes it possible to study the constitution of the domestic water consumption system, and also enables us to watch over the whole system in a dynamic way. As apparently population and Gross Domestic Product (GDP) best give a picture of the current stage an economy is in, and industrial structure shows how water consumption is distributed in different industries, we put all these variables into the System Dynamic Model, and keep other variables constant throughout the period we study. A simplified overview of the System Dynamic Model is displayed in figure 1.

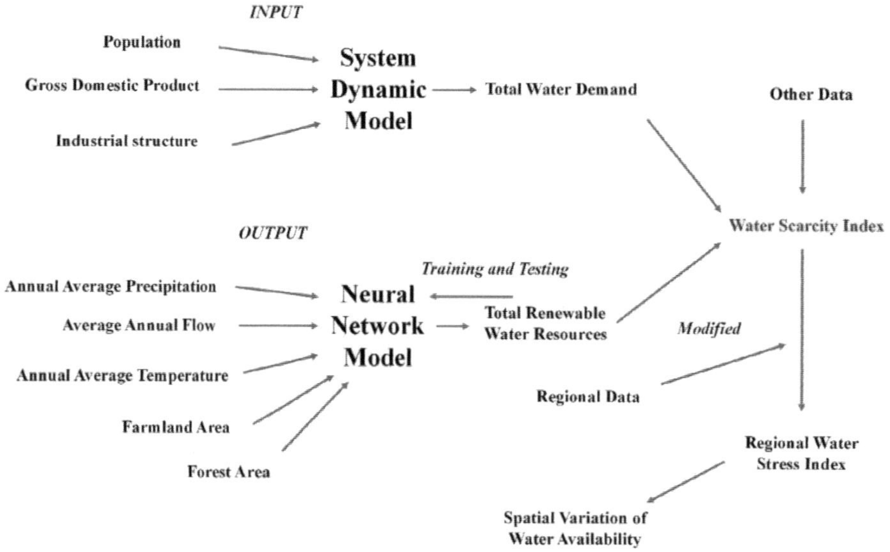

Figure 1. Overview of the water availability model.

When understanding water scarcity in a region, water distribution among areas is indispensable, for it exactly tells us how water is scarce in every corner of this country, instead in the whole country. This is important, for water is still scarce in a country, when some in this country are provided with so much water that they even squander it, while others are faced with terrible water scarcity, too terrible for them to survive, even if the country is seemingly water abundant on the surface. Therefore, Water Scarcity Indices in all regions in a country are also calculated. To keep simple, we assume that water consumption in every region only depends on population and GDP in that area, and only make effort to give an accurate estimation of water supply in an area, based on Neural Network mentioned above. While precipitation in specific region has been already observed and data are available, how citizens in different regions benefit from the main river in this country remains unknown, but critical. Thus, we establish formula below to speculate the extent to which citizens benefit from the main river:

$$benefit_i = \ln\left[ce^{-d_i}\right] \cdot runoff \tag{3}$$

where is c calculated by solving $\sum benefit_i = runoff$.
One-step-forward, prediction of water scarcity is also conducted. For relationships among variables keep stable and fast, only variables are predicted in this passage, and in various ways. Population is believed to fit a logistic progress, and increasing rate of which is described as below.

$$\frac{dx}{dt} = rx(1 - \frac{x}{x_m} - a), \qquad x(0) = x_0 \tag{4}$$

where x is population at time t, x_m is environment population capacity, r is natural growth rate of population; a is net migration rate.
Besides, GDP is predicted with grey forecasting model, while precipitation and annual flow are predicted based on normal distribution, since they greatly fit a normal distribution according to what observed data in last decades show.

4. Application in Sudan

Sudan, located in South-Africa, is one of the countries that suffer from severest water scarcity, in accordance with water scarcity map released by UN. Water in Sudan is heavily overloaded as a result of both environmental and social constraints: for one thing, though the body of Nile river system is right located in Sudan, surface water in Sudan is subject to utter evaporation and discharge; for another thing, while water resources may be abundant, renewable and available water resources are still in dearth due to a lack of advanced water purification technology, and therefore citizens in Sudan find it difficult to obtain enough water without disease carried: as data shows, only 69% of rural dwellers and 86% of city dwellers in Sudan have an access to purified drinking water. To make things worse, population in Sudan keeps growing at a relatively high rate, generating an increasingly greater demand of clean water [5].

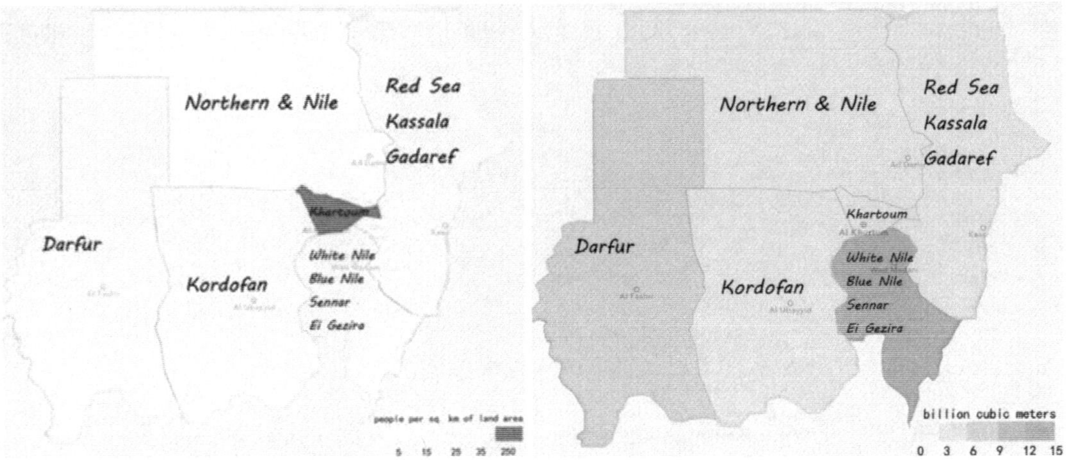

Figure 2. Uneven WSI distribution in Sudan.

Moreover, regarding water distribution and population distribution, the ordeal faced by Sudan seems tougher. Khartoum, for instance, is of highest population density and relatively low water resources in Sudan, which makes it the least water abundant city in Sudan. Figure shows a rough estimation of population density and renewable water resources in every region in Sudan, from which we can draw a precise picture of a country with water utterly unevenly distributed, as in figure 2.

Table 1 presents overview of Sudan in 2014 regarding water supply and demand:

Table 1. Data of key variables of Sudan in 2014.

Precipitation (km^3)	Annual Flow (billion km^3)	Population	GDP (dollar)
4.32×10^{11}	6.08×10^{10}	3.94×10^7	7.38×10^{11}

With System Dynamic Modeling and Neural Network Model, aggregate water supply and demand, as well as Water Scarcity Index, are stated in Table 2.

Table 2. Estimated water resource situation in 2014.

Water Supply (billion cubic meters)	Water Demand (billion cubic meters)	Water Scarcity Index
39.26	27.18	1

Water Scarcity Index is equal to 1, which means water resources are greatly overloaded in 2014. In contrast, Water Scarcity Index in Europe is generally below 0.5 in 2014, and estimation of Water Scarcity Index of Sudan in 2004 is 0.763, far below its current value, which shows that Sudan is not only faced with severe water scarcity, but also exhausting its water resources at a high speed.
Further estimation of the extent to which water is scarce in all regions is then conducted, and outputs are interpreted as left in figure 3, suggesting a more concerning water problem in Sudan. The result of water supply and demand prediction in 15 years is calculated by inputting date into the Demand-Supply Model as indicated in figure 3 at the right.

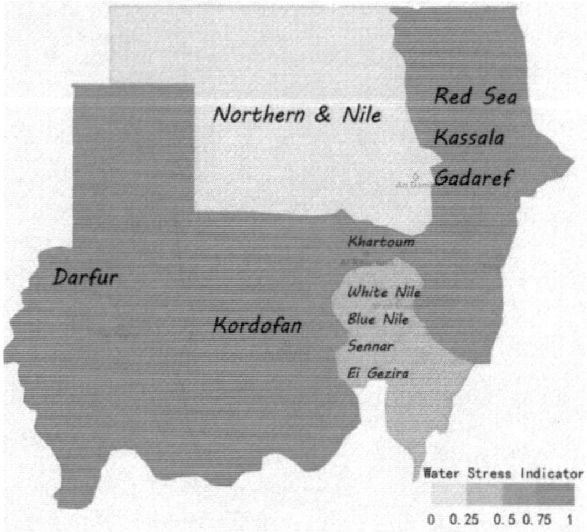

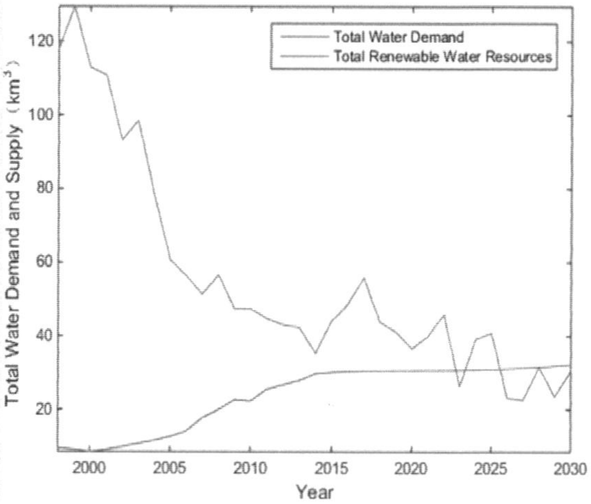

Figure 3. WSI in different states and water prediction in Sudan.

The results show that total renewable water resource has decreased sharply in the past decade while the water demand is increasing because of the accelerated increase of population. Without any intervention, Sudan's water supply will not be able to fulfil its demand after year 2025.
Rainwater should be collected in Sudan in order to get more water supplies. Besides, as the agriculture irrigation technology is applied and population control is enacted, the water demand would decreases accordingly. We put all parameters related to these interventions into our model and get a new prediction of demand of water resources for the following years. Figure show that our intervention plan will make Sudan less susceptible to water scarcity. However, the water shortage will reappear in the year 2027, although we put it off by 15 years compared to no intervention condition.
Water distribution can be optimized by redirect water courses in Sudan. We input modified data of water supply in various states in our model, and obtain a new WSI distribution figure. Figure 4 shows prediction of WSI distribution and water supply and demand after the intervention plan.

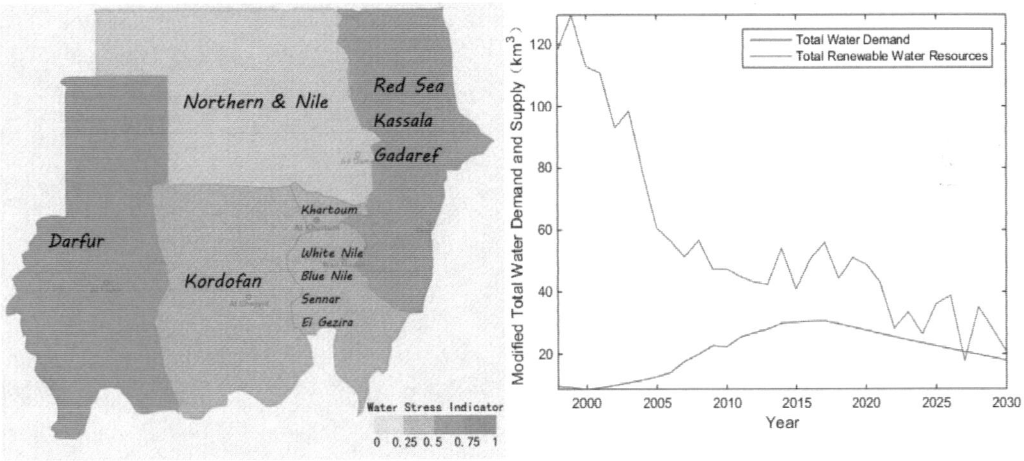

Figure 4. WSI distribution and water supply and demand after the intervention plan.
The results are quite important since we they give us a view of the effectiveness of possible intervention plans in a specific country and predict the reoccurring "bad time" so as to help policy makers greatly.

5. Conclusion
Estimation and Prediction of water scarcity in Sudan are conducted in this passage, and its outputs lead to some important conclusions.
Aggregate water demand and supply in Sudan were 27.18 and 39.26 million cubic meters in 2014, respectively, which shows that renewable water resource in Sudan was able to cover water consumption. However, as predicted in Demand-supply model developed in this passage, the country will exhaust its water resources in 2025, and as a result, leave citizens in an unprecedented water crisis.
Water Scarcity Index is calculated both domestically and endemically. While Water Scarcity Index of Sudan remains 1 from now on, that of regions in Sudan vary significantly, ranging from 0.541 to 1, implying that water distribution in Sudan is so uneven, that some dwellers have already suffered from a dearth of renewable water, while others may own more than necessity.
Therefore, intervention plans are needed. With some viable intervention plans induced in this case, depletion of water resources delayed to 2027 as predicted using our model. Though it takes time and effort, certain effective measures should be taken, for the sake of people's living.

IOP Conf. Series: Earth and Environmental Science **73** (2017) 012013 doi:10.1088/1755-1315/73/1/012013

Acknowledgments
Authors wishing to acknowledge assistance or encouragement from colleagues, special work by technical staff or financial support from organizations should do so in an unnumbered Acknowledgments section immediately following the last numbered section of the paper.

References
[1] Forrester J 1961 *Industrial dynamics* (Waltham: Pegasus Communications)
[2] Sušnik J, Vamvakeridou-Lyroudia L, Savić D and Kapelan Z 2012 Integrated System Dynamics Modelling for water scarcity assessment: case study of the Kairouan region *Sci. Total. Environ.* **440** 290–306
[3] Pfister S, Koehler A and Hellweg S 2009 Assessing the environmental impacts of freshwater consumption in LCA *Environ. Sci. Technol.* **43** 4098–104
[4] Alcamo J and Henrichs T 2002 Critical regions: a model-based estimation of world water resources sensitive to global changes *Aquat. Sci.* **64** 352-62
[5] Omer A. 2001 Solar water pumping clean water for Sudan rural areas *Renew. Energ.* **24** 245-58

IOP Conf. Series: Earth and Environmental Science **73** (2017) 012014 doi:10.1088/1755-1315/73/1/012014

Study on the Preparation Process and Influential Factors of Large Area Environment-friendly Molten Carbonate Fuel Cell Matrix

Ruiyun Zhang, Shisen Xu, Jian Cheng, Hongjian Wang and Yongqiang Ren
China Huaneng Clean Energy Research Institute, Beijing 102209, China

E-mail: zhangruiyun@hnceri.com

Abstract. Low-cost and high-performance matrix materials used in mass production of molten carbonate fuel cell (MCFC) were prepared by automatic casting machine with α-LiAlO2 powder material synthesized by gel-solid method, and distilled water as solvent. The single cell was assembled for generating test, and the good performance of the matrix was verified. The paper analyzed the factors affecting aqueous tape casting matrix preparation, such as solvent content, dispersant content, milling time, blade height and casting machine running speed, providing a solid basis for the mass production of large area environment-friendly matrix used in molten carbonate fuel cell.

1. Introduction

The fuel cell is a kind of power generation device which directly converts the chemical energy of the fuel into electric energy [1]. Unlike the traditional coal-fired power generation technology, the fuel cell adopts electrochemical catalysis to convert the chemical energy in the fuel directly into electric energy, which go beyond the heat engine Carnot cycle efficiency limit because there's no thermodynamic cycle, and the current total efficiency of fuel cell is between 45% to 60%. The molten carbonate fuel cell (MCFC) is a high temperature fuel cell working at 650℃ developed in the late 1950s. The waste gas generated by the cell stack can be combined with the gas turbine or steam turbine to further recover the heat, and the total efficiency is more than 80%, and the power generation efficiency is higher than that of the thermal power plant, and its environmental performance is better than conventional thermal power plants. With many advantages, fuel cell is considered to be the fourth largest energy conversion power generation system after the thermal power generation, hydropower and nuclear power generation, and can be an alternative power plant of the internal combustion engine[1]. It will be the most attractive green energy known as the 21st century.

On December 2, 2015, delegates at the Ninth China-Japan Energy Conservation and Environmental Protection Forum enthusiastically discussed all of the clean coal power generation technologies from now until 2030 as shown in Fig. 1, including ultra supercritical thermal power technology (A-USC) (power generation efficiency about 46%), 1700 ℃ level IGCC (power generation efficiency about 46 ~ 50%), the overall coal gasification fuel cell power generation technology (IGFC) (power generation efficiency of 55%) , as well as thc innovative IGFC. As of all those power generation methods, fuel cell was regarded as the most ideal distributed power generation for its advantages of good modular, and higher power generation efficiency at lower power.

Figure 1. The route of advanced clean coal power generation technology

Among the many fuel cells, MCFC is believed to have good practical prospects, and is the mainstream of the current study. MCFC has porous nickel as an anode, porous nickel oxide as a cathode, an eutectic mixture consisting of Li_2CO_3/K_2CO_3 or Li_2CO_3/Na_2CO_3 as an electrolyte, and the molten electrolyte is dispersed in matrix prepared by α-$LiAlO_2$ or γ-$LiAlO_2$ powder. Matrix is one of the core components of MCFC with the function of electronic insulation, ion conduction, gas seal and so on. Therefore, matrix must have high strength, be able to bear high temperature molten salt corrosion, could block the gas through when immersed in molten salt electrolyte, and could have good ionic conductivity.

The performance of matrix is closely related to its porosity and average pore size. The larger porosity and average pore size is, the more electrolyte immersed in the matrix, then the matrix resistance becomes small. But the average pore size is large, so it may have the risk of channeling gas between cathode and anode. Therefore, it is required that the matrix has a reasonable porosity and pore size distribution, the general porosity required is of 50% to 70%, pore size is less than 1μm, and has a good distribution.

This paper studied a simple and environment-friendly matrix preparation method, α-$LiAlO_2$ powder synthesized by solid-state reaction method and polyvinyl alcohol (PVA) are blended together, with distilled water as solvent, to form a slurry which is put into the automatic casting machine to make water - based matrix.

2. Experiments

2.1. Preparation of α-LiAlO$_2$ Powder by Solid State Reaction

After many experiments, the $LiAlO_2$ was chosen as a matrix material. $LiAlO_2$ has many kinds of crystal form, the most common are α, β and γ type, usually α-$LiAlO_2$ is more stable under high pressure, γ-$LiAlO_2$ is more stable at high temperature, and β-$LiAlO_2$ is a transition state and is not stable[1]. α-$LiAlO_2$ was prepared by sol-gel solid-phase reaction method that alkaline alumina and lithium carbonate were mixed by equimolar ratio and calcined at 750℃. The reaction process was as follows:

$$Al_2O_3 + Li_2CO_3 = 2LiAlO_2 + CO_2$$

Fig. 2 shows the XRD patterns of the prepared α-$LiAlO_2$ powder samples. Fig. 3 shows the particle size distribution of the α-$LiAlO_2$ powder.

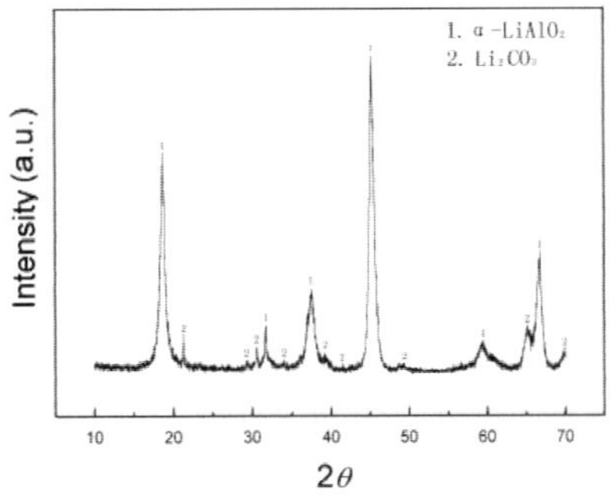

Figure 2. The XRD patterns of the prepared $LiAlO_2$ powder samples

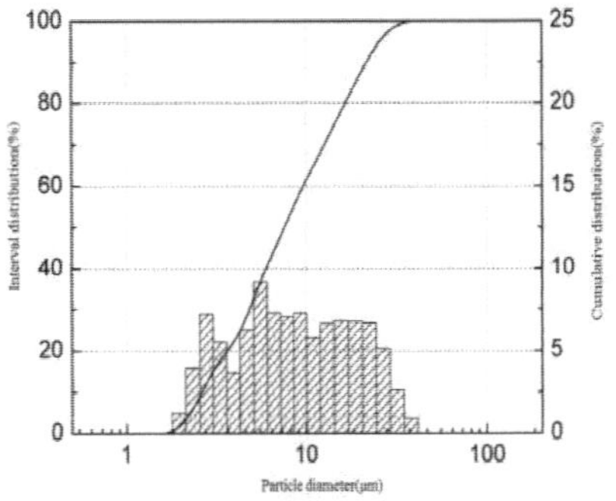

Figure 3. The particle size distribution of the α-powder prepared by solid phase method

It can be seen from Fig. 2 and Fig. 3 that the α-$LiAlO_2$ powder prepared by the solid-phase reaction of the gel has a high purity and whose relative content reaches 96.58%, the equivalent particle diameter D10 is 2.6μm, the average particle diameter D50 is 7.53μm, the maximum particle size is 21.23μm, which has reached the requirements of MCFC matrix [2].

2.2. Preparation of MCFC Matrix

There are many methods for preparing MCFC electrolyte matrix: hot pressing, electrophoretic deposition, vacuum casting, hot and cold rolling and strip casting (casting method) and so on, the key is how to prepare high quality slurry. Large experiments show that large-scale water-soluble and environmentally friendly MCFC matrix can be prepared by using automatic casting machine, the preparation process shown in Fig. 4.

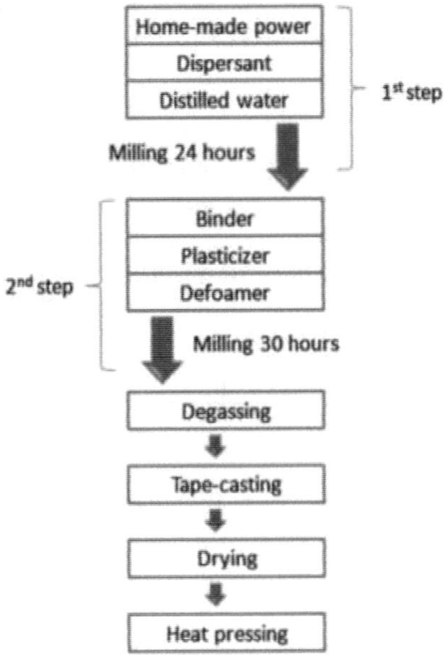

Figure 4. The preparation process of water - based matrix

Fig. 5 shows the prepared large area of MCFC water-soluble matrix, the average thickness is of 0.8mm, the porosity is of 66.2% measured by mercury intrusion, the average pore size is of 0.2μm, the overall performance is ideal.

(a) Automatic Tape-Casting (b) Single matrix (c) 400t precision hot press (d) Large area matrix (900mm*600mm)

Figure 5. The tape-casting preparation process of water - based matrix

2.3. Matrix performance test

The prepared matrix, the sintered nickel electrode and the carbonate electrolyte salt are assembled to be a molten carbonate fuel cell unit to test the electrochemical performance of the prepared matrix. The assembly diagram and the physical drawing are shown in Fig. 6.

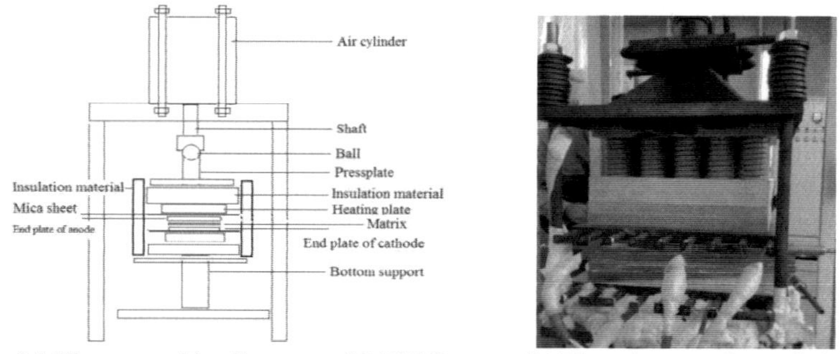

(a) The assembly diagram of MCFC (b) The picture of test field

Figure 6. The molten carbonate fuel cell unit assembly

The assembled molten carbonated fuel cell was heated by electric heating at room temperature to 923K according to a certain temperature rise curve. In the 813K, H_2 was introduced into the MCFC anode, and the air and CO_2 were introduced into cathode, and electrochemical reaction occurs, resulting in open circuit voltage. After 923 K, electronic load was accessed to test cell discharge performance [3], [4].

Using the electrochemical workstation produced by German Zanner company to test its performance and long-term stability, the test results shown in Fig. 7 and Fig. 8.

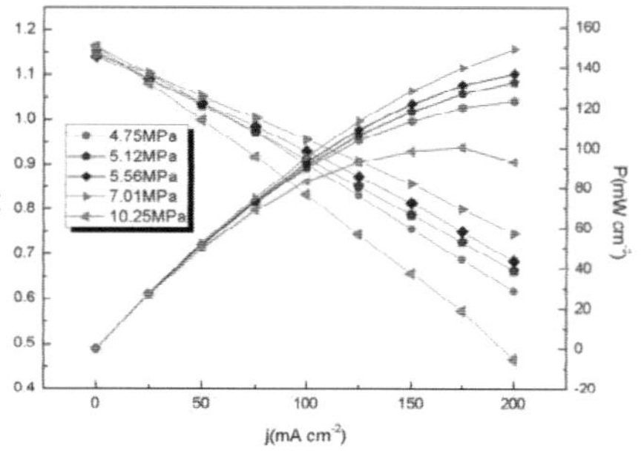

Figure 7. The I-V curve test of MCFC single cell

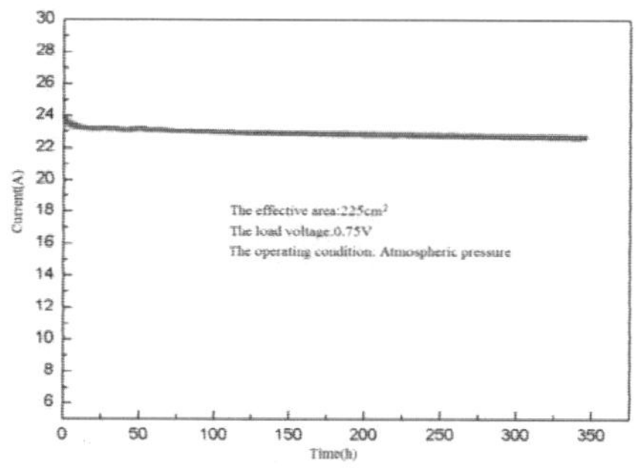

Figure 8. The diagram of MCFC long discharge

It can be seen from Fig. 7 and Fig. 8, at the discharge voltage of 0.75V, the discharge current density of the assembled single cell reached $120mA/cm^2$, after nearly 350 hours of continuous operation, the performance is basically stable, indicating that the prepared matrix has good electrochemical properties and could fulfill the requirements of MCFC power generation.

3. Influencing Factors of Preparing matrix by Water - based Tape Casting

3.1. Effect of solvent

Water-based casting was carried out using demonized water as the solvent during the experiment. The amount of water content, not only affects the viscosity of cast slurry, but also seriously affects the sedimentation properties and the thickness control of the later casting process. At the same time, the solvent content should also take into account with the dispersant, binder, plasticizer solubility relationship. Table 1 shows the effect of different solvent contents on the casting effect [5], [6].

Table 1. The effect of different solvent content on the casting

Solvent content	Casting effect
Less than 20%	The slurry is not uniform and is dry.
20~30%	Cast film cracking, and the green body is difficult to remove from the polyester film.
30~35%	The slurry is relatively uniform, cast paste thickness is better control.
More than 35%	The slurry is not uniform, casting process can not control the thickness of the matrix

As can be seen from Table 1, when the solvent content is in the range of 30~35%, it doesn't occur the deposition phenomenon, and the slurry is more uniform, it could be better control the thickness of the green body to achieve experimental purposes. So the amount of adding demonized water should be 30 to 35% during the experiments [7]-[11].

3.2. The effect of amount of dispersant added and milling time

In the slurry system, the function of dispersant is to disperse the powder particles in the slurry, so that the binder, plasticizer could better wrap particles to achieve the desired effect. Dispersant not only can enhance the solid capacity, but also has the function of weakening the viscosity of the slurry. In the experiments, lactic acid and citric acid were used as dispersants. The results showed that the milling time required was controlled between 45-48 hours when lactic acid as the dispersant, while the milling time was controlled between 20-24 hours when the citric acid as the dispersant [12], [13].

3.3. Scraper height and tape speed on the film formation

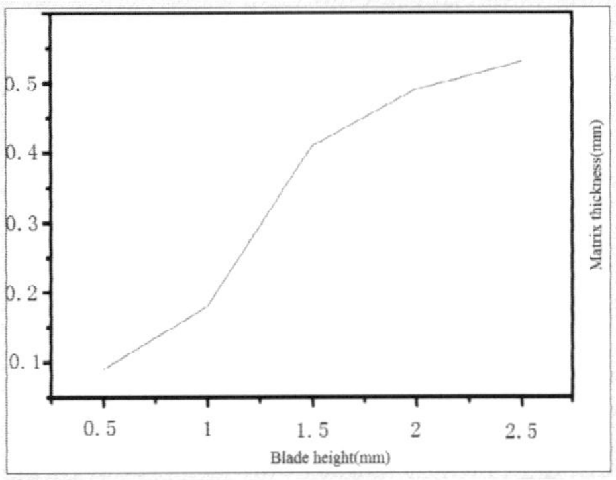

Figure 9. The relationship between blade height and matrix thickness

In the preparation of large area MCFC matrix, the scraper height and tape speed determines the thickness of the film after drying. Experiments shows that when the transport speed was 0.15m / s , the drying temperature was 50 ° C. The results are shown in Fig. 9.
From the fig. 9 we can see that although the thickness of the film is increased with the height of the scraper, it fluctuated when the blade thickness is in the range of 1.0~1.5mm. We hope that the prepared single matrix be thinner so that good gas barrier could still be kept after hot pressed together of many single films, which is beneficial to the open circuit voltage and power. In this case, the height of the scraper is controlled at 1 mm, and the thickness of the dried film is about 0.2 mm. We have 5 to 6 sheets of film heated at 90℃ and pressed at 50kg/cm^2 ,and the final matrix is produced whose thickness is of about 1.0mm, with a strong gas barrier properties[14], [15].

4. Conclusions

The large area environment-friendly Molten Carbonate Fuel Cell matrix was prepared by the automatic casting machine and the overall performance is ideal. The molten carbonate fuel cell unit was assembled with the prepared water-based matrix and the discharge test was carried out. The test results showed that the prepared matrix can be used for MCFC power generation. It is believed that the MCFC has good practical prospects and further optimizing the porosity and long lifetime test of larger area matrix are needed to promote China's MCFC industry.

Acknowledgments

This work was financially supported by Beijing Municipal Science & Technology Commission in the Future Science & Technology Park (Development of 20kW molten carbonate fuel cell power generation system) (TZ-16-BJKW01).

References

[1] Yi Baolian. Fuel Cell-theory technology application [M]. Beijing: Chemical Industry Press, 2004(in Chinese).
[2] Cheng Jian, Guo Liejin, Xu Shisen, et al. Preparation of α-LiAlO2 powder for MCFC by gel-solid method and its' characteristics [J]. Proceedings of the CSEE, 2013, 33(s1): 148-152(in Chinese) .
[3] Cheng Jian, Guo Liejin, Xu Shisen, et al. Key components of molten carbonate fuel cell matches and its performance diagnostic [J]. Proceedings of the CSEE, 2012, 32(s1): 101-107(in Chinese) .
[4] U. S. Department of Energy Office of Fossil Energy, Morgantown Energy Technology Center, Fuel Cell Handbook, January 1994.
[5] Antolini E. The stability of LiAlO2 powders and electrolyte matrices in molten carbonate fuel cell environment [J]. Ceramics International, 2013, 39(4): 3463-3478.
[6] Choi H J, Lee J J, Hyun S H, et al. Fabrication and performance evaluation of electrolyte-combined α-LiAlO2 matrices for molten carbonate fuel cells[J]. International Journal of Hydrogen Energy, 2011, 36(17): 11048-11055.
[7] Zhang J, Zhang X, Tu M, et al. Preparation of core(Ni base)-shell(Silicalite-1) catalysts and their application for alkali resistance in direct internal reforming molten carbonate fuel cell[J]. Journal of Power Sources 2012, 198(1):14-22.
[8] Cheng Jian, Xu Shisen, Shu Kai. Test study on power generation from molten carbonate fuel cells[J]. Thermal Power Generation, 2010, 39(6):11-13(in Chinese) .
[9] Huang Bo, Zhu Xinjian, Yu qingchun, et al. kW- class molten carbonate fuel cell stack[J]. Scientia Sinica Chimica, 2011,41(12):1884-1888(in Chinese) .
[10] Lin Huaxin, Zhou Li, Ming Pingwen, et al. Influence of MCFC stacking pressure on matrix performance [J]. Battery Bimonthly, 2007, 37(1):6-8(in Chinese) .
[11] Zhou Li, Yi Baolian, Cheng Mojie, et al. Study on kW-scale molten carbonate fuel cell stacks[J]. Battery Bimonthly, 2002, 32(3):138-141(in Chinese) .

[12] Patil K Y, Yoon S P, Han J, et al. Phase stabilities in molten Li-K carbonate of efficient matrix materials for molten carbonate fuel cells thermodynamic calculations and experimental investigations[J]. Journal of Material Science, 2011, 46(8):2557-2567.

[13] Zhou L, Lin H, Yi B, et al. Study on a new structure of the separator plate assembly for molten carbonate fuel cell(MCFC)stacks[J]. Chemical Engineering Journal, 2007, 125(3):187-192.

[14] Wang Pengjie, Cheng Jian, Zhang Ruiyun, et al. Preparation and test of electrolyte film in molten carbonate fuel cell[J]. Thermal Power Generation, 2015, 44(12):42-45(in Chinese) .

[15] Zhou L, Lin H, Yi B, Zhang H, Shao Z, Ming P, Cheng M. A study on the start-up and performance of a kW-class molten carbonate fuel cell (MCFC) stack [J]. Electrochimica Acta 2006; 51:5698.

Dependence of Optical and Thermal Properties on Substrate of Solar Thermal Collectors

Bhim Kafle, Rishi Lamichhane, and Sandesh Basnet

Kathmandu University

Email: bhim@ku.edu.np; rishilamichhane17@gmail.com;basnetjack.sb@gmail.com;

Abstract. The optical and thermal properties of the black crome based Solar thermal collectors (STCs) deposited on three different substrates (aluminium, Al; galvanized iron, GI; and stainless steel, SS) were investigated. The devices were prepared by two different methods: electro-deposition and dip coating and were heat treated at 300 ^{O}C. Each STC's performance was evaluated by measuring optical and thermal properties: Optical properties were measured with UV-Vis, Raman and IR Spectroscopy. For later measurements, all the STC samples were kept inside an air tight glass box and are exposed to the solar radiation over all the sunshine hours in summer (from 7:30 am – 5 pm, August). Then, the instantaneous temperature was recorded, simultaneously, of all samples with IR-temperature sensor. Among all the samples, the STC with black chrome coated on Al substrate showed the highest temperature, reaching the maximum value of ca. 95 ^{O}C at about 1 pm. Moreover, the STC samples fabricated by dip coating found to possess as equal optical and thermal properties as samples prepared by electro-deposition.

1. Introduction

As in the other parts of world, solar water heating systems (SWHSs) are commonly used by many households in south Asian nations as well, owing to their relatively affordable cost (around 300 USD). These systems have acceptable performance due to the relatively high solar radiation in the region. For example, in Nepal, the average daily solar radiation is around 5.5 kW h/m2 and over around 300 yearly sunny days [1].

The most crucial component of any SWHS is the solar thermal collector (STC), a device that absorbs the incoming solar radiation, converts it into heat, and transfers this heat to a fluid (e. g., water) flowing through the collector [2,3]. The solar energy thus collected is carried from the circulating fluid either directly to the hot water or a thermal energy storage tank from which can be drawn for use at night and/or cloudy days.

A preliminary survey on SWHSs in Nepalese market shows that there are, mainly, two kinds of STCs: One utilizes commercial black paint coated flat plate collector, which is attached with the galvanized iron pipes underneath. The STC exchanges heat with the GI pipes through which water is heated as it circulates by thermo-gravity. While another kind of SWHS comprises glass covered steel boxes encompassing, usually, glass tubes coated with cupper based nanoparticles. The SWHSs with STC of later kind are imported from abroad, while the former kind has been commercially fabricated in Nepal. However, both kinds of SWHSs have own problems: For e.g., former kind is less efficient and requires large surface areas to meet customers' requirement, while on the other hand, second kind is expensive and are unaffordable unless government provides heavy subsidies for purchasing them.

Therefore, for harnessing of solar energy efficiently, the traditional material for STC (black paint) must be replaced with an efficient absorber material with low fabrication process.
Specifically, efficient conversion of solar energy into thermal energy requires collector panels covered with special coatings that absorb across the solar spectrum while emitting poorly in the infrared region [4-6]. Surfaces with black paint are perfect absorbers (in the solar radiation range, 0.3 – 2.5 μm), however, have high emissive values in the infra-red range (> 2.5 μm) [7]. By contrast, so-called "selective" absorber surfaces absorb all incident radiation but show reduced emission in the infra-red region. Hence, advanced absorber substrates are typically covered with a coating that strongly absorbs in the solar radiation wavelength range in order to obtain high values of absorptivity, while yielding reduced emittance in the infrared region [8].
Electroplated black chromium is one of the most widely used selective absorbers [9], mainly due to the possibility of large area processing. While other materials have also been investigated: For e.g., semiconductor-dielectric composites (lead sulphide, PbS, on aluminum substrate) and metal-dielectric composite selective coatings (black nickel, NiS-ZnS, on a metal substrate, stable nickel, Ni–pigmented alumina, Al_2O_3, cermet coating on aluminum substrates, TiN_xO_y cermet coating on copper substrate, black nickel-tin coating on Cu substrate, NiO_x cermet, CrNxOy and CrAlO-based solar selective absorbing nanocomposite coating [5-7,10-14]. Among these examples, as indicated above, black crome selective coating possesses superior optical and thermal properties [7]. Its high solar-absorbing characteristics, selectivity factor (ratio of absorbance to emittance equals to 0.92 - 0.98/0.08-0.25) and a thermal stability up to 300 ^{O}C, are a result of the micro surface light-trapping morphology [9].
However, when attempts are made to reproduce any of the published formulations, it is difficult to obtain the desired features of the coatings [15]. In particular, method of fabrication of STCs and the substrates on which they are deposited has great influence over the thermal emittance values of the solar collector due to their direct physical contact with absorber [16]. Recently, Khamlich *et al.*[17,18] demonstrated that preparation of STCs with "dip coating" (with a low cost chemical bath) could be one of viable alternatives to conventional electroplating. Authors showed that STCs produced with the dip coating possess comparable optical and thermal properties [17, 18]. In this work, we performed the comparative study of performance of black crome STCs prepared by electrochemical deposition and dip coating of chromium onto Al, GI and SS substrates. Also for comparison, STCs with commercial black paint as selective coating materials were also evaluated.

2. Methodology

2.1. Preparation of Cr-Cr_2O_3 cermet films

The chromium based STCs presented in this work were fabricated by two different methods: (1) Electrochemical deposition and (2) dip coating of black chrome onto Al, GI and SS substrates with the size 6.5cm x 5.5cm x 0.1cm. For former method, the firstly a nickel under coat layer was deposited on each substrate using a Watt-type warm bath [16]. That was followed by the fabrication of a thin layer of chromium (Cr) selective absorber coating from a near-neutral electrolyte at room temperature. Details of each step are given below.

- Before electro-deposition, each substrate's roughness was enhanced by employing mechanical polishing with emery paper and was cleaned with the following scheme: Alkaline solution of KOH → distilled water → aqueous solution of 10% H_2SO_4 → distilled water. After cleaning process the substrates were dried in hot air oven at 100 ^{O}C for one hour.
- Nickel plating to thickness of about 10 microns.
- Cr plating was performed with the parameters 216 mA/cm2, 24 volts at 24 ^{O}C for 2 - 4 minutes.
- Then samples were rinsed with water, alcohol and dried (at room temp.) which was followed by heat treatment at 300 ^{O}C.

All the experiments for electro-plating the substrates were conducted in a conventional two-electrode cell. The substrate was used as the working electrode (cathode) and a Sn-Pb as anode (with Sn/Pb ratio

= 3/2). The relationship between film thickness and plating time was determined experimentally and thus the time during step (2) and (3) was adjusted to a desired thickness.

For coating Ni undercoat layer, 310 g/L Nickel sulphate hexahydrate ($NiSO_4.6H_2O$), 50 g/L Nickel chloride hexahydrate ($NiCl_2.6H_2O$) and 40 g/L boric acid (H_3BO_3) was used. The employed DC current for carrying out electroplating was ca. 14 A/dm2 and the pH was maintained at ca. 4. Also, a replicate of each sample was prepared for reproducibility issues. Then, thus formed nickel layer was thereafter cleaned with acetone, alcohol and distilled water and stored in a dust free environment before deposition of Cr layer. Following the deposition of the nickel layer, a thin layer of Cr selective absorber was deposited. To obtain an effective black selective absorber coating, a Cr-C Cr_2O_3 cermet layer of thickness ca. 0.2 micron was fabricated, using the electro-chemical bath of 125 g/L Chromic acid (CrO_3), 20g/L Barium carbonate (Ba_2CO_3), 10 g/L H_3BO_3, 0.5 g/L Ammonium fluoride (NH_4F) and g/L 5 g/L Nitric acid.

While with the dip coating, samples were prepared by following the method prescribed in Ref. [17]. Briefly, the cleaned substrates were immersed into the bath of only CrO_3 solution for 6 hrs at temperature 70 ^oC. Then the as prepared sample was air dried, before thermal treatment.

2.2. Thermal treatment of the electrodeposited Ni-Cr films

In order to submit the fresh Cr-Cr_2O_3 cermet film to a similar thermal treatment as that for a solar collector, samples were heated in a muffle furnace to 300 ^oC for 6 h. Then the selective absorber coated samples were analyzed to characterize their effectiveness in harnessing the incident radiation by measuring it optical and thermal properties. Thermal treatment of the electrodeposited Ni-Cr films
In order to submit the fresh Cr-Cr_2O_3 cermet film to a similar thermal treatment as that for a solar collector, samples were heated in a muffle furnace to 300 ^oC for 6 h. Then the selective absorber coated samples were analyzed to characterize their effectiveness in harnessing the incident radiation by measuring it optical and thermal properties.

2.3. Optical and photo-thermal response analysis of STC

Tests were conducted to evaluate STCs' optical and photo-thermal responses for their potential in solar thermal energy scavenging. This section describes the testing and test setups for these experiments.

Spectral reflectance measurements in UV-Vis and Near IR regions were carried out to investigate the optical selectivity of each sample. Reflectance measurements for all the samples in the wavelength range 0.28 – 1.1 microns were performed on a Reflectometric Spectrophotometer (Amstec., US). All the reflectance measurements were taken at near normal incidence at room temperature (ca. 20 ^oC). Raman spectra were also recorded with Raman Spectrometer (Model RIRM, RI Instruments and Innovation, India).

The study of optical properties was followed by analysis of thermal performance of the collector samples. That represented a key element of the STC characterization effort. The ability of the collector to achieve high temperature is critical for SWHSs. Hence, the temperature profile of each STC plate was characterized to determine the ability of the collector to transform solar radiation into heat. For this measurement, the experimental setup consisted of three kinds identical STC films. The first system (C1), a substrate (Al) of size 6.5cm x 5.5cm coated with commercial black paint as STC. The second system (C2), substrate (Al) of same size coated with black crome layer as STC (prepared by electro-deposition), and the third system (C3), substrates (Al, GI and SS) coated with black crome layer as STC (prepared with deep coating). Each solar collector plate (C1, C2, C3) has total surface area of about 35.75 cm2. The temperature measurements were conducted using a IR thermometer which was calibrated using a calibrated digital thermometer. All temperatures were recorded every hour from 7:30 am to 5 pm for consecutive two days on the first week of August, 2016 to ensure the reproducibility and reliability of the experimental output. Measured maximum radiation flux in the spot was about 900 W/m^2.

3. Results

3.1. X-Ray analysis

The X-Ray diffraction (XRD) patterns of the synthesized and annealed chromium/chromium oxide, Cr/Cr_2O_3 particles deposited on Al and GI substrates using electroplating and dip coating are demonstrated in Fig. 1. In between, black crome and substrate a Ni undercoat layer was coated for better adsorption of chromium STC. It can be seen that even though the samples were prepared with different techniques and at different substrates, all the samples show intense peaks in the planes [110], [024], [300] and [311] at $2\theta = 45.24^O$, 52.58^O, 65.75^O and 78.82^O, respectively. Observation of these peaks is in line with in earlier contributions [17,18]. However, the intensities of all the peaks observed with the samples prepared with electroplating found to be significantly higher (For. e.g., the intensity peak [110] was higher by the factor of ca. 8).

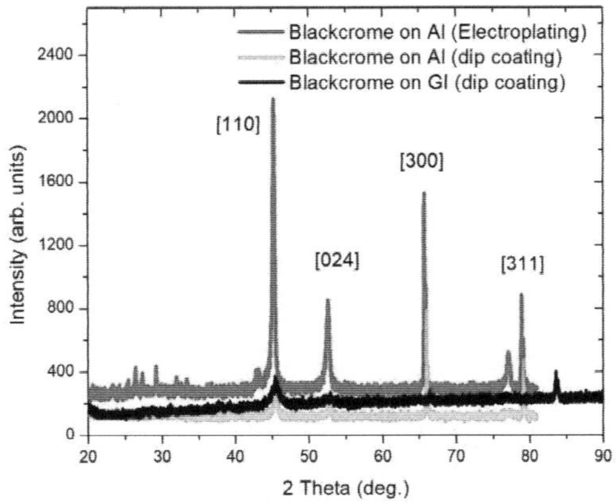

Figure 1. X-Ray diffraction patterns of the Cr/Cr_2O_3 particles deposited on Al and GI substrate.

3.2. Optical Properties of STC

Fig. 2 shows the spectral reflectance from UV to near IR region (wavelength ≈ 0.2 μm – 1.2 μm) for Al substrate and for thin film of black crome fabricated by electro-deposition and dip coating (represented with red blue and black solid lines, respectively). As expected, one can clearly observe magnificent reduction in the reflectance of the thin film of black crome compared to that of Al substrate. Moreover, the reflectance curve observed from the sample C2 (with electroplating process) almost overlapped with that of C3 (with dip coating process), which confirms that the film prepared by later process is also of good quality.

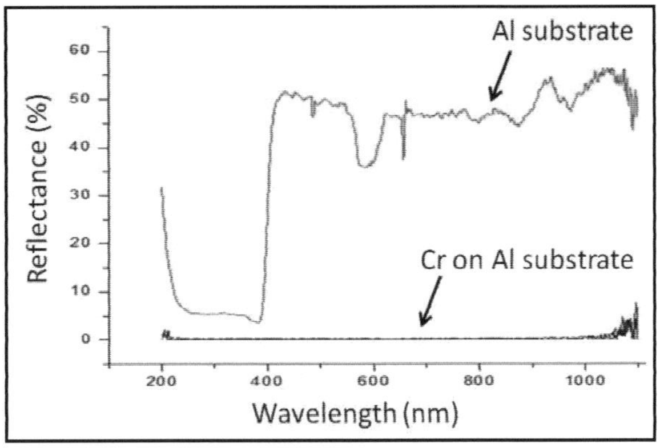

Figure 2. Reflectance from bare Al substrate and black crome (deposited on Al substrate by both the electroplating and dip coating).

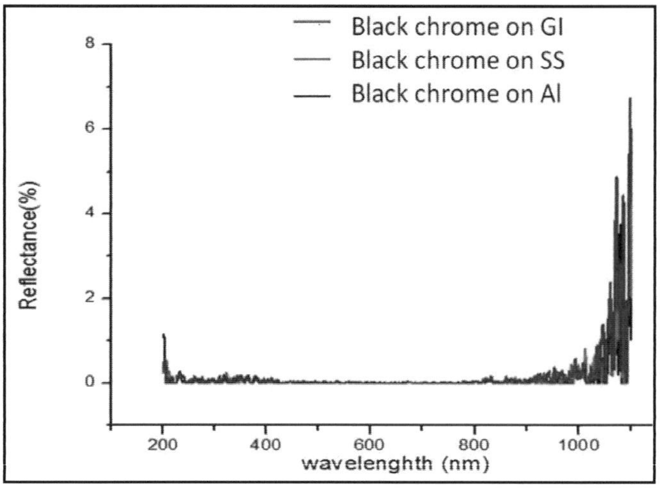

Figure 3. Reflectance for C3 samples (black crome films coated on Al, SS and GI substrates prepared by dip coating). All three spectrums found to overlap each other.

Fig. 3 shows the spectral reflectance from UV - near IR regions for sample set C3: Black crome films deposited by dip coating on substrates Al, SS and GI. All three samples showed less than 2% reflectance for all the scanned wavelengths below 1 μm, above which one would expect gradual rise in the spectrum (noise in the upper edge is due to upper limit of our spectrophotometer). The findings from present study are consistent with that the results of Khamlich et al. [18].

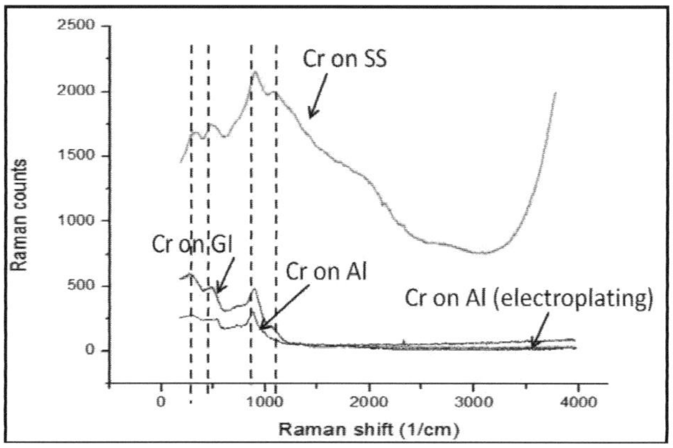

Figure 4. Raman spectra for black crome films coated on Al, SS and GI substrates (prepared by dip coating).

Fig. 4 displays the Raman spectra for black crome film coated on Al, SS and GI substrates. The Raman spectra are due to presence of chromium, as the contribution of the substrate has been eliminated in base line correction for each peak. Furthermore, a peak ca. 475 cm^{-1} confirms the presence of Cr_2O_3.

3.3. Photo-Thermal response Analysis of STC

Fig. 5 shows the setup for thermal behaviours measurement of our STCs, which comprises an air tight transparent glass box with wooden base on which samples are placed horizontally. The temperature profile of each STC was monitored in real sun condition. To verify the increase in collectors' temperature is only due to absorption of solar radiation by the STC, the temperature of the surrounding (which serves as baseline), substrate and of all STC samples was simultaneously monitored.

In Fig. 6, temperature as a function of time duration (hr) of radiation exposure on the surface of black chrome coated STCs samples C1, C2 and C3 are demonstrated. Also, included are the temperature profiles of Ni undercoat layer and of Al substrate. Temperature measurements were carried out in every 5 minutes time interval, in order to monitor gradual change in temperature with the radiation exposure time. As displayed in the Fig. 5, monotonic increase of the temperature was observed for both the STC and for Ni coated plate and substrate until 12:30 pm, but decreasing trend was observed thereafter. Notice, the large difference in temperature during mid day between STC and Al substrate (differ by a factor of ca. 4).

Figure 5. Setup for thermal measurement which comprises an air tight transparent glass box with wooden base on which samples are placed horizontally with the surface.

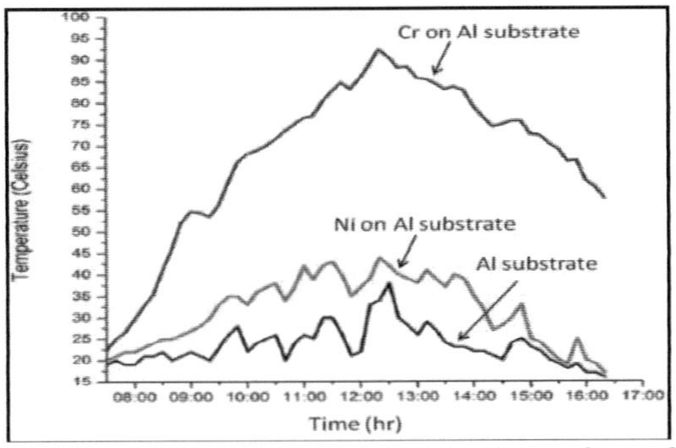

Figure 6. Temperature as a function of time (hr) recorded on the surface of black chrome coated STC. Also, included are the temperature profiles of Ni undercoat layer and of Al substrate.

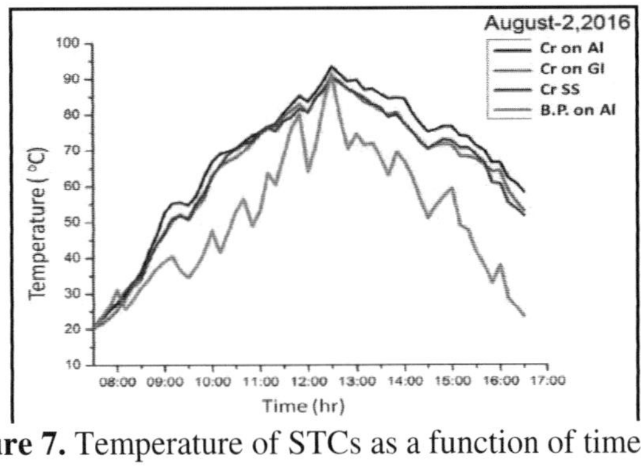

Figure 7. Temperature of STCs as a function of time (hr).

Fig. 7 displays the temperature as a function of time (hr) recorded on the surface of samples C3. Also, included is the temperature profile of sample C1, STC prepared by coating of commercial black paint (B.P.) on the Al substrate. One can apparently notice higher temperature for all the time duration of radiation exposure for STCs with black crome selective coating compared to that of black paint. Also, very interestingly, the fluctuation in the later spectrum is due to cloud cover in the sky (decrease in temperature directly reflects the higher emissivity of the film fabricated with black paint). Although very nominal, the STC with Al substrate always showed higher temperature compared to that of STCs prepared on GI and SS substrates.

4. Conclusion

The STCs deposited by "dip coating" showed similar optical and thermal properties to that of STCs prepared by "electro-deposition". Results on optical properties from present study are consistent with that the results of Khamlich *et al.* [17] and provide a further evidence (basis) for employing this low cost technique to fabricate commercial solar collectors. Moreover, about 20 % higher temperature observed with STCs with Cr-Cr_2O_3 than that of STCs with black paint likely to influence the performance of solar water heater system (and also of solar dryers) by the same factor. However, the films fabricated with dip coating were not as homogeneous as of "electro-deposited" one. Therefore, further study on this matter is required and is in progress.

Acknowledgement

Dr. B. Kafle would to thank "The World Academy of Science (TWAS), for developing countries, Italy, for the grant support [Grant No. 15-114 RG/CHE/AS-G FR3240287025]".

References

[1] Zahnd A and Kimber H 1992 Benefits from a renewable energy village electrification system. *Renewable Energy* **34** 362

[2] Kalogirou S 2004 Solar thermal collectors and applications. *Progress in Energy and Combustion Science* **30** 231

[3] Colangelo G, Favale E, Miglietta P and Risi A 2016 Innovation in flat solar thermal collectors: A review of the last ten years experimental results. *Renewable and Sustainable Energy Reviews* **57** 1141

[4] Chaurasia P and Twidell J 2001 Collector cum storage solar water heaters with and without transparent insulation material. *Solar Energy* **70** 403

[5] AlShamaileh E 2010 Testing of a new solar coating for solar water heating applications. *Solar Energy* **84** 1637

[6] Kennedy C and Price H 2005 Progress in development of high-temperature solar–selective coating. in *Proceedings of ISEC2005 International Solar Energy Conference, Orlando, Florida, USA.*

[7] Ogbonnaya E, Gunasekaran A and Weiss L 2011 Fabrication of a MEMS- based solar thermal collector for energy sustainability. *Proceedings of the ASME 2011 International Mechanical Engineering Congress & Exposition IMECE2011*, Denver, Colorado, USA

[8] Duffie J and Beckman W 1991 Solar Engineering of Thermal Processes. *Wiley-Interscience, 2nd ed.*

[9] Sukhatme S and Nayak J 2009 Solar Energy: Principles of Thermal Collection and Storage. *McGraw- Hill Education (India), 3rd ed.*

[10] Kirilov R, Stefchev P, Alexieva Z and Dikov H 2010 A High Effective Selective Absorbing Coating for Solar Thermal Collectors. *Solid State Phenomena*, **159** 97

[11] Shanmugam N, Selvam M, Srinivasan K, John S and Shenoi B 1984 Black Nickel-Tin Selective Coating for Solar Thermal Energy Conversion. *Metal Finishing* **82** 91

[12] Kennedy C 2002 Review of Mid-to High-Temperature Solar Selective Absorber Materials, *Technical Report, NREL/TP-520-31267, Colorado, USA*

[13] Wu L, Gao J, Liu Z, Liang L, Xia F and Cao H 2013 Thermal aging characteristics of CrN_xO_y solar selective absorber coating for flat plate solar thermal collector applications. *Solar Energy Materials & Solar Cells* **114** 186

[14] Liu H, Wan Q, Lin B, Wang L, Yang X, Wang R, Gong D, Wang Y, Ren F, Chen Y, Cheng X, Yang B 2014 The spectral properties and thermal stability of CrAlO-based solar selective absorbing nanocomposite coating. *Solar Energy Materials & Solar Cells* **122** 226

[15] Aguilar M, Barrera E, Palomar-Pardave M, Huerta L, Muhl S 2003 Characterization of black and white chromium electrodeposition films: surface and optical properties. *Journal of Non-Crystalline Solids* **329** 31

[16] Lee Y, Ruby D, Peters D, McKenzie B and Hsu J 2008 ZnO Nanostructures as Efficient Antireflection Layers in Solar Cells. *Nano Lett.* **8**, 1501

[17] Khamlich S, Nemraoui O, Mongwaketsi N, McCrindle R, Cingo N and Maaza M 2011 Black Cr/a-Cr_2O_3 nanoparticles based solar absorbers. *Physica B: Physics of Condensed Matter* **407** 1509

[18] Khamlich S, McCrindle R, Cingo N and Maaza M 2013 Annealing effect on the structural and optical properties of Cr/Cr_2O_3 monodispersed particles based solar absorbers. *Applied Surface Science* **265** 745

IOP Conf. Series: Earth and Environmental Science **73** (2017) 012016 doi:10.1088/1755-1315/73/1/012016

The Design and Construction of a Battery Electric Vehicle Propulsion System - High Performance Electric Kart Application

Mark Burridge[1] and Sanath Alahakoon[2]

[1] School of Engineering & Technology, Central Queensland University, Rockhampton, Australia

[2] School of Engineering & Technology, Central Queensland University, Gladstone, Australia

Email: s.alahakoon@cqu.edu.au

Abstract. This paper presents an electric propulsion system designed specifically to meet the performance specification for a competition racing kart application. The paper presents the procedure for the engineering design, construction and testing of the electric powertrain of the vehicle. High performance electric Go-Kart is not an established technology within Australia. It is expected that this work will provide design guidelines for a high performance electric propulsion system with the capability of forming the basis of a competitive electric kart racing formula for Australian conditions.

1. Introduction

Electric propulsion systems for competition racing karts have not yet reached the point in their development cycle whereby a recognised technical formula has been achieved for their introduction into sanctioned competitive electric kart racing. This is particularly true in Australia where there is no existing or proposed category for electric kart racing under the control of the governing body of this sport.

This paper attempts to provide an insight into the design and construction procedure that is necessary when engineering a high performance electric powertrain specifically for a competition karting application. In doing so it will provide a working example of a high performance orientated electric propulsion system with the capability of forming the basis of a competitive electric kart racing formula for Australian conditions.

The procedure requires the documentation of the design, construction and testing of an electric powertrain to determine if indeed it has the capacity to at least match the performance characteristics of a comparable competition Internal Combustion Engine (ICE) powered vehicle. This will be accomplished with the installation of the designed and constructed powertrain into an Australian Karting Association (AKA) level, competition rolling kart chassis for testing and evaluation.

At this introductory level, in order to minimise costs of the project, only proven Electric Vehicle (EV) technology will be incorporated in the design and construction of the propulsion system. However it is envisaged that this research project will form a platform for further development into more advanced EV traction systems ideally suited to competition applications.

The overall research project will deliver the following main outcomes;

- A specification for an EV propulsion system commensurate with an electric kart platform.
- A complete and working EV propulsion system manufactured and installed into an AKA level kart rolling chassis.
- A set of test results confirming the dynamic performance of the electric powered kart as a complete vehicle system.
- A basic operator's manual.

2. Methodology

There are three main work functions required for this assignment: (1) design which includes modelling and simulation, (2) construction and assembly of the powertrain, (3) testing and evaluation.

The design phase follows the sequence outlined in Figure 1 below.

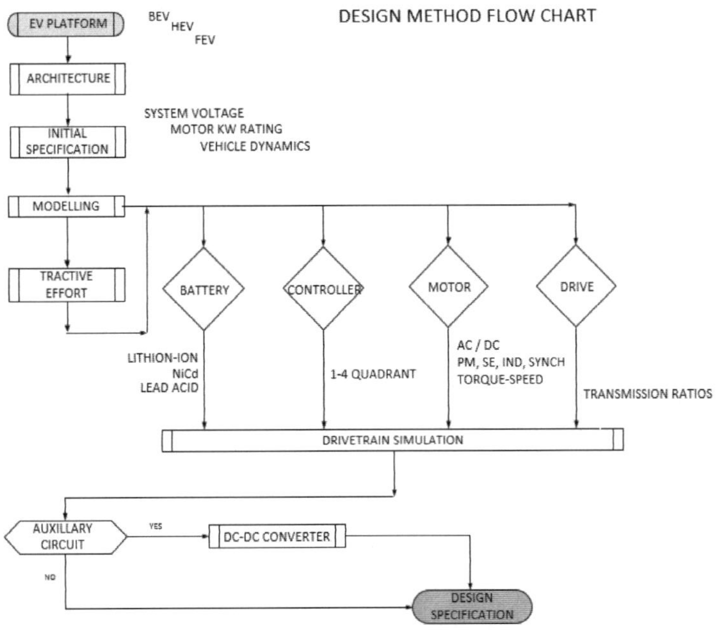

Figure 1. Design method flow chart.

3. Electric vehicle platform & architecture

The platform adopted for the kart will be battery electric vehicle (BEV) [1], [2], with a standard DC electric powertrain architecture as shown in Figure 2.

The propulsion system consists of a battery pack for the power source, motor controller for torque, and speed regulation, an electric traction motor to generate the torque and angular velocity, and a final drive to transfer the motor's output mechanical power to the rear axle to drive the vehicle forward. An auxiliary circuit to control any vehicle accessories not associated with the operation of the powertrain is an optional extra.

The DC architecture has been chosen because of its mature technology, simplicity in its operation and construction and availability of suitable light weight and compact EV components. An AC system, though providing superior torque speed characteristics, is more suited to full size production passenger road vehicles.

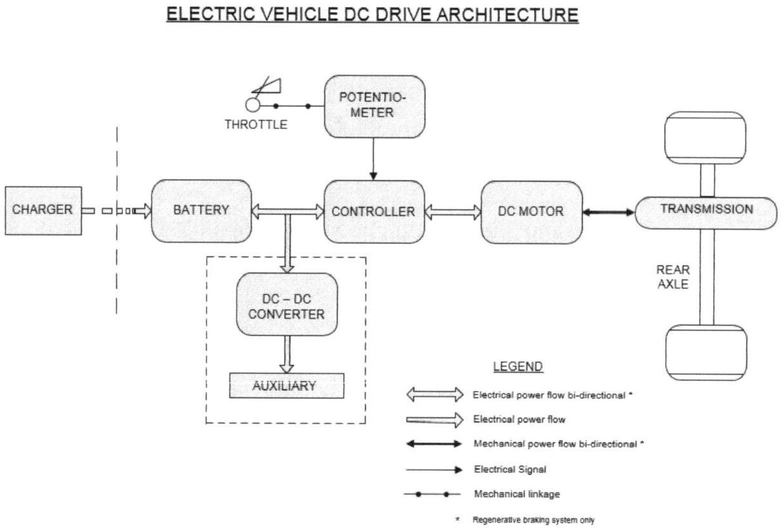

Figure 2. Electric kart powertrain architecture

4. Initial design specification

As this research project is primarily intended at the introductory level of the sport, it is only necessary to produce a machine which has the performance characteristics comparable to an equivalent ICE powered vehicle currently competing in the senior entry level class of the sport. This corresponds to the existing AKA class of karting, known as National or KA4. This class has the following performance characteristics:

- Power plant: IAME or Yamaha 100 cc air cooled ICE, restricted to 13 h.p. (9.7kW), or unrestricted to 22 h.p. (16.4kW) [3].
- Max speed: 77 to 104 kph depending on port restrictions [4].
- Acceleration: ICE acceleration figures unavailable. Therefore figures based on 50 to 75% of the recently released Bosch e-kart which has a claimed figure of 5.36 ms^{-2} with a 20kW motor [5].

The remainder of the specification relating to the major powertrain components has been determined from a literature review of the available technology applicable to the application. Hence the initial design specification can be summarised by the following criteria:

Initial Vehicle Dynamic Performance Specification

- Maximum velocity: 75 – 85 kph
- Acceleration: 2.68 – 4.02 ms^{-2}
- Range: 10 km minimum

Initial Powertrain Specification

- EV platform: DC BEV battery electric vehicle
- EV architecture: refer to Figure 2
- System voltage: 48V to 72V
- Rated motor power: 10 to 15 kW
- Motor type: DC brushed/brushless
- Controller: Electronic semi-conductor type with either 1 to 2 quadrant operation

- Battery pack: lead acid, nickel metal based or lithium metal based chemistries
- Final drive: Chain drive (gear ratio determined from modelling)

5. Traction Forces

In order for the design of an EV powertrain to satisfy a design specification, it must first be evaluated in the context of its intended operating environment. Any electric vehicle by virtue of its application is subjected to a number of resistive forces which oppose its forward motion. These same forces have a direct impact on the ability of the powertrain to overcome those resistive forces and in turn produce enough forward momentum to drive the vehicle forward to meet a predetermined rate of acceleration and final velocity [6]. Hence it is imperative to determine which acceleration and final velocity performance characteristics are feasible with a particular powertrain configuration relative to the vehicle to which it is providing a traction drive capability. Table 1 below summarises the traction forces applicable to the electric kart application researched in this project.

Table 1. Total tractive effort.

Total Tractive Effort $\mathbf{F_{te} = F_{rr} + F_{ad} + F_{he} + F_{la} + F_{wa}}$ (1)

Total Tractive Effort $F_{te}(N)$	*Rolling Resistance* F_{rr} *(N)*	*Aerodynamic Drag* $F_{ad}(N)$	*Hill Climbing* F_{hc} *(N)*	*Linear Acc.* F_{la} *(N)*	*Angular Acc.* $F_{\omega a}$ *(N)*
619.88	27.95	172.39	refer note 1	419.54	refer note 2

Notes:
1. Calculations based on flat ground hence $F_{hc} = 0$
2. Angular acceleration has been included into linear acceleration with an additional 5% mass to allow for indeterminate rotational inertia quantities.

The instantaneous traction power required to move the vehicle forward at a given velocity is given by the equation,

$$P_{te} = (F_{te} \times v) / n_g \quad (2)$$

P_{te} = motor output power?
v = vehicle velocity
n_g = drivetrain efficiency ~ 0.9

Assuming a maximum velocity of 85 kph or 22.2 m/s is required from the vehicle, the total motor power requirements P_{te}= 15.29 kW. It must be noted here that more advanced differential drive applications have also been reported in scientific literature [7].

6. Battery pack

When selecting a battery power source for a performance orientated EV drive, there are a number of criteria which require an assessment before any attempt at battery selection can be undertaken. There is sufficient previously published work also in the area of selecting various battery technologies for electric kart applications [8]. The following items make up the assessment criteria and include,

- Commercial availability and associated cost.

- Physical properties of weight and volume.
- Static electrical energy properties.
- Dynamic electrical energy properties.
    - Constant current discharge rate – the effect on battery voltage from a number of current discharge C rates which determines the corresponding amp hour (Ah) rating of the battery cell.
    - Constant power discharge rate to meet motor electrical power requirements.
    - Peukert Coefficient – the mathematical relationship describing the effect which an increasing discharge rate will have on a corresponding decrease in the available battery capacity. As a general rule, the lower the value the more efficient the battery.
- Battery charging and balancing requirements.

It must be stipulated that when evaluating battery behaviour, any results obtained must be taken as a guide only. The indeterminate nature of the chemical make-up of most battery types leads to a wide ranging disparity in their performance characteristics from battery manufacturers and suppliers alike. Hence any modelling or predication of a battery's behaviour is only an approximation and should be treated as an indication of their performance potential [6].

At the completion of the assessment process there is only one real contender for this application – the lithium chemistry based batteries. There are a number of different subtypes in this chemistry, however the most popular and readily available is Lithium-Iron Phosphate LiFePO4. This battery type exhibits a very flat discharge curve Figure. 3, which contributes to its high energy density characteristic.

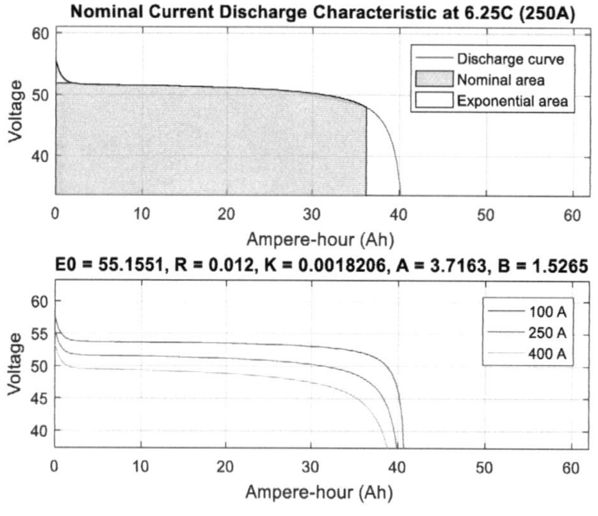

Figure 3. Typical 48V 40Ah LiFePO4 Current Discharge Curve.

The $LiFePO_4$ in a 40 to 60 Ah rating is the preferred battery chemistry and rating for the EV kart application.

7. Controller

The rpm speed and torque control of a DC motor can be undertaken by varying the motor's supply voltage or by flux weakening in the case of the separately excited DC motor. Either method results in a variation of the motor's armature current by means of a device more commonly known as a DC-DC converter or a DC "chopper". The chopper is so called because it takes a constant input DC source voltage and combined with high speed semiconducting switching, transforms the output voltage into a

chopped or segmented waveform. The variation in the time duration of the switching is the means of providing the variable DC voltage for motor control. Motor speed is proportional to voltage while torque is proportional to current [6].
The features of a ready-made commercial EV power electronics controller include:

- MOSFET, IGBT or GTO based high speed switching.
- Microprocessor control.
- Requires and external 5 K-Ohm throttle potentiometer.
- Automatic motor current limiting.
- Thermal protection.
- Under/over voltage protection.
- Programmable control of throttle response to motor speed control.

The controller is required to operate in either one or two quadrant mode of operation to obtain the desired direction of motion and or generator capabilities from the traction motor. In order to obtain both forward motoring and forward regenerative braking, two-quadrant operation is required. Figure 4 below details the basic circuitry and waveforms for a class C two quadrant chopper.

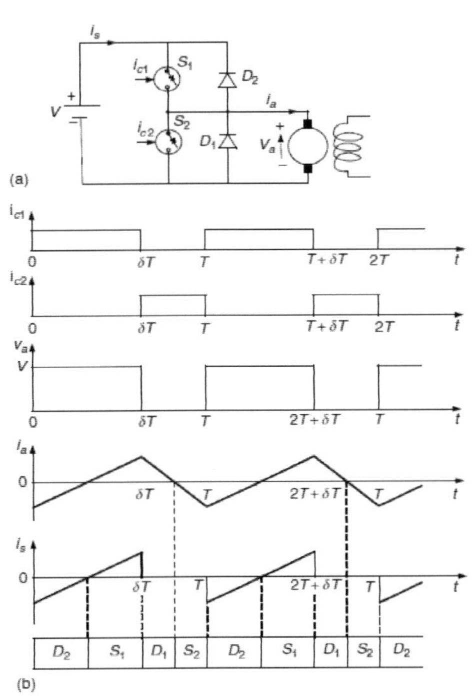

Figure 4. Two quadrant "Chopper" (a) circuit & (b) waveforms, source: [3]

The selection of a motor controller is therefore quite straight forward with a number of manufacturers such as Alltrax®, Curtis® and Sevcon® providing the latest in power electronic controllers for both AC and DC motors. In this instance, the controller is simply matched to the system voltage and max motor current ratings of the corresponding DC motor in the powertrain. A comparison of recognised DC controllers designed for traction applications is shown in Table 2 below. The preferred controller is the Sevcon due to its high current ratings and quadrant 2 forward regenerative braking capability.

IOP Conf. Series: Earth and Environmental Science **73** (2017) 012016 doi:10.1088/1755-1315/73/1/012016

Table 2. DC controller comparison

	Voltage range	Current limit 1 min rating	Current limit 1 hr.	Plug braking	Quadrant
Alltrax	12 - 72	300-450	125-250	optional	1
Curtis	36 - 144	250-550	-	standard	1
Sevcon	36-72	300-600	400-550	n/a	4

8. Motor and final drive

The parameters for choosing an electric drive for an electric kart application can be found in [9]. The circuitry of DC motors which are well suited to EV traction drive applications are shown in Figure 5 below. These include the brushed separately excited, permanent magnet and the brushless BLDC motors depicted in the circuit diagrams (a) (b) (c). There suitability is primarily due to the ease at which the speed and torque of each motor can be controlled to obtain a drive system which is very flexible and ideally suited to the demands of a modern motor vehicle.

The selected motors all exhibit the same inverse linear torque-rotational speed relationship as shown in Figure 5 (d). These include the characteristic high stall torque decreasing linearly to a no load value at maximum rpm. It is this simple straight line torque-speed characteristic which lends itself to ease of control and which makes it so attractive to a traction motor application over other motor types.

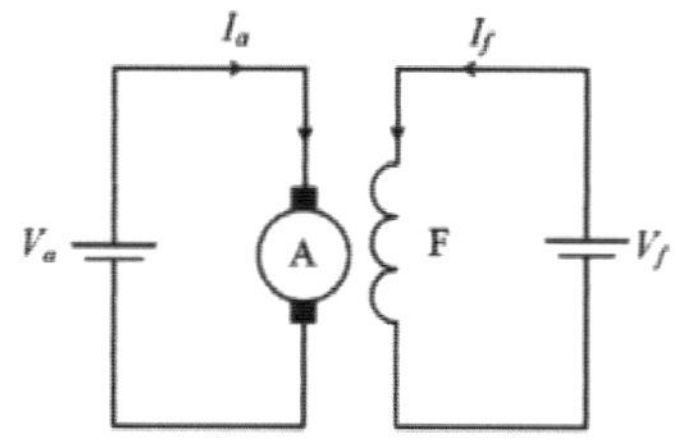

(a) Separately Excited
Source: [1]

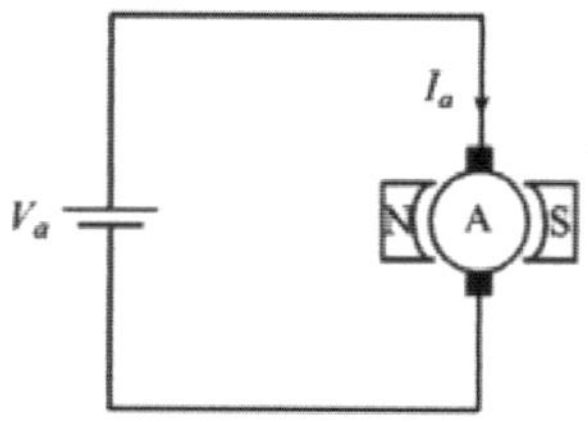

(b) Permanent Magnet
Source: [1]

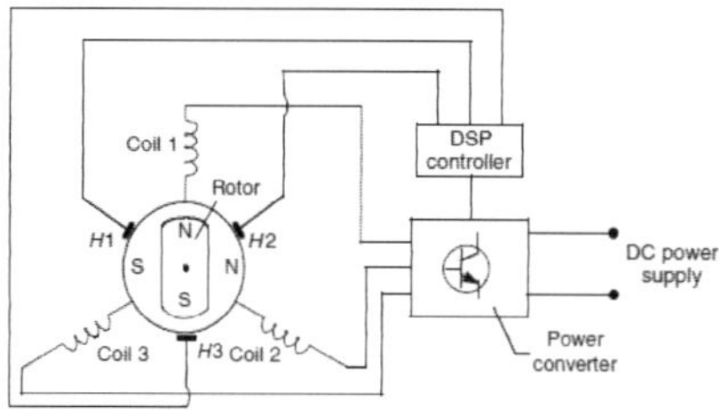

(c) Brushless DC (BLDC)
Source: [3]

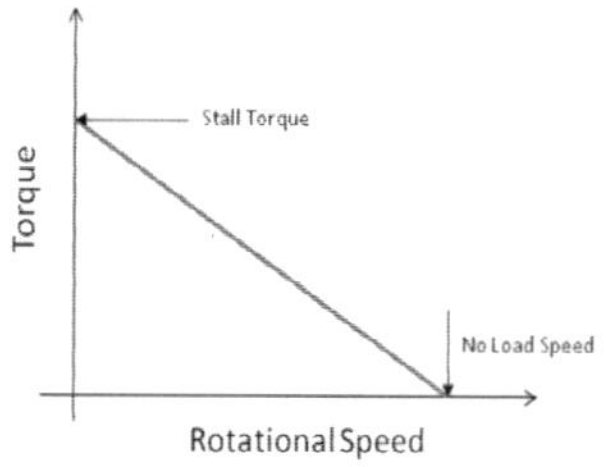

(d) Torque Speed Curve
Source: NI Instruments

Figure 5. EV DC traction motor options

A Permanent Magnet (PM) motor has been selected on the basis of its high kW and current ratings providing superior torque speed, combined with light weight, compact dimensions and commercial availability.

For a single gear transmission, a 2.6:1 final drive gear ratio has been selected based on the relationship between motor rpm, desired top speed (85 kph) and driven wheel diameter.

9. Modelling and simulation

Ideally a complete vehicle model containing drive cycles, control systems, powertrain components, mechanical drivetrain and longitudinal vehicle dynamics is the preferred simulation method for determining the vehicle's performance capabilities. In the absence of such a complex model, a rather more basic approach has been followed.

The basic steps include modelling the main propulsion system components to compare various options as undertaken in previous sections 6, 7 and 8, followed by the modelling of the vehicle dynamics to evaluate performance capability. Finally a complete powertrain model is employed to confirm the motor's response to an applied torque at the rear wheels.

The vehicle dynamics must be evaluated to ensure the drive system has sufficient capacity to deliver the mechanical power required to meet the initial performance specification listed in section 4. Hence the tractive forces opposing forward motion explained in section 5 are incorporated into a model to determine acceleration, velocity and range potential for a given electric motor, gear ratio and total vehicle mass configuration. The torque speed characteristics of the motor must also be included into the calculation using the first order differential equation (3) of the following form [10].

$$\frac{G}{r}T_{\max}\eta_g = \mu_{rr}mg + \frac{1}{2}\rho AC_d v^2 + \left(m + I\frac{G^2}{\eta_g r^2}\right)\frac{dv}{dt} \tag{3}$$

$$G = \text{gear ratio}$$

$$r = \text{radius of drive wheel}$$

$$T = \frac{K_m \Phi E_s}{R_a} - \frac{(K_m \Phi)^2}{R_a}\omega = \text{motor torque}$$

$$\mu_{rr}mg = \text{rolling resistance}$$

$$\frac{1}{2}\rho AC_d v^2 = \text{aerodynamic drag}$$

The application of (3) results in the Figure 6 Matlab® plots of velocity and acceleration capability of the vehicle.

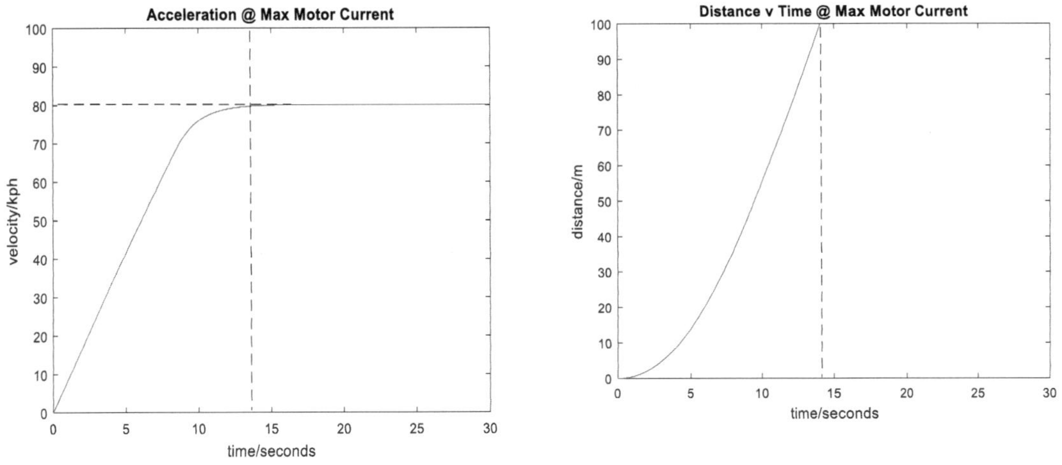

Figure 6. Vehicle dynamic modelling

The Matlab® Simscape® powertrain model as depicted in Figure 7 below has been utilised to simulate the traction motor's power, input current and rpm response to various torque applications at the rear wheels.

10. Final design specification

Table 3 below provides a summary of the final design specification resulting from the previous modelling simulations, with adjustments carried out during the procurement of components. The adjustments were necessary for reasons of cost and availability in order to fulfil the practical requirement of building a working prototype of the EV powertrain.

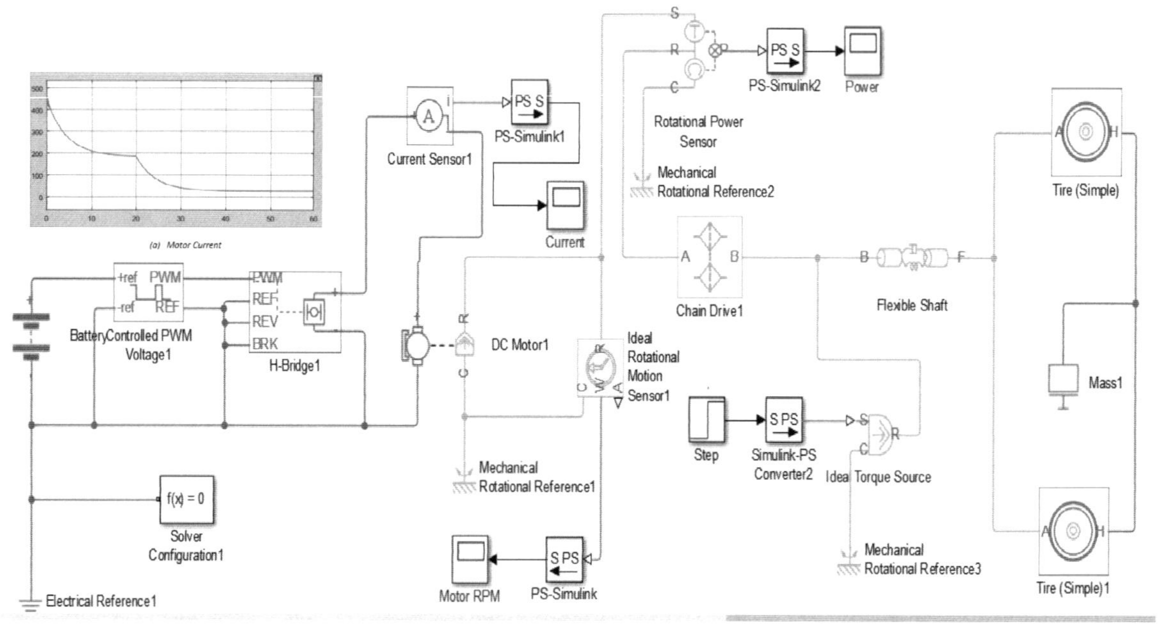

Figure 7. Matlab powertrain simulation

Table 3. Final design specification

	Initial Design Specification	**Final Design Specification**	Comments
EV platform & Architecture	BEV - DC	**BEV - DC**	
Performance.			
Max velocity	75 to 85 kph	**80 kph**	Flat gnd.
Acceleration	2.7 to 4.0 ms^{-2}	**2.33** ms^{-2}	Flat gnd
Range	10 km	**12 km**	Full throttle
Powertrain			
Sys. voltage	48 to 72 V	**48 V**	60V optional
Battery pack	Lithium 40Ah	**$LiFePO_4$ 20Ah**	Cost restriction
Controller	Sevcon 2Q	**Alltrax 1Q**	Controller is specific to the motor type.
Motor	PM 10 to 15kW	**Lynch PM Brushed 10kW**	
Final drive	2.6:1	**2.3:1**	Availability

IOP Conf. Series: Earth and Environmental Science **73** (2017) 012016 doi:10.1088/1755-1315/73/1/012016

11. Construction testing and evaluation

The following photographs Figure 8 depict the construction phase culminating with the complete propulsion system installed in the kart Figure 9.

(a)

(b)

(c)

(d)

Figure 8. Construction (a) Contactor & controller, (b) Lynch pm motor & final drive, (c) LiFePO4 battery pack , (d) Installed powertrain

On road testing was undertaken on a short track of approximately 500 metres in total length which included a straight section of 225 metres. Within this limited distance the vehicle was able to accelerate to an average maximum recorded speed of 76 kph (and still accelerating) before the limited straight distance precluded any attempt to reach the design top speed of 80 kph.
Maximum acceleration times have been calculated at 1.95 ms^{-2} from recorded data.

Figure 9. Complete vehicle with functioning electric powertrain.

IOP Conf. Series: Earth and Environmental Science **73** (2017) 012016 doi:10.1088/1755-1315/73/1/012016

Though not previously discussed or depicted in Figures 8 (c) and 9 above, a battery management system (BMS) has since been installed adjacent to the battery pack on the vehicle. This is a necessary piece of equipment to ensure individual $LiFePO_4$ cell voltages are maintained within limits and balanced across the battery pack. Without a BMS the lithium based cells will suffer from premature failure resulting in a severe reduction in the battery pack's life cycle.

12. Conclusion

This research project has demonstrated that by using an engineering systems approach, an all-electric propulsion system is a viable alternative to the ICE, in so far as outright performance is concerned, for a karting application. This has been achieved using commercially available current state, electro-mechanical, power electronics and energy storage devices.

When evaluated on purely technical considerations, the electric powertrain has proven to be capable of achieving the performance goals set in the beginning of the project. Further development of a complete system model combined with the refinement of the powertrain componentry can only further enhance the performance potential of the above propulsion system.

Acknowledgement

Authors would like to acknowledge the support received through Central Queensland University for the software simulations and access to research databases.

References

[1] Chau K T 2015 *Electric Vehicle Machines & Drives Design Analysis & Applications* (Singapore, John Wiley & Sons) pp.18–19.

[2] Neudorfer H 2007 *Comparison of Three Different Electric Powertrains for the use in High Performance Electric Go-Kart* (IEEE International Aegean Conference on Electrical Machines and Power Electronics ACEMP'07) pp. 21-26.

[3] Http://www.karting.net.au/summary-of-2016-class-names-and-tyres.

[4] Https://www.youtube.com/watch?v=Y_lhrfPQp-g.

[5] Http://www.motoring.com.au/bosch-reveals-e-kart-102493/.

[6] Larminie J Lowry J 2003 *Electric Vehicle Technology Explained* (John Wiley & Sons Ltd Chichester West Sussex England) pp. 183–187.

[7] Vitols K Galkin I 2012 *Analysis of Electronic Differential for Electric Kart* (15th International Power Electronics and Motion Control Conference EPE-PEMC 2012 ECCE Europe Novi Sad Serbia)

[8] Stepanek J Bednar B *et al* 2013 *Electric Kart "FeLis" with LiFeYPO4 batteries* (Clean Electrical Power ICCEP International Conference Alghero) pp. 151-154.

[9] Vitols K Reinberg N, *et al* 2010 *Drive Selection for Electric Kart* (Power Electronics and Motion Control Conference EPE/PEMC 2010 14th International Ohrid) pp. T9-15-T9- 18.

[10] Ehsani M Gao Y *et al* 2005 *Modern Electric, Hybrid Electric, and Fuel Cell Vehicles - Fundamentals, Theory and Design*. (Boca Raton Florida: CRC Press) pp 142-154.

A Bi-Level Particle Swarm Optimization Algorithm for Solving Unit Commitment Problems with Wind-EVs Coordinated Dispatch

Lei Song, Bo Zhang
School of Electrical Engineering, Shandong University, Jinan 250061, China

E-mail: songlei19920812@foxmail.com

Abstract. Nowadays, the grid faces much more challenges caused by wind power and the accessing of electric vehicles (EVs). Based on the potentiality of coordinated dispatch, a model of wind-EVs coordinated dispatch was developed. Then, A bi-level particle swarm optimization algorithm for solving the model was proposed in this paper. The application of this algorithm to 10-unit test system carried out that coordinated dispatch can benefit the power system from the following aspects: (1) Reducing operating costs; (2) Improving the utilization of wind power; (3) Stabilizing the peak-valley difference.

1. Introduction

With the improvement of public awareness of environmental protection, the penetration level of wind power generation and electric vehicles (EVs) have kept rapidly growing in recent years. According to the study [1], a high penetration of wind power may bring many challenges to the grid; besides, the unordered access to grid by a large amount of EVs could significantly impact the power system [2].
In this work, a model of coordinated dispatch was proposed. Then added it into the conventional Unit Commitment (UC) problem and solve it by bi-level particle swarm optimization algorithm. The outer level is based on Quantum-inspired Binary Particle Swarm Optimization (QBPSO) [3] to solve the UC problem; The inner level adopts conventional Particle Swarm Optimization (PSO) [4] to distribute the load between thermal units, simultaneously, calculate the accepted wind power and schedule EVs charging demand. According to the characteristics of QBPSO, this paper also proposed a novel mutation strategy to prevent the algorithm from falling into premature too early.

2. Model of Wind-EVs Coordinated Dispatch

2.1. Forecasting of EVs charging demand and wind power

(a) (b)

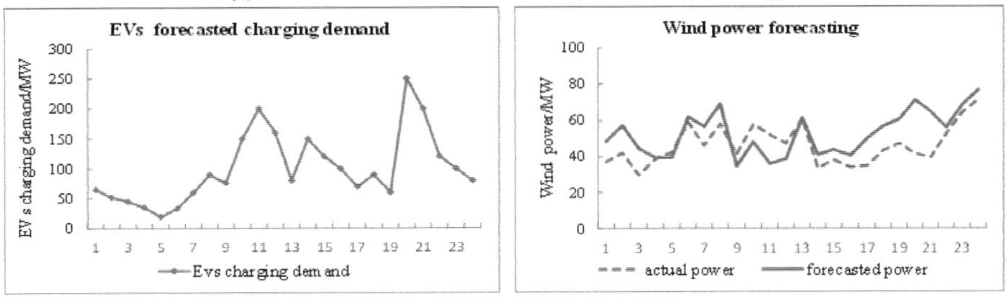

Figure 1. Forecasting of EVs charging demand and wind power

Figure 1 (a) shows the forecasted demand of EVs based on [5]. And according to Ouammi's work [6], taking the operating data of a certain wind farm in Northern China Power Grid as an example, the forecasted wind power curve is shown in figure 1 (b).

2.2. *Wind-EVs Coordinated Dispatch*

In this work, the forecasted wind power is regarded as a variable that can be scheduled. When discussing the grid's ability of accepting wind power, first of all, it is necessary to examine whether the thermal unit ramp-rate can cover the fluctuation of the load and wind power. Second, research the relationship between the economic benefit that brought by wind power and the amount of the wind power that the grid accepted. The wind power acceptance region is bound to following constrains:

$$(\sum_{g}^{G} -r_{d,g} \cdot u_g^t)\Delta T \le \Delta P_W^t + \Delta P_L^t \le (\sum_{g}^{G} r_{u,g} \cdot u_g^t)\Delta T \tag{1}$$

$$\Delta C_G - \Delta C_W < 0 \tag{2}$$

where $-r_{d,g}$ and $r_{u,g}$ are decent rate and rising rate respectively of gth unit; ΔP_W^t and ΔP_L^t are the variation of the wind power and load respectively in tth interval; u_g^t is the status of ith unit in tth interval; ΔC_G means the variation of coal consumption caused by wind power; ΔC_W is the coal conserved by wind power. Constraint (2) narrows the region of the wind-power-acceptance which determined by constraint (1). When the accepted wind power exceed this region, the increased ΔC_G will counteract the ΔC_W, because the wind power will cause many uneconomical problems such as increasing reserve capacity and making the thermal units deviate from economic operation point [7].

The wind power acceptance region can be extended by scheduling EVs charging demand. In order to ensure a basic charging demand can be satisfied, for each time interval, we set 30% of the forecasted charging demand as basic demand, and the rest 70% as variable that can be scheduled. The strategy can be expressed as follows:

$$P_{EV}^t = P_{EV,base}^t + \Delta P_{EV,ed}^t \tag{3}$$

where P_{EV}^t is the EVs charging demand after scheduling; $\Delta P_{EV,ed}^t$ is the variation according to the coordinated dispatch; EVs basic charging demand $P_{EV,base}^t = 30\% \times P_{EV,pre}^t$; $P_{EV,pre}^t$ is the forecasted demand of EVs.

Then, the total demand of EVs charging should obey constraint (4):

$$\sum_{t=1}^{T} P_{EV}^t = \sum_{t=1}^{T} P_{EV,pre}^t \tag{4}$$

The total EVs charging demand after scheduling of all intervals shouled be equal to the total forecasted demand.

3. A bi-level particle swarm optimization algorithm for solving unit commitment problems with wind-EVs coordinated dispatch

3.1. *Application of QBPSO to UC problem*

QBPSO adopts quantum-bit (Q-bit) to contain information, kind of like the chromosome encode mode in Genetic Algorithm (GA). And "Quantum Rotation Gate", which is a more efficient way for particle to update itself, is introduced into QBPSO. Given that, all possible combinations of decision variables can be derived from a single representation.

For a UC problem with T intervals and G thermal units, the outer-level particles should be initialized as a $1 \times T$ matrix as:

$$x = [x_1^1 x_2^1 \cdots x_N^1, x_1^2 x_2^2 \cdots x_N^2, \cdots, x_1^T x_2^T \cdots x_G^T] \tag{5}$$

where $(t\text{-}1) \times G + g$ is the index of gth unit in tth interval.

The generation for probability matrix and rotation angle adjustment is mainly based on [8].

When to decide the value of the rotation angle, a linear-decreasing-weighted method is proposed by this work, this method can expand the searching space in prophase, and accelerate the convergence in final phase, just as follow:

$$\theta = \theta_{max} - (\theta_{max} - \theta_{min}) \times \frac{k}{k_{max}} \tag{6}$$

where k is number of iterations; θ_{max} and θ_{min} can be found in table 1.

3.2. *Application of PSO to economic dispatch and coordinated dispatch*

3.2.1. *Initialization.*

Corresponding to the outer level, the inner level particles should also be initialized as a $1 \times T$ matrix. For thermal units, their particles are defined as:

$$p = [p_1^1 p_2^1 \cdots p_N^1, p_1^2 p_2^2 \cdots p_N^2, \cdots, p_1^T p_2^T \cdots p_N^T] \tag{7}$$

The Initialization of the initial position for thermal-unit particles is mentioned in equation (18).

$$[P_{wind}^1, P_{wind}^2, \cdots, P_{wind}^T] \tag{8}$$

$$P_{wind}^t = \min(P_{wind,forecasted}^t, P_{wind,accepted}^t) \tag{9}$$

Equation (8) is the initialization for wind power particles. The $P_{wind,accepted}^t$ in equation (9) is determined by equation (1) and (2). The initial positions for wind-power particles are set as the minimum value between the forecasted wind power and the margin of acceptance region in every interval, considered that the wind power has the highest priority to be put into the grid.

The particles of EVs charging demand are initialized according to (10). The initial positions for EVs-charging-demand particles are set as the basic charging demand as mentioned in equation (3).

$$[P_{EV}^1, P_{EV}^2, \cdots, P_{EV}^T] \tag{10}$$

3.2.2. *Update.*

The inner-level particles share the same pattern of update. The equation is as follow:

$$\begin{cases} v_{id} = \omega \times v_{id} + c_1 \times r_1 \times (P^{id} - P) + c_2 \times r_2 \times (P^{gd} - P) \\ P = P + v_{id} \end{cases} \tag{11}$$

where ω is inertia weight; c_1 and c_2 are acceleration constants; r_1 and r_2 are acceleration-weighted coefficient; v_{id} is the velocity of the particle, as its value gets smaller, the solution gets more accurate; P^{id}, P^{gd} and P can be replaced according to the type of particles.

3.3. *Objective function and constraints for UC problem with wind-EVs coordinated dispatch.*

- Objective function:

$$\min \sum_{t=1}^{T} [\sum_{i=1}^{N} (u_{i,t} F_i(P_{i,t}) + S_i u_{i,t} (1 - u_{i,t-1}))] \tag{12}$$

where $u_{i,t}$ is the ON/OFF status of ith unit in tth interval; F_i is the cost function of ith unit; S_i is the start-up cost of ith unit.

- Constraints:
  - Generation limit constraint:

$$P_{i,min} \le P_{i,t} \le P_{i,max} \tag{13}$$

  - Ramp-rate constraint:

$$-r_{di} \Delta T \le P_{i,t} - P_{i,t-1} \le r_{ui} \Delta T \tag{14}$$

  - Minimum up-time/down-time constraints:

$$\begin{cases}(T_{i,t\text{-}1}^{on}-T_{\min,i}^{on})(u_{i,t\text{-}1}-u_{i,t})\geq 0\\(T_{i,t\text{-}1}^{off}-T_{\min,i}^{off})(u_{i,t}-u_{i,t\text{-}1})\geq 0\end{cases}\tag{15}$$

- Spanning reserve constraints:

$$\begin{cases}\sum_{i=1}^{N}P_{i,\max}-\sum_{i=1}^{N}P_{i,t}\geq\alpha P_W+\beta(P_D+P_{EV})\\\sum_{i=1}^{N}P_{i,t}-\sum_{i=1}^{N}P_{i,\min}\geq\alpha P_W+\beta(P_D+P_{EV})\end{cases}\tag{16}$$

- Load balance constraint:

$$\sum_{i=1}^{G}P_{i,t}+P_W^t=P_{EV}^t+P_L^t\tag{17}$$

where $P_{i,\min}$ and $P_{i,\max}$ are the minimum power and maximum power of ith unit respectively; $P_{i,t}$ is the output power of ith unit in tth interval; $T_{i,t-1}^{on}$ and $T_{i,t-1}^{off}$ are the continuous time of ON/OFF in t-1th interval respectively; $T_{\min,i}^{on}$ and $T_{\min,i}^{off}$ are minimum continuous time of ON/OFF of ith unit respectively; $u_{i,t\text{-}1}$ is the status of ith unit in t-1th interval; α/β is the coefficient of reserve for wind / load.

3.4. Some improvements on the algorithm

- The advanced initialization for particles of thermal units

The thermal units with a higher priority should bear the load as much as possible, and the unit should remain above its minimum rated power once it is decided to be put into operation. So, for all the units, its initial position can be initialized by:

$$p_g^t=x_g^t\times[P_{\min}^g+(1-\frac{g-rand()}{G})\times(P_{\max}^g-P_{\min}^g)]\tag{18}$$

g is unit number after priority-ordering, and as g gets smaller, the priority of this unit gets higher, so the value of $(1-\frac{g-rand()}{G})$ is more close to 1, then p_g^t is close to its maximum rated power, and vice versa. After that, the particles of unit can be generated in a proper position which closed to the final solution. Obviously, this method can accelerate the convergence of the algorithm.

- The Mutation Strategy in outer-level algorithm

Applied with the concept of "mutation" in GA, a new evolution strategy can be modeled. According to the features of the 10-unit test system, unit 9 and 10 are regarded as base-load units and they are operated at full load in the most of intervals. So the "mutation" can be applied only to from unit 1 to unit 8. The procedures are demonstrated as follows:

- To select a Q-bit with a certain probability. For example, if the 5th bit is selected, then calculate the total power can be provided according to the status of units from 5th bit to the final bit.
- If the total power calculated by step (1) exceed the demand, then retrace step (1); otherwise, calculate the power shortage.
- Selecting a certain Q-bit or few Q-bits before the 5th bit, set its/their value to 1, others are set to 0. Examine if the units selected in step (3) can meet the shortage calculated in step (2).
- Putting the new particle into inner algorithm.

The mutation can not only change the value of a certain bit, but also change the evolution path for this bit. Different from the mutation in GA, the mutation here keeps the population diversity. The mutation can operated as many times as needed. If the fitness after mutation is better than before, then save this individual and put it into the next iteration; otherwise, try again or quit the mutation. If the algorithm gets no better results beyond continuous 100 iterations, the mutation strategy will be introduced automatically.

IOP Conf. Series: Earth and Environmental Science **73** (2017) 012017 doi:10.1088/1755-1315/73/1/012017

4. Case studies

The proposed algorithm was programmed with Matlab and tested with 10-unit test system in 24-hour horizon. The parameters were set as table 1:

Table 1. Parameters for the algorithm.

Population size	Maximum iteration	θ_{max}	θ_{min}	ω	c_1	c_2	r_1	r_2
60	1000	0.05π	0.01π	1	[1,2]	[1,2]	[0,1]	[0,1]

According to the forecasted data, the penetration of wind power is 4.32%, and the penetration of EVs charging load is 8.15%.

Figure 2 shows that the algorithm fell into premature since around 100th iteration. After the "mutation" proposed by this work, the algorithm continued the search in solution-space. The mutation procedure was introduced into the algorithm for the second time in around 300th iteration. The result proved that the "mutation" procedure can prevent the algorithm from falling into premature and improve the accuracy of the solution.

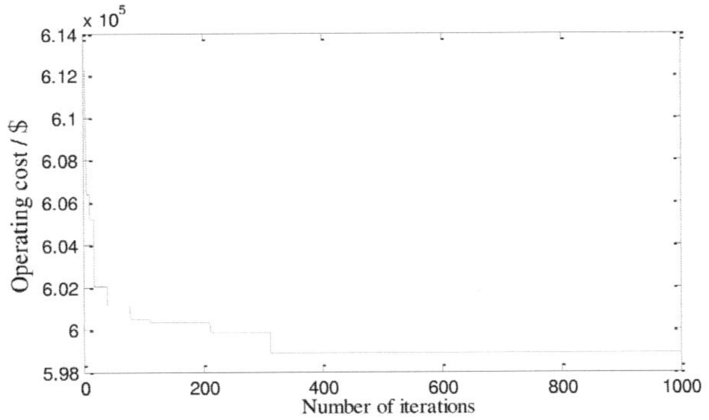

Figure 2. The convergence curve of the algorithm

The two vallies of wind acceptance region in figure 3 (a) are mainly because that the load has reached the peak, which led to the insufficient capacity for covering the fluctuation. After the coordinated dispatch, not only the acceptance region has been expanded, but also the actual wind power follow closely with the forecasted power, in other words, the utilization of wind power is improved, just as figure 3 (b) demonstrated.

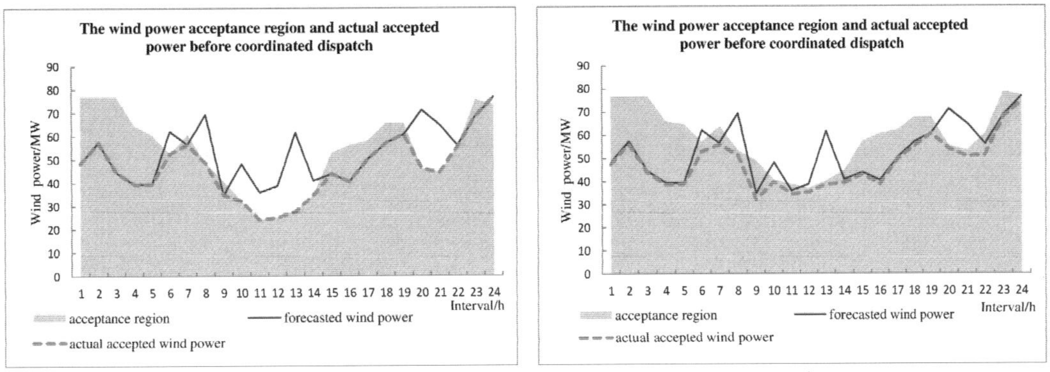

(a) (b)

Figure 3. Wind-acceptance region and accepted wind before/after coordinated dispatch

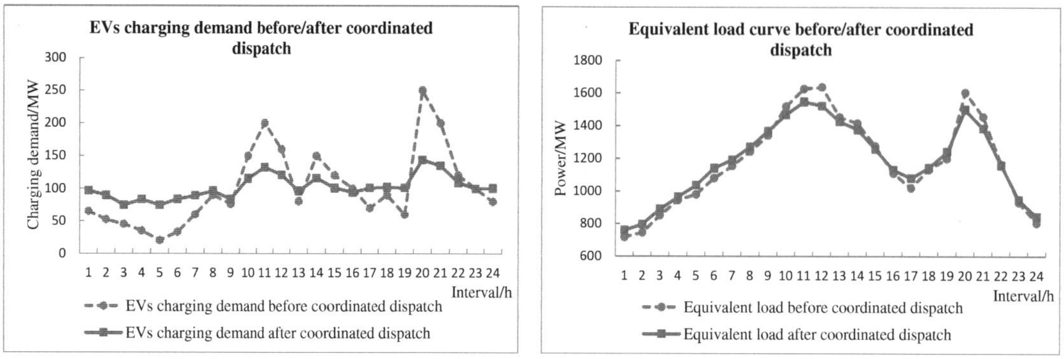

(a) (b)

Figure 4. EVs charging demand and equivalent load before/after coordinated dispatch

Moreover, the superposition of power demand on account of the participation of EVs charging is no longer severe as before, because that the coordinated dispatch can smooth the load curve to reduce frequent output-adjustments between thermal units. The curve of EVs charging demand and equivalent load before/after coordinated dispatch are shown in figure 4 .

Figure 5 demonstrates the results of unit commitment problem and the generation plan.

Interval \ Unit number	1	2	3	4	5	6	7	8	9	10
1	0	0	0	0	0	0	0	0	1	1
2	0	0	0	0	0	0	0	0	1	1
3	0	0	0	0	0	0	0	1	1	1
4	0	0	0	0	0	0	0	1	1	1
5	0	0	0	0	0	0	1	1	1	1
6	0	0	0	0	0	0	1	1	1	1
7	0	0	0	0	0	1	1	1	1	1
8	0	0	0	0	1	1	1	1	1	1
9	0	0	0	0	1	1	1	1	1	1
10	0	0	1	1	1	1	1	1	1	1
11	1	0	1	1	1	1	1	1	1	1
12	1	0	1	1	1	1	1	1	1	1
13	0	0	0	1	1	1	1	1	1	1
14	0	0	0	0	1	1	1	1	1	1
15	0	0	0	0	1	1	1	1	1	1
16	0	0	0	0	0	1	1	1	1	1
17	0	0	0	0	0	1	1	1	1	1
18	0	0	0	0	0	1	1	1	1	1
19	0	0	0	1	0	1	1	1	1	1
20	0	1	1	1	0	1	1	1	1	1
21	0	0	0	1	0	1	1	1	1	1
22	0	0	0	0	0	1	0	1	1	1
23	0	0	0	0	0	0	0	1	1	1
24	0	0	0	0	0	0	0	0	1	1

Interval \ Unit number	1	2	3	4	5	6	7	8	9	10	Wind
1	0	0	0	0	0	0	0	0	294	455	47
2	0	0	0	0	0	0	0	0	328	455	56
3	0	0	0	0	0	0	0	25	400	455	44
4	0	0	0	0	0	0	0	85	455	455	39
5	0	0	0	0	0	0	35	90	455	455	38
6	0	0	0	0	0	0	91	130	455	455	53
7	0	0	0	0	0	38	105	130	455	455	56
8	0	0	0	0	30	70	105	130	455	455	51
9	0	0	0	0	40	140	130	130	455	455	32
10	0	0	30	35	80	162	130	130	455	455	40
11	16	0	35	85	80	162	130	130	455	455	34
12	34	0	55	85	80	162	130	130	455	455	35
13	0	0	0	45	80	162	130	130	455	455	39
14	0	0	0	0	45	162	130	130	455	455	39
15	0	0	0	0	33	75	110	130	455	455	43
16	0	0	0	0	0	55	60	82	455	455	39
17	0	0	0	0	0	50	60	60	425	455	50
18	0	0	0	0	0	60	60	115	455	455	55
19	0	0	0	32	0	75	93	130	455	455	61
20	0	18	55	85	0	162	130	130	455	455	54
21	0	0	0	52	0	162	130	130	455	455	51
22	0	0	0	0	0	116	0	130	455	455	51
23	0	0	0	0	0	0	0	52	425	455	68
24	0	0	0	0	0	0	0	0	370	455	74

Figure 5. Results of UC problem and generation plan

5. Conclusion

The data demonstrated in table 2 can be obtained by analyzing the information in figure 3 and figure 4.

Table 2. The benefits to grid in contrast with the data before coordinated dispatch

	Total operation cost /$	Peak-valley difference / MW	Utilization of wind	EVs utilization of electricity at night
Before coordinated dispatch	612761	918.26	86.8%	16.5%
After coordinated dispatch	598846	787.7	90.7%	25.8%

where the night period is defined as 23:00 to 5:00 the next day.

By compared the results after coordinated dispatch with before, the conclusion can be drawn that the total operation cost has reduced by 2.27%, the peak-valley difference of equivalent load has decreased by 14.22%, and the utilization of wind power has risen nearly 4%. Besides, the EVs charging demand can be partly transferred to the night-off-peak period to let EVs use more power at this period which can benefit both the grid and the EVs users.

References

[1] Lei Y, Wang W and Yin Y 2002 Analysis of wind power value to power system operation *Power Syst. Tech.* **26**(5) 10-14

[2] Clement-Nyns K, Haesen E and Driesen J 2008 The impact of charging plug-in hybrid electric vehicles on a residential distribution grid *IEEE Trans. on Power Syst.* **25**(1) 371-380

[3] Wei M, Li Y, Jiang D, He Y, Huang X and Xu X 2008 A new evolutionary algorithm based on quantum statistical mechanics *IEEE Congress on Evolutionary Computation* pp 1722-1727

[4] Yu X and Zhang X 2014 Unit commitment using lagrangian relaxation and particle swarm optimization *Int. J. of Electr. Power & Energy Syst.* **61** 510-522.

[5] Yi F and Li F 2012 An exploration of a probabilistic model for electric vehicles residential demand profile modeling *Power and Energy Society General Meeting* pp 1-6

[6] Ouammi A, Dagdougui H, and Sacile R 2012 Short term forecast of wind power by an artificial neural network approach *Syst. Conf.* Vol 544 pp 1-5

[7] Zhi L I, Han X, Yang M, Lei Y and Kang K 2010 Power system dispatch considering wind power grid integration *Automation of Electr. Power Syst.* **34**(19) 15-19

[8] Jeong Y W, Park J B, Jang S H and Lee K Y 2009 A New Quantum-Inspired Binary PSO for Thermal Unit Commitment Problems *Int. Conf. on Intelligent Syst. Applications To Power Syst.* PP 1-6

Technical and Economic Assessment of the Implementation of Measures for Reducing Energy Losses in Distribution Systems

Alexander Aguila[1] and Jorge Wilson[2]
1. Universidad Politécnica Salesiana, Ecuador; 2. Universida Pontificia Bolivariana, Colombia

Email: aaguila@ups.edu.ec; jorgew.gonzalez@upb.edu.co;

Abstract. This paper develops a methodology to assess a group of measures of electrical improvements in distribution systems, starting from the complementation of technical and economic criteria. In order to solve the problem of energy losses in distribution systems, technical and economic analysis was performed based on a mathematical model to establish a direct relationship between the energy saved by way of minimized losses and the costs of implementing the proposed measures. This paper aims at analysing the feasibility of reducing energy losses in distribution systems, by changing existing network conductors by larger cross-section conductors and distribution voltage change at higher levels. The impact of this methodology provides a highly efficient mathematical tool for analysing the feasibility of implementing improvement projects based on their costs which is a very useful tool for the distribution companies that will serve as a starting point to the analysis for this type of projects in distribution systems.

1. Introduction

Until nowadays ninety percent of the energy needs of our planet are satisfied with the use of fossil fuels (oil, gas, coal), all extinguishable, heavily polluting and inefficiently used due to the predominant interest on energy production, rather than on its ecological effect. The importance of reducing the consumption of these primary sources has transformed from an economic problem to a vital problem and even worse, from a vital problem for the future it has transformed to one of the major accidents suffered in the development of mankind. The acidity of rainfall, natural disasters, the consequences of the greenhouse effect and the reduction of the ozone layer, are consequences that must be cured with a new way of energy production based mainly on efficiency [1]. The ability we have today in the energy distribution companies to produce and transmit electricity is due to the continuous action in the preparation of electrical workers, who install, operate and give maintenance to generating plants and the thousands of kilometers lines, towers, cables and poles that distribute electricity. For that reason, improving processes, optimizing network management and minimizing losses in transmission and distribution are key issues to reduce energy waste [2].

To achieve energy efficiency in a systematic way, the proper application of a body of knowledge and methods to ensure this practice is necessary. They constitute a technology and they are applied to the means of labor, human resources, processes and methods of management, control and planning [3].

Electrical power losses are inherent to real physical systems, in the electro - energy systems they are classified into two groups: technical and non-technical. Technical losses are largely related to the state of the physical characteristics of the system and its parts, which are a consequence of planning and designing criteria used in the past, initially in the materials used to construct the electric machines, conductors, insulators and other devices, and then, in the constructive characteristics and operative conditions in the analysed period [4]. Moreover, non-technical losses are the reflection of the activity of billing systems, the accuracy of the measuring equipment, the difficulties of the company to collect the service, the level of tariffs and the service quality.

The design and the construction of networks have a decisive influence on its subsequent operation, and even though technical losses will always be present, we should project adequately, so that they can be minimum [5].

Losses in electrical systems, reported in studies and measurements of electrical circuits, are not due to a sudden change in the conditions previously cited. Electrical losses are mostly caused by the natural aging of the systems which are not renewed promptly or it is not yet profitable to replace them, and they gradually reveal the consequences of design criteria that could be economic in the past or erroneously determined as economic.

The solution of these and other problems should be holistically evaluated to reduce the losses levels to minimum values [1][6].

This paper proposes a technical and economic assessment of measures to minimize energy losses in distribution systems, making emphasis on improving the resistance of the conductors and the exchange or conversion of distribution voltage at a higher level [7][8]. It also makes a planning model for estimation and classification of electrical losses in distribution networks [9] [10].

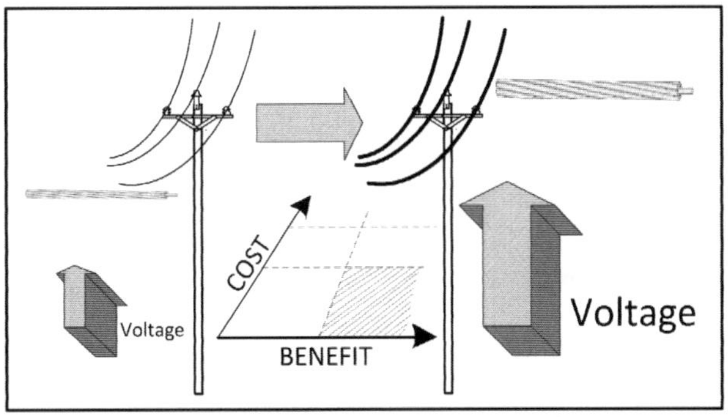

Figure 1. Planning diagram of the technical and economic improvements in distribution networks.

Figure 1 shows the purpose of the paper, which is to improve the energetic efficiency applying technical measures minimizing the investment costs.

2. General methods of economic analysis.

The most common electrical power losses are evaluated in association with heat dissipation in conductors (in the lines and the windings of electric machines), and they are linked in quadratic form with the current. For this reason, it should be treated by all means that in none of the parts of a network, current values exceed the nominal [5][10]. They are also related to the electrical resistance of the conductive materials, so it is absolutely necessary to properly calculate the circuits to choose wire sizes to satisfy the current and voltage values that will exist in them. The losses which depend on the voltage in distribution channels with voltage levels below 35 kV are small compared with those that

depend on current and mainly occur by the transformation to heat energy, from the electrical energy transmitted by drivers due to its electrical resistance[11]. The greater the amount of energy to be transmitted (consumer demand), and the greater the resistance of the conductors, the greater the losses would be [12]. But current and resistance values can be limited to optimal levels, so that the only losses which would occur are the strictly necessary in the process of transmission [5].
In many cases the power systems are in unwanted regimes, as a product of problems or deficiencies in them, and in order to solve them, we must implement measures aiming at achieving technical and economic improvements [5][13] [14], That is why for many years, methods of economic balance that serve to demonstrate whether a measure is applicable or not, have been studied. In this chapter different economic methods will be analyzed, to prove whether the technical measures applied in this paper, are feasible or not[1].

General methods of economic analysis:
Net Present Value (NPV)

$$NPV = -K + \sum_{J=1}^{N} \frac{FC_J}{(1+TD)^J} \tag{1}$$

Where:

K - Cost of the initial investment
TD - Discount rate
FC_J - Cash flow

$$FC_J = E_J - S_J$$

E_J - Inputs
S_J - Operating expenses

Si: $NPV > 0$ the measure is economically feasible.
$NPV \leq 0$ The measure is not applied.

The best variant is the highest NPV.
Net present value cost (NPVC)
The NPVC in an electric network can be expressed as follows:

$$NPVc = K + \sum_{j=1}^{N} \frac{(G_E + G_a + \Delta E \cdot \beta(1+r)^J)_J}{(1+TD)^j} \tag{2}$$

Where:
K- Equipment cost or initial investment as corresponds.
$GE = pE*K$; $Ga = pa*K$ Annual operating expenses and by amortization concept, which can be expressed in percent of the investment K.
pE, pa - Coefficients of annual expenses (ordinary maintenance and repairs); amortization coefficients, including the renewal discount of the equipment and those meant to the capitalized or general reparations, the latter are not included nowadays in the planned fixed expenses, and they are made according to the needs of the net.

ΔE - Energy losses through the electrical net.
β - Cost of kV of losses.
TD - Annual discount rate.
N- Number of years of exploitation evaluated.
r - Coefficient of vegetative annual load growth.

Here the best option is the one with the least Vanc.
Let´s evaluate the application of a determined technical measure, from the analysis of the Net Present Value Cost, this may arise as follows:
The introduction of a technical measure, aiming at reducing energy losses, will be considered economic if:

$$NPVc1 \leq NPVc \tag{3}$$

Where:
NPVc1- Net Present Value of the costs of the new variant (variant or improvement measure whose application is object to analysis)
NPVc- Net Present Value of the costs of the previous variant.

Substituting equation (2) in (3) and making some mathematical arrangements we have:

$$\frac{\Delta E - \Delta E1}{\Delta K} \geq \frac{1 + p_{\Sigma}\sum_{j=1}^{N}\frac{1}{(1+TD)^{j}}}{\beta\sum_{j=1}^{N}\frac{(1+r)^{J}}{(1+TD)^{j}}} \tag{4}$$

Where:

$pS = pe + pa$
$\Delta K = K1 - K$ – Additional costs required to implement the technical measure.
If after applying the new measure aiming at reducing energy losses, transmitted energy increases, the expression

(4) can be written as:

$$\frac{\Delta E - \frac{\Delta E1}{h}}{K(1-h)+K} \geq \frac{1 + p_{\Sigma}\sum_{j=1}^{N}\frac{1}{(1+TD)^{j}}}{h\beta\sum_{j=1}^{N}\frac{(1+r)^{J}}{(1+TD)^{j}}} \tag{5}$$

Where:
$\frac{E1}{E} = h$ - Coefficient that shows how many times the transmitted energy has increased.

Internal Rate of Return (IRR)

$$0 = -K + \sum_{J=1}^{N}\frac{FC_J}{(1+IRR)^J} \tag{6}$$

Here the best variant is the highest IRR.

2.1. Voltage change at a higher level[1][12].
For the voltage change in the expression (4) the term ΔK is expressed as:

$$\Delta K = Kd + Km1 + Keq + Kr - K \quad (7)$$

Where:

Kd – Cost of dismantling transformers (if necessary) and the old structures.

Km1 – Cost of installation of new transformers and new structures.

Keq - Cost of equipment needed for the new voltage level.

Kr – Unamortized costs of the initial investment (structure and equipment to be dismantled).

$$Kr = K(1 - pr * tj)$$

pr - Coefficient of renewal. Reverse of the regulatory duration of equipment for the previous voltage level.
tj – Equipment operating time when the change is applied
As it is already deduced from the application of the term:

$$\frac{\Delta E - \Delta E1}{\Delta K}$$

It's the relationship that expresses the energy savings invested additional monetary value which is actually achieved with the implementation of the measure, which must be greater than the term of the right of expression (4).

2.2. Change of conductors to superior transversal section [1][12].
In this case ΔK expression takes the following form:

$$\Delta K = Kd + (a + b * F1) - pr * tj(a + b * F) \quad (8)$$

Where:

F1, F – New and old section respectively

- Coefficient expressing the part of the investment that doesn´t depend on the caliber or cross section. For lines up to 13kV it is calculated by the following expression:
- Coefficient expressing the part of the investment that depends on the caliber or cross section.

$a + b * F1 = Km1 + Keq1$ - Represents the cost of mounting plus the equipment (structures, insulators, etc.) of the new caliber.
$a + b * F = Km + Keq$ - Represents the cost of mounting plus the equipment (structures, insulators, etc.) of the initial caliber.

As, $Kd = 0.2 * (Km1 + Keq1)$ then the expression ΔK is:

$$\Delta K = 1.2 * (a + b * F1) - pr * tj * (a + b * F) \quad (9)$$

Substituting the terms of ΔE, ΔE1 y ΔK in the expression (4), and clearing I, we obtain the equation to determine the value of the maximum current (Im).

$$I_m = \sqrt{\frac{\left[1 + p_\Sigma \sum_{j=1}^{N} \frac{1}{(1+TD)^j}\right]\left[a(1.2 - p_r t_j) + b(1.2F_1 - F \cdot p_r t_j)\right]}{3\rho\beta\left(\frac{1}{F} - \frac{1}{F_1}\right)\sum_{j=1}^{N} \frac{1}{(1+TD)^j}}} \quad (10)$$

The maximum current is defined as the current value, from which it becomes feasible. From an economic point of view, change the section F by the F1, since from that value; the Net Present Value Cost (VANC) will be lower with the new section than with the previous one.

If the caliber change, aiming at reducing the losses, increases power transmission capacity of the line, we introduce the coefficient h, expressing the increase of transmitted energy, in the equation. Then the expression to maximum current is:

$$I_m = \sqrt{\frac{\left[1 + p_{\Sigma} \sum_{j=1}^{N} \frac{1}{(1+TD)^j}\right]\left[a\left(2.2 - h - p_r t_j\right) + b\left(F - Fh + 1.2F_1 - Fp_r t_j\right)\right]}{3\rho\beta\left(\frac{1}{F} - \frac{1}{F_1 h}\right)\sum_{j=1}^{N} \frac{1}{(1+TD)^j}}} \quad (11)$$

Notice that h is a coefficient greater than 1, in the case h=1 (when there is not an increase of the energy that will be transmitted), the expression (11) transforms in (10).

3. Necessary information for estimating losses in distribution systems [1].

In any study of losses it is advisable to start the analysis with adequate knowledge of the system under study. The necessary information is related to:

Technical characteristics:

- Line Diagram.
- Voltage level.
- Length of conductors.
- Types of conductors.
- Features of conductors.
- Geometric setting of structures.
- Phases by circuits.
- Routes of circuits.
- Location of transformers.
- Electrical characteristics of transformers.
- Location of other equipment (capacitors for example)

Charge information [15] [16].

- Charge factor.
- Hourly demand of power plants, distribution substations, primary feeders, distribution transformers, etc. According to the scope of the study.
- Power factor.
- Sales of energy according to the type of user.
- Users associated with each distribution transformer.

Necessary information for the study of electric losses [1][17].

- Zoning of the city circuit.
- Visual inspection of the networks to classify the status of existing materials (poles, fittings, transformers, conductors, etc.) and the percentage use in the event of remodelling networks.
- Statistical data analysis and estimate global demand for global growth rates into the future.
- Measurement of losses in typical existing transformers (study protocol or test) to determine the appropriateness of their use in case of remodelling, economic chargeability, etc [18].

- Selection of sample size for measuring transformers secondary circuits, representative for the sectors in which the city was divided.
- Update, if necessary with lifting planes of primary networks, number of phases, size, drivers, length of sections, location, capacity and power factor, sectioning points and interconnection with neighboring circuits.
- Collection and analysis of information related to the future growth of the city in the short and medium term (roads, residential, commercial, industrial plans, etc.) [19].
- Projections of growth of new consumers by consumption category and geographic location.
- Estimate of number of current and future consumers of the city circuit.
- Data collection and analysis of data on existing circuits operation, including location, type and duration of interruptions, cost of materials and labor, etc [20].
- Data collection and analysis of information on business practices according to the design, use of type and wire size, type of primary and secondary distribution, chargeability of transformers, chargeability of primary circuits , connections, etc[18].
- Collection and analysis of information on unit costs of materials and labor for later use in the preparation of budget and economic analysis.

General methodology for estimating losses in distribution circuits [1].
To estimate power and energy losses, the following considerations must be taken into account:
For distribution substations losses can be evaluated from the value of losses in the peak demand and the loss factor of the system, then:

$$\Delta E = F_p \Delta PT$$

Calculating the loss factor depends on each particular system. In first approximation, its value can be estimated from the charge factor. It can be shown that the loss factor is bounded by the following values:

$$F_C^2 < F_p < F_C$$

An empirical relationship between the loss factor and the charge factor (developed by Bullery Woodrow) is:

$$F_p = X * F_C + (1 - X) * F_C^2$$

Where the variable $X \leq 1$ and depends on the characteristics of each system.
Usually, the power supplied by the substation and the value of the voltage are known, but as the considered points are taken away from it, the voltage value decreases [21].
This is due to the voltage drops occurring in the elements (feeders, transformers), by the effect of the current.

4. Conclusions and future works

The model proposed allows us to verify the technical and economic feasibility for the implementation of measures of electrical improvements in distribution nets, by posing different scenarios for future energy in the study circuits.
Specific cases were modelled, such as voltage conversion at a higher level and the replacement of low caliber conductors or crossed section, where all implementation activities in each of the improvements with their associated costs were related, using tools of economic calculation, which allows us to

visualize the coefficients that define the feasibility in terms of possible future increase in the transmitted energy.

For changing primary voltage to a superior level, it was established a relation expressing the energy savings by invested additional monetary value, which is actually achieved with the application of the measure.

To replace low caliber conductors by others with larger cross section, it was defined a value of current from which it becomes feasible, from an economic point of view, to change the section F of an existing circuit by the new proposed section. If the caliber change, aiming at minimizing losses, increases the capacity of power transmission of the line, we introduce a coefficient h in the equation, expressing the increase of the transmitted energy, from which a result of final calculation is obtained, verifiable for any situation of the project, whether current or future.

It was made a collection of necessary information and basic activities of analysis in study of electrical losses.

References

[1] A. Aguila, "Medidas para minimizar las pérdidas de energía en circuitos de distribución." .

[2] T. M. Krishna, N. V. Ramana, and S. Kamakshaiah, "A novel algorithm for the loss estimation and minimization of radial distribution system with distributed generation," 2013, pp. 1289–1293.

[3] P. Pezzini, O. Gomis-Bellmunt, and A. Sudrià-Andreu, "Optimization techniques to improve energy efficiency in power systems," *Renew. Sustain. Energy Rev.*, vol. 15, no. 4, pp. 2028–2041, May 2011.

[4] R. Viral and D. K. Khatod, "Optimal planning of distributed generation systems in distribution system: A review," *Renew. Sustain. Energy Rev.*, vol. 16, no. 7, pp. 5146–5165, Sep. 2012.

[5] S. Kalambe and G. Agnihotri, "Loss minimization techniques used in distribution network: bibliographical survey," *Renew. Sustain. Energy Rev.*, vol. 29, pp. 184–200, Jan. 2014.

[6] L. William E., R. Luke J., W. Liam D., A. Colin F., and S. Anthony R., "An economic evaluation of the potential for distributed energy in Australia," *Energy Policy*, vol. 51, pp. 277–289, Dec. 2012.

[7] T. J. Hammons, L. L. Lai, and K. P. Wong, "International practices for methods and techniques for energy loss minimization worldwide," 2008, pp. 1–4.

[8] A. Augugliaro, L. Dusonchet, M. G. Ippolito, and E. R. Sanseverino, "Minimum losses reconfiguration of MV distribution networks through local control of tie-switches," *IEEE Trans. Power Deliv.*, vol. 18, no. 3, pp. 762–771, Jul. 2003.

[9] M. A. Kashem, V. Ganapathy, G. B. Jasmon, and M. I. Buhari, "A novel method for loss minimization in distribution networks," 2000, pp. 251–256.

[10] P. S. Georgilakis and N. D. Hatziargyriou, "A review of power distribution planning in the modern power systems era: Models, methods and future research," *Electr. Power Syst. Res.*, vol. 121, pp. 89–100, Apr. 2015.

[11] R. Dashti, S. Yousefi, and M. Parsa Moghaddam, "Comprehensive efficiency evaluation model for electrical distribution system considering social and urban factors," *Energy*, vol. 60, pp. 53–61, Oct. 2013.

[12] J. A. Lobão, T. Devezas, and J. P. S. Catalão, "Influence of cable losses on the economic analysis of efficient and sustainable electrical equipment," *Energy*, vol. 65, pp. 145–151, Feb. 2014.

[13] R. Weron, "Electricity price forecasting: A review of the state-of-the-art with a look into the future," *Int. J. Forecast.*, vol. 30, no. 4, pp. 1030–1081, Oct. 2014.

[14] K. Qian, C. Zhou, M. Allan, and Y. Yuan, "Effect of load models on assessment of energy losses in distributed generation planning," *Int. J. Electr. Power Energy Syst.*, vol. 33, no. 6, pp. 1243–1250, Jul. 2011.

[15] F. Tanrisever, K. Derinkuyu, and M. Heeren, "Forecasting electricity infeed for distribution system networks: An analysis of the Dutch case," *Energy*, vol. 58, pp. 247–257, Sep. 2013.

[16] A. S. Bouhouras and D. P. Labridis, "Influence of load alterations to optimal network configuration for loss reduction," *Electr. Power Syst. Res.*, vol. 86, pp. 17–27, May 2012.

[17] M. M. Rahman, J. V. Paatero, and R. Lahdelma, "Evaluation of choices for sustainable rural electrification in developing countries: A multicriteria approach," *Energy Policy*, vol. 59, pp. 589–599, Aug. 2013.

[18] A. Aguila, D. Carrión, and L. Ortiz, "Analysis_of_power_losses_in_the_asymmetr," IEEE Lat. Am. Trans., vol. 13, no. 7, pp. 2190 – 2194, 2015.

[19] V. Orejuela, D. Arias, and A. Aguila, "Response of Residential Electricity Demand Against Price Signals in Ecuador," in PROCEEDINGS of the 2015 IEEE Thirty Fifth Central American and Panama Convention, 2015, no. Concapan XXXV, pp. 373 – 378.

[20] J. Vasco and A. Aguila, "Design of Automatic Reclosing scheme in distribution feeders," IEEE Lat. Am. Trans., vol. 13, no. 8, pp. 2587–2593, 2015.

[21] R. Pérez, A. Aguila, and C. Vásquez, "Classification of the Status of the Voltage Supply in Induction Motors Using Support Vector Machines," in 2016 IEEE/PES Transmission and Distribution Conference and Exposition (T&D), 2016, pp. 1 – 5.

IOP Conf. Series: Earth and Environmental Science **73** (2017) 012019 doi:10.1088/1755-1315/73/1/012019

Model Predictive Controlled Active NPC Inverter for Voltage Stress Balancing Among the Semiconductor Power Switches

Md. Parvez Akter[1] and Dylan Dah-Chuan Lu[1, 2]
[1]School of Electrical and Information Engineering, the University of Sydney, NSW 2006, Australia
[2]School of Electrical, Mechanical & Mechatronic Systems, University of Technology Sydney, NSW 2007, Australia

Email: md.akter@sydney.edu.au; dylan.lu@uts.edu.au;

Abstract. This paper presents a model predictive controlled three-level three-phase active neutral-point-clamped (ANPC) inverter for distributing the voltage stress among the semiconductor power switches as well as balancing the neutral-point voltage. The model predictive control (MPC) concept uses the discrete variables and effectively operates the ANPC inverter by avoiding any linear controller or modulation techniques. A 4.0 kW three-level three-phase ANPC inverter is developed in the MATLAB/Simulink environment to verify the effectiveness of the proposed MPC scheme. The results confirm that the proposed model predictive controlled ANPC inverter equally distributes the voltage across all the semiconductor power switches and provides lowest THD (0.99%) compared with the traditional NPC inverter. Moreover, the neutral-point voltage balancing is accurately maintained by the proposed MPC algorithm. Furthermore, this MPC concept shows the robustness capability against the parameter uncertainties of the system which is also analyzed by MATLAB/Simulink.

1. Introduction

The neutral-point-clamped (NPC) inverter introduced in 1981 by Nabae [1], is the most widely used topology in widespread applications in various fields such as renewable energy generation [2], medium voltage drives [3] and electric vehicle traction [4]. This NPC inverter provides significant advantages over the two-level voltage source inverter (VSC), such as lower THD, higher power capability and smaller *dv/dt* and output inverter common mode voltages [5]. Aside from these several advantages, the NPC inverter faces some major limitations including the uneven voltage stress across the semiconductor power switches, neutral point voltage balancing and robustness against the parameter uncertainties. The uneven voltage stress across the power switches limits the switching frequency, maximize the entire phase current and reduce the output power of the NPC inverter [6]. This voltage stress problem is solved by using the active neutral-point-clamped (ANPC) which offers additional zero switching states by employing two active switches in every phase of the NPC inverter [7]. Moreover, the neutral point voltage balancing and the robustness problems are solved by using the model predictive control (MPC).

The MPC is a multidimensional constrained based control technique which solves the constrained finite time optimal control problems of the nonlinear systems at every sample time instant. MPC has gained significant attention in power converters and electrical drives research due to its intuitive approach, elimination of linear controllers and modulators, and easy inclusion of nonlinearities and restrictions [8-10]. Due to these various advantages, MPC has been implemented in different power converter topologies such as voltage source inverter [11], active-front-end rectifier [12-14], rectifier and inverter (ac-dc-ac) system [15], bidirectional ac-dc inverter [16, 17], direct matrix converter [18] and indirect matrix converter [19], etc. Although, some research has been done on using of MPC in NPC inverter [20, 21], complete investigation on the model predictive controlled three-level ANPC inverter, namely, ensuring the voltage stress distribution across the power switches, neutral-point voltage balancing and robust operation, is still yet to be done in the existing state-of-art.

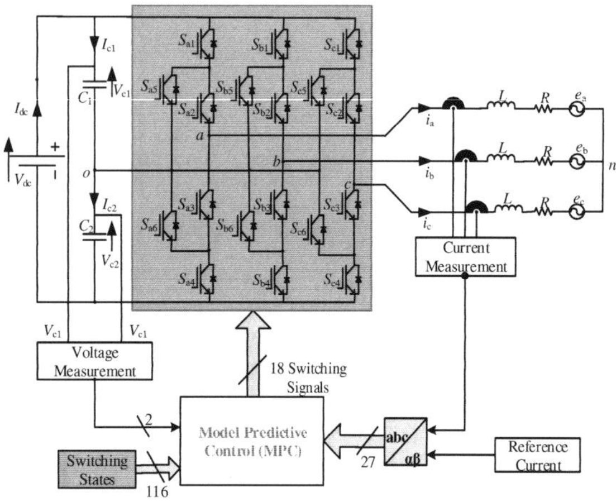

Figure 1. The ANPC inverter and its proposed MPC scheme.

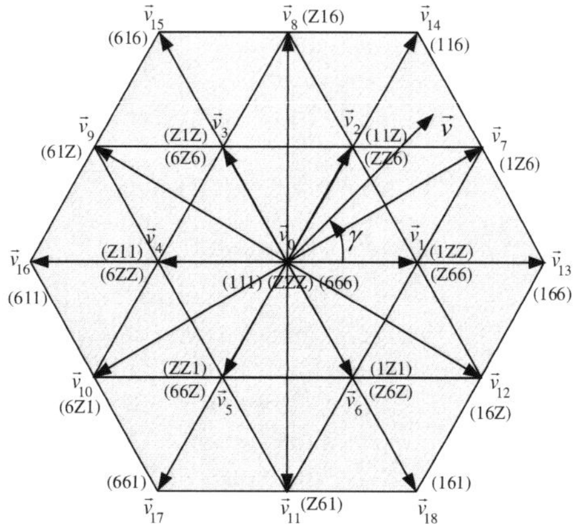

Figure 2. The space vector diagram of ANPC inverter.

Table1. Each Phase Switching States of the ANPC Inverter

State	Switching Sequence						Voltage vector ($\vec{v}_{ANPC}$)	No	
	S_{x1}	S_{x2}	S_{x3}	S_{x4}	S_{x5}	S_{x6}			
+	1	1	0	0	0	1	$+V_{dc}/2$	**1**	
0U2	0	1	0	0	1	0	0	**2**	
0U1	0	1	0	1	1	0	0	**3**	**Z**
0L1	1	0	1	0	0	1	0	**4**	
0L2	0	0	1	0	0	1	0	**5**	
-	0	0	1	1	1	0	$-V_{dc}/2$	**6**	

A model predictive controlled three-level three-phase active neutral-point-clamped (ANPC) inverter for distributing the voltage stress among the semiconductor power switches as well as balancing the neutral-point voltage and improving the system performance is presented in this paper. The rest of the paper is structured as follows. The dynamic modeling of the ANPC inverter with MPC scheme is introduced in section II. Section II presents the simulation results and discussion in order to validate the proposed model predictive control scheme for three-level three-phase ANPC inverter. Moreover, the comparative evaluation between the proposed ANPC and the traditional NPC inverter and the

IOP Conf. Series: Earth and Environmental Science **73** (2017) 012019 doi:10.1088/1755-1315/73/1/012019

robustness of the proposed MPC concept are analyzed in section IV. Finally, fruitful conclusions are drawn in section VIII.

2. Modelling of the ANPC inverter with MPC

Figure 1 shows the schematic system configuration which includes a three-level three-phase active neutral-point-clamped (ANPC) inverter and an inductive-resistive-active load. The ANPC inverter consists of 18 anti-parallel diode-clamped IGBT switches. Each phase-leg consists of six switches (S_{x1}-S_{x6}), where x is phase (a, b, or c). In this converter, the active switches S_{x5} and S_{x6}, and their associated anti-parallel diodes are clamped to neutral-point (o) which generates six switching states as shown in Table 1, these are capable of generating three different voltage levels, namely, $+V_{dc/2}$, 0 (0U2, 0U1, 0L1, 0L2) and $-V_{dc/2}$. The space vector block diagram of the three-level ANPC inverter is presented in Fig. 2, in which all the zero states are replaced by Z. Therefore, for a three-level ANPC inverter, one can still utilize 27 switching states to express all the possible switching states, such as [1 Z Z].

2.1. Mathematical Modelling of the ANPC Inverter

The power circuit of the ANPC inverter transfers the electrical power from dc voltage supply to three-phase inductive-resistive-active load as shown in the figure 1. The relation between the output voltage space vector ($\vec{v}_{\text{ANPC}}$), switching state function vector ($\vec{S}$) and the dc-link voltage (V_{dc}) of the NPC inverter can be presented as:

$$\vec{v}_{\text{ANPC}} = \vec{S} \times V_{\text{dc}} \tag{1}$$

Otherwise, the voltage vector ($\vec{v}_{\text{ANPC}}$) generated by the ANPC inverter is described with the phase to neutral voltages ($v_{\text{ao}}, v_{\text{bo}}$, and v_{co}) as:

$$\vec{v}_{\text{ANPC}} = \frac{2}{3}(v_{\text{ao}} + \vec{\omega} v_{\text{bo}} + \vec{\omega}^2 v_{\text{co}}) \tag{2}$$

where, $\vec{\omega} = e^{j2\pi/3}$, denotes 120° phase shift between the phases.

The three-level ANPC inverter is connected to the three-phase inductive-resistive-active ac load as shown in figure 1. Hence, the output space voltage vector dynamics generated by the inverter can be obtained as:

$$\vec{v}_{\text{ANPC}} = L\frac{d\vec{i}}{dt} + R\vec{i} + \frac{2}{3}\left(e_{\text{a}} + \vec{\omega} e_{\text{b}} + \vec{\omega}^2 e_{\text{c}}\right) + \frac{2}{3}\left(v_{\text{no}} + \vec{\omega} v_{\text{no}} + \vec{\omega}^2 v_{\text{no}}\right) \tag{3}$$

The ac load current ($\vec{i}$) and back-emf ($\vec{e}$) space vector model can be expressed as:

$$\vec{i} = \frac{2}{3}(i_{\text{a}} + \vec{\omega} i_{\text{b}} + \vec{\omega}^2 i_{\text{c}}) \tag{4}$$

and

$$\vec{e} = \frac{2}{3}(e_{\text{a}} + \vec{\omega} e_{\text{b}} + \vec{\omega}^2 e_{\text{c}}) \tag{5}$$

where i_a, i_b, and i_c are the ac load phase currents and e_a, e_b and e_c are the back-emf phase voltages.

Then the load current dynamics of the three-level three-phase ANPC inverter can be presented from (3), (4) and (5) as the last term of (3) is zero. Hence the three-phase load current dynamics is:

$$L\frac{d\vec{i}}{dt} = \left(\vec{v}_{\text{ANPC}} - \vec{e}\right) - R\vec{i} \tag{6}$$

Finally, dynamics of the dc-link capacitor voltages (V_{c1} and V_{c2}) can be described with the capacitor current (i_{c1} and i_{c2}) as:

$$dV_{c1}/dt = (1/C)i_{c1} \text{ and } dV_{c2}/dt = (1/C)i_{c2} \tag{7}$$

where, C is the capacitor value.

2.2. *Dynamic discrete time model of the ANPC inverter*

The continuous time based dynamic system of the three-level three-phase ANPC inverter, as represented in (6) and (7), needs to transform into discrete time domain as the MPC is formulated in discrete time domain. Then, an appropriate cost function is derived from these discrete variables to achieve the controllability of the MPC technique for the ANPC inverter.

The dynamic discrete time model of the ANPC inverter is approximated with Euler equation $(dx/dt \approx \{x(k) - x(k-1)\}/T_S)$ at a sampling time T_s. By utilizing one step ahead of this Euler approximation, the discrete future predictive current ($\vec{i}(k+1)$) vector of the three-phase ANPC inverter can be derived from the continuous time current dynamic of (6) as:

$$\vec{i}(k+1) = \frac{1}{L+RT_s}\left\{L\vec{i}(k) + T_s\left[\vec{v}_{ANPC}(k+1) - \vec{e}(k+1)\right]\right\} \tag{8}$$

On the other hand, the discrete time model of the future dc-link capacitor voltages (k+1) are:

$$V_{c1}(k+1) = V_{c1}(k) + \frac{1}{C}i_{c1}(k)T_s \tag{9}$$

and

$$V_{c2}(k+1) = V_{c2}(k) + \frac{1}{C}i_{c2}(k)T_s \tag{10}$$

There are 27 future value of the three-phase load current ($\vec{i}(k+1)$), and the dc-link capacitor voltages (V_{c1} and V_{c2}) are predicted by the switching states of the ANPC inverter. Therefore, an appropriate cost function is essential for the MPC algorithm for the proper selection of the switching states. The cost function ($\vec{c}$) is evaluated for every switching states. The switching state that gives minimum value of the cost function is applied to the three-level three-phase ANPC inverter for operation and also uses during the next sampling instant. The cost function ($\vec{c}$) of the MPC scheme is defined with future measured reference load current ($\vec{i}_{ref}(k+1)$), predicted load current ($(\vec{i}(k+1))$) and dc-link capacitor voltages ($V_{c1}(k+1) - V_{c2}(k+1)$) as:

$$\vec{c} = \left|\vec{i}_{ref}(k+1) - \vec{i}(k+1)\right| + \lambda_{dc}\left|V_{c1}(k+1) - V_{c2}(k+1)\right| \tag{11}$$

where, λ_{dc} is the weighting factor which maintains the voltage balance of the two dc-link capacitor voltages.

2.3. *Control scheme*

Figure 1 shows the MPC strategy for three-level ANPC inverter. The load current ($\vec{i}(k)$) is measured and the future predicted current($\vec{i}(k+1)$) is calculated by utilizing (9) for each one of 116 possible switching state function vector ($\vec{S}$) of the converter. This future value ($\vec{i}(k+1)$) is compared with reference load current ($\vec{i}_{ref}(k+1)$) of the NPC inverter by using cost function ($\vec{c}$) of equation (11).

Finally, the switching states ($\vec{S}$) of the three-level three-phase ANPC inverter which minimizes the cost function ($\vec{c}$), is selected for next sampling time. Moreover, the voltage balancing between the two capacitor voltages $(V_{c1}(k+1)-V_{c2}(k+1))$ is also achieved by utilizing the cost function in (11).

3. Simulation results

The proposed model predictive controlled ANPC scheme is simulated in MATLAB/Simulink to validate efficacy of the system. Parameters used in simulation are listed in Table 2 and the simulation result is presented in figure 3. The three-phase load current drawn by the MPC controlled ANPC inverter is depicted in figure 3(a). The reference load current is set at a value of 5 A (peak) for each phase and the figure 3(a) demonstrates that, each phase load current $\vec{i}(k+1)$ is accurately tracking the reference value $(\vec{i}_{ref}(k+1))$, which verifies the feasibility of the model predictive control method. Furthermore, the MPC algorithm effectively reduces the THD of load current by accurately tracking the sinusoidal reference current. Figure 3(c) confirms that the total harmonic distortion (THD) of the model predictive controlled ANPC inverter is 0.99%, which can be accepted as a very good performance.

Additionally, another important outcome of the proposed MPC algorithm is to balance the neutral-point voltage of the ANPC inverter. This neutral-point voltage can be denoted as the difference between the two dc-link capacitor voltages. The figure 3(b) shows two dc-link capacitor voltages (V_{c1} and V_{c2}), which confirms that the neutral-point voltage of the ANPC inverter is accurately balanced. The value of the weighting factor is fixed at $\lambda_{dc}=0.01$, to achieve the best performance.

Table 2. Simulation Parameters

Variables and Parameters	Values	Units
Power Rating (P)	4.0	kW
DC-link Voltage (V_{dc})	533	V
Load inductance (L_s)	30	mH
Load resistance (Rs)	10	Ω
Capacitor (C_1, C_2)	750	μF
Sampling time (T_s)	50	μs

IOP Conf. Series: Earth and Environmental Science **73** (2017) 012019 doi:10.1088/1755-1315/73/1/012019

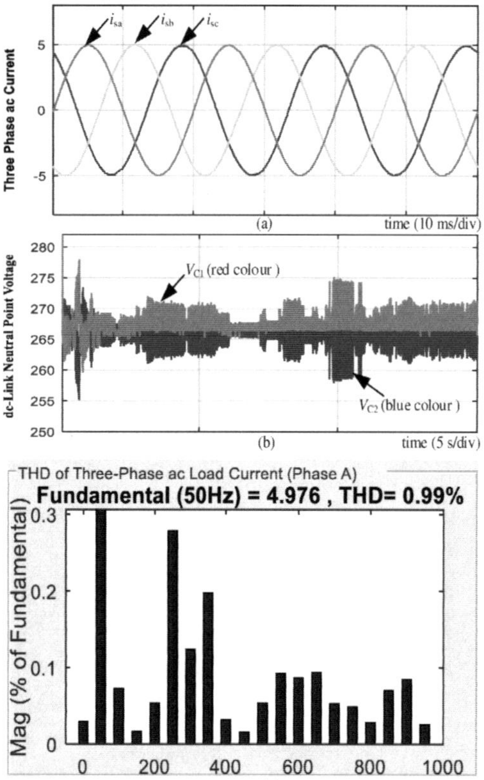

Figure 3. Simulation results of model predictive controlled ANPC inverter: (a) three-phase ac load current, (b) neutral point voltage and (c) THD of the load current (phase A).

4. Comparative Evaluation and Robustness Analysis

4.1. Comparative Evaluation

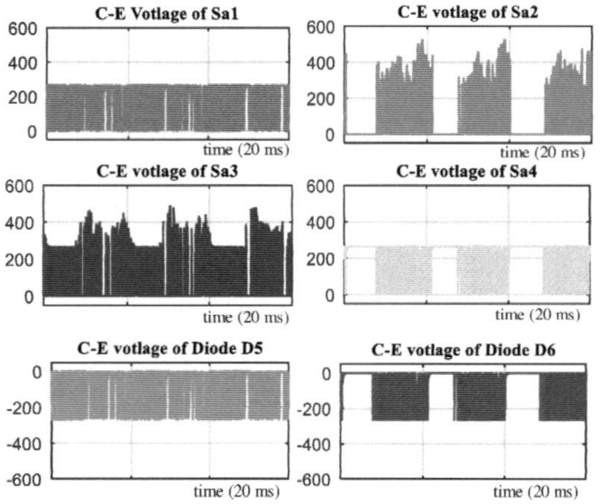

Figure 4. Simulation results of voltages across the semiconductor power switches and the diodes of the model predictive controlled NPC inverter (Phase A).

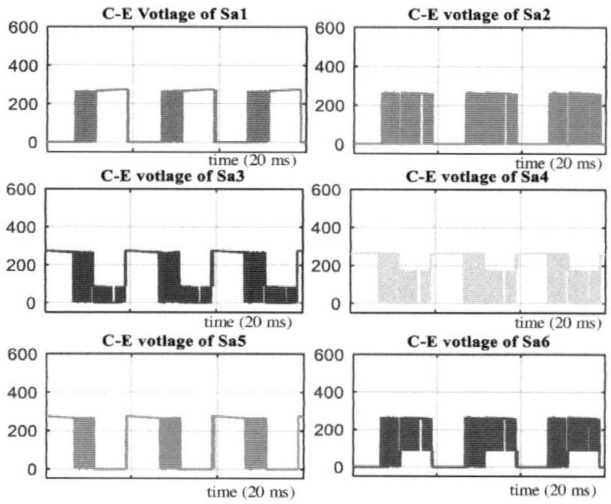

Figure 5. Simulation results of voltages across the semiconductor power switches of the model predictive controlled ANPC inverter (Phase A).

A comparative evaluation is performed in this section to verify the efficacy of the proposed MPC controlled ANPC inverter over its traditional NPC inverter. One of the major aim of this research is to balance and equalize the voltage stress among all the semiconductor power switches of the inverter. Figure 4 shows the voltages across the semiconductor power switches $(S_{a1} - S_{a4})$ and the diodes $(D_5 - D_6)$ of phase A of the model predictive controlled NPC inverter. The simulation result shows that there is uneven voltage stress among the power switches of the NPC inverter which further cause the unequal power loss and failure of the devices. In order to overcome this problem, MPC controlled ANPC inverter is proposed. Figure 5 presents the collector-emitter voltages of the IGBT power switches $(S_{a1} - S_{a6})$ of phase A, which confirms that all the power switches of the ANPC inverter face equal voltage stress.

4.2. Robustness Analysis

The robustness of the proposed MPC algorithm is analyzed against the filter parameters uncertainties of the three-level ANPC inverter. Figure 6 shows the current error variation $(\vec{i}_e = \vec{i} - \vec{i}_{ref})$ with the variation of filter parameters. The maximum deviation of the current error $(\vec{i}_e)$ is limited within 0.2 A (less than 4% which is negligible) which confirms the robustness of the proposed scheme. This current deviation against the filter parameter variation is verified in MATLAB/Simulink software. The equation for the calculation of current deviation used in this section is as follows:

$$\vec{i}_e = \{(\vec{i} - \vec{i}_{ref}) / \vec{i}_{ref}\} \times 100\% \qquad (12)$$

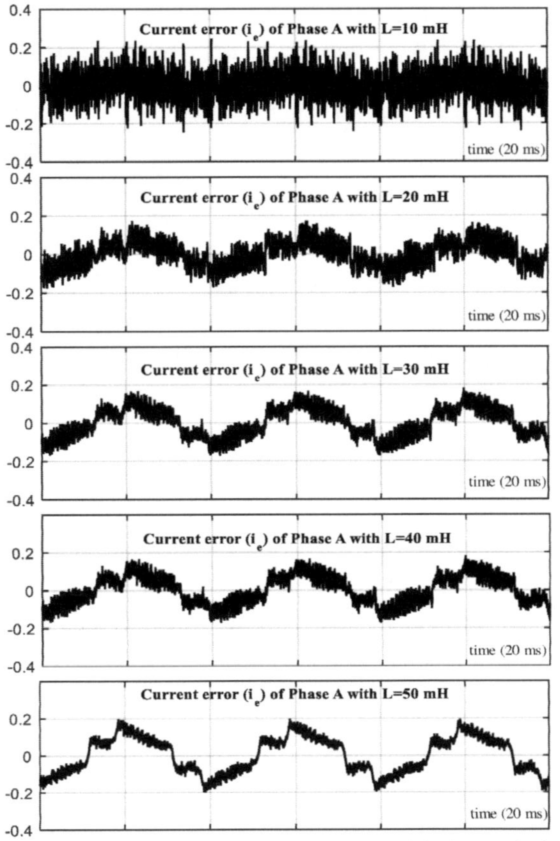

Figure 6. Robustness analysis of the proposed MPC scheme with the variation of current error in case of parameter uncertainties.

5. Conclusions

A model predictive controlled three-level three-phase active neutral-point-clamped (ANPC) inverter is presented in this paper to equally distribute the voltage stress among all the semiconductor power switches. The proposed concept is enhanced by deriving a cost function which provides very good system performance as well as ensures the robustness. Moreover, the dc-link capacitor voltage balance is also accurately maintained by the proposed controller at all time. The simulation results of three-phase ANPC inverter strongly agree and show very good system performance. Again, the MPC also shows the robustness capability with the parameter uncertainties, which is verified by the robustness analysis in against filter parameter variation. The results associated in this investigation are very encouraging and will continue to play a strategic role in the equal loss distribution among all the power switches and fault-tolerant capability of the modern high-performance active neutral-point-clamped (ANPC) inverters.

Acknowledgments
The authors would like to thank the Australian Government and the University of Sydney for providing financial support through the 2016 International Postgraduate Research Scholarships (IPRS).

References
[1] A. Nabae, I. Takahashi, and H. Akagi 1981 "A new neutral-point-clamped PWM inverter," *IEEE Trans. Ind. Appl.,* pp. 518-523.
[2] S. Kouro, J. I. Leon, D. Vinnikov, and L. G. Franquelo 2015 "Grid-connected photovoltaic systems: An overview of recent research and emerging PV converter technology," *IEEE Ind. Electron. Magazine,* vol. 9, pp. 47-61.
[3] J. Rodríguez, S. Bernet, B. Wu, J. O. Pontt, and S. Kouro, "Multilevel voltage-source-converter topologies for industrial medium-voltage drives 2007" *IEEE Trans. Ind. Electron.,* vol. 54, pp. 2930-2945.
[4] A. Choudhury, P. Pillay, and S. S. Williamson 2016 "Modified DC-Bus Voltage-Balancing Algorithm Based Three-Level Neutral-Point-Clamped IPMSM Drive for Electric Vehicle Applications," *IEEE Trans. Ind. Electron.,* vol. 63, pp. 761-772.
[5] X. Jing, J. He, and N. A. Demerdash 2014 "Loss balancing SVPWM for active NPC converters," in *Proc. IEEE Applied Power Electron. Conf. and Expo. (APEC-2014)* pp. 281-288.
[6] T. Bruckner, S. Bernet, and H. Guldner 2005 "The active NPC converter and its loss-balancing control," *IEEE Trans. Ind. Electron.,* vol. 52, pp. 855-868.
[7] S. R. Pulikanti, M. S. Dahidah, and V. G. Agelidis 2011 "Voltage balancing control of three-level active NPC converter using SHE-PWM," *IEEE Trans. Power Delivery,* vol. 26, pp. 258-267.
[8] M. P. Akter, S. Mekhilef, N. M. L. Tan, and H. Akagi 2016 "Modified Model Predictive Control of a Bidirectional ac-dc Converter Based on Lyapunov Function for Energy Storage Systems," *IEEE Trans. Ind. Electron.,* vol. 63, pp. 704-715.
[9] M. Rivera, J. Rodriguez, and S. Vazquez 2016 "Predictive control in power converters and electrical drives- part ii," *IEEE Trans. Ind. Electron.,* vol. 63, pp. 4472-4474.
[10] J. Rodriguez, M. P. Kazmierkowski, J. R. Espinoza, P. Zanchetta, H. Abu-Rub, H. A. Young, *et al. 2013* "State of the Art of Finite Control Set Model Predictive Control in Power Electronics," *IEEE Trans. Ind. Info.,* vol. 9, pp. 1003-1016.

[11] S. Kwak and S. k. Mun 2015 "Model Predictive Control Methods to Reduce Common-Mode Voltage for Three-Phase Voltage Source Inverters," *IEEE Trans. Power Electron.,* vol. 30, pp. 5019-5035.

[12] M. P. Akter, S. Mekhilef, N. M. L. Tan, and H. Akagi 2015 "Stability and performance investigations of model predictive controlled active-front-end (AFE) rectifiers for energy storage systems," *Journal of Power Electron.,* vol. 15, pp. 202-215.

[13] M. Parvez, S. Mekhilef, N. M. Tan, and H. Akagi 2015 "An improved active-front-end rectifier using model predictive control," in *Proc. IEEE Applied Power Electron. Conf. and Expo. (APEC-2015)*, 2015, pp. 122-127.

[14] M. Parvez, S. Mekhilef, N. M. Tan, and H. Akagi 2016 "A robust modified model predictive control (MMPC) based on Lyapunov function for three-phase active-front-end (AFE) rectifier," in *Proc. IEEE Applied Power Electron. Conf. and Expo. (APEC-2016)*, 2016, pp. 1163-1168.

[15] D. Zhou, J. Zhao, and Y. Li 2016 "Model-Predictive Control Scheme of Five-Leg AC-DC-AC Converter-Fed Induction Motor Drive," *IEEE Trans. Ind. Electron.,* vol. 63, pp. 4517-4526.

[16] M. P. Akter, S. Mekhilef, N. M. L. Tan, and H. Akagi 2015 "Model predictive control of bidirectional AC-DC converter for energy storage system," *J. Electr. Eng. and Tech.,* vol. 10, pp. 165-175.

[17] M. Parvez, S. Mekhilef, N. M. Tan, and H. Akagi 2014 "Model predictive control of a bidirectional AC-DC converter for V2G and G2V applications in electric vehicle battery charger," in *Proc. IEEE Transportation Electrification Conf. and Expo. (ITEC-2014)*, pp. 1-6.

[18] M. Rivera, C. Rojas, J. Rodr, x00Ed, guez, P. Wheeler, *et al. 2011* "Predictive Current Control With Input Filter Resonance Mitigation for a Direct Matrix Converter," *IEEE Trans. Power Electron.,* vol. 26, pp. 2794-2803.

[19] J. L. Elizondo, A. Olloqui, M. Rivera, M. E. Macias, O. Probst, O. M. Micheloud, *et al.* 2014 "Model-Based Predictive Rotor Current Control for Grid Synchronization of a DFIG Driven by an Indirect Matrix Converter," *IEEE Journal of Emerging and Selected Topics in Power Electron.,* vol. 2, pp. 715-726.

[20] A. Calle-Prado, S. Alepuz, J. Bordonau, J. Nicolas-Apruzzese, P. Cortés, and J. Rodriguez 2015 "Model predictive current control of grid-connected neutral-point-clamped converters to meet low-voltage ride-through requirements," *IEEE Trans. Ind. Electron.,* vol. 62, pp. 1503-1514.

[21] M. Narimani, B. Wu, V. Yaramasu, and N. R. Zargari 2015 "Finite Control-Set Model Predictive Control (FCS-MPC) of Nested Neutral Point-Clamped (NNPC) Converter," *IEEE Trans. Power Electron.,* vol. 30, pp. 7262-7269.

IOP Conf. Series: Earth and Environmental Science **73** (2017) 012020 doi:10.1088/1755-1315/73/1/012020

Multi-Criteria Optimization of the Deployment of a Grid for Rural Electrification Based on a Heuristic Method

L. Ortiz-Matos[1], A. Aguila-Tellez[1], R.C. Hincapié-Reyes[2] and J.W González-Sanchez[2]
[1] Universidad Politécnica Salesiana and research GIREI Group, Quito, Ecuador.
[2] Universidad Pontificia Bolivariana, Medellín, Colombia.

Email: {lortizm, aaguila}@ups.edu.ec; {roberto.hincapie, jorgew.gonzalez}@upb.edu.co

Abstract. In order to design electrification systems, recent mathematical models solve the problem of location, type of electrification components, and the design of possible distribution microgrids. However, due to the amount of points to be electrified increases, the solution to these models require high computational times, thereby becoming unviable practice models. This study posed a new heuristic method for the electrification of rural areas in order to solve the problem. This heuristic algorithm presents the deployment of rural electrification microgrids in the world, by finding routes for optimal placement lines and transformers in transmission and distribution microgrids. The challenge is to obtain a display with equity in losses, considering the capacity constraints of the devices and topology of the land at minimal economic cost. An optimal scenario ensures the electrification of all neighbourhoods to a minimum investment cost in terms of the distance between electric conductors and the amount of transformation devices.

1. Introduction

The design of electrification systems, involve recent mathematical models which solve location problems, types of electrification components, and the design of possible distribution microgrids. However, as the amount of demand points increases, the solution to these models require high computational times, which at the same time, become unviable practice models [1], [2]. In order to find the best possible solution to this kind of problem described as NP-Complete, it is necessary to consider the use of heuristics in search of the optimal solution. Heuristics are an algorithm that approximates the solution or a subset of the problem [3]. There are general forms of heuristic algorithms such as Gross Furze, the Random Search algorithms and Smart Search. Combination type problems that are NP- Complete or NP-Hard, have an exponential search space that becomes very large; therefore, it is necessary to approximate a heuristic that is finite, scalable and that seeks an optimal solution [3].

The problem that needs to be solved in this study is an NP- Complete problem that has no defined single optimal solution and therefore we will rely on a heuristic method to find an approximation to the solution. This case study seeks to solve a complex problem of electrification at minimal cost, and reduce losses in electrical circuits in rural areas. The problem has a set of rural developments, in which their location is known and will be electrified from known transmission points. The objective is to trace the exact location of the transformation centers in the transmission Grid and its connection with the subtransmission Grid, transformation centers of subtransmission to distribution. All these centers are connected by an optimal route for optimal Grid deployment, considering further losses therein. The system locates the general connection of the entire circuit with one of the candidate points of the transmission grid. Among the restrictions are the following: the grid topology, the ability to locate transformers, location, and losses; finally electrifying the proposed rural areas. The algorithm design allows for a general electrical circuit comprising of transformation centers, and the deployment of transmission and distribution grids, with a location and routing that meets the minimum investment cost and losses in the electricity grid [4], [5].

2. State of the art: Electrification of rural areas

The figure 1 shows the electrification problem raised in the case study, where a group of rural developments with their location will be electrified from an existing transmission point. The goal is to locate a multi-criteria analysis that includes loss and costs, an exact location of the transformation centers, the optimal deployment of the transmission network, its connection with the subtransmission network, and transformation centers of subtransmission to distribution.

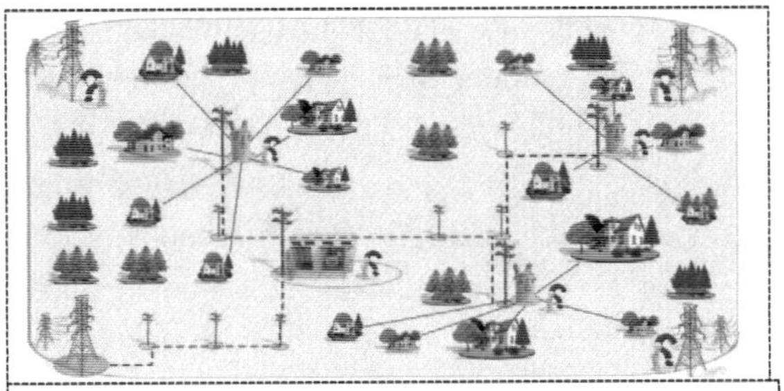

Figure 1. Electric grid for Rural Electrification.

This is due to the dispersal of loads, low expected consumption, lack of road infrastructure and accessibility, topography and environmental setting with many limitations and complications to deploy an air grid, distant electrified points either by transmission or distribution, or inaccessibility to their villages, among many other causes that may arise in the individual analysis of each issue of electrification. For these reasons, it is of vital importance the development of an optimal technical-economic strategy in order to carry out the electrification in rural areas [6], [5].

Most electrification projects of rural areas are based on a design in regards to an economic scope or coverage. Based from this preliminary analysis the design of routing grids and location of processing centers is carried out in compliance with the minimum quality requirements and energy efficiency [7]. Achieving the electrification of these areas at minimal cost is an optimization problem NP Complete the due to the amount of restrictions mainly related to accessibility for the construction of Grids [8]. The best and only solution, is quite difficult to find due to the designs that are based on the experience of the professionals involved, hence, the analysis is done by creating scenarios and comparing them to each other without exploring other possible solutions, leaving the design without a mathematical analysis of feasibility and with more search space [9], [10].

IOP Conf. Series: Earth and Environmental Science **73** (2017) 012020 doi:10.1088/1755-1315/73/1/012020

3. Problem Formulation

The biggest problems in electrification are posed in rural areas but with much less impact in urban peripheries. The electrification of rural areas in many countries has reduced levels of coverage and low service quality [10]. Substations must be placed in any cell, including the same cell with an existing neighborhood. The cost of wiring is 1M pesos per km. Distances are measured using a straight line between the cells, which are each 1km apart. For example, the distance between cell A1 and B2 is 1.42 km, losses in distribution circuits is considered one of the biggest problems that electric companies face that should be another important part of the analysis.
Whereas the active power losses are given by [4]:

$$\Delta P = \sum_{i=1}^{n} I_i^2 * R_i \tag{1}$$

where: n is the number of nodes in the system; I_i corresponds to the value of the current at node I; and R_i resistance is seen in the same node

3.1. Description of Heuristics

The heuristic model for optimal deployment of a new electric grid that electrifies rural neighborhoods considers cost variables associated with necessary material resources, load capacity and losses. The model is based on a set of algorithms that seek an optimal solution that meet the need for electrification of the grid under the criteria of loss reduction, efficiency, and profitability.
In order to define a first instance of the solution, the k-means algorithm is used, in which the clustering of the sites are defined as houses or neighborhoods. This grouping is based on the basis of Euclidean distances and number of centroids (clusters) proposed that are also defined, based on the allowable capacity that will be projected from transformers in order to assume the charge required to electrify. Once the centroids are defined for neighborhoods with the above characteristics, it is necessary to move them to an appropriate location in an exact coordinate defined by the candidate sites, which must comply with the restriction of not having a home or neighborhood in that location.
This movement is performed through a smart search in candidate sites closest to the centroid point making a comparison between the distances from the centroid to the sites. Once a first approximation of the location of the centroid is defined, as it would be the position of the substation of subtransmission to distribution, a comparison of distances of these centroids is done against the termination sites of the transmission grid that are the candidates to carry out the connection. In this case, those sites are defined with the coordinates of each corner of the map.
By comparing the measurements obtained, the k-means algorithm is again used to select a new centroid between the connection point and the above defined centroids. The new centroid is the transformer transmission to subtransmission which delimits routing grids as which will allow the addition of various functions of cost electric conductor used in transmission, subtransmission and distribution. The position of this centroid is relocated to a candidate site under the same criteria taken for subtransmission substations to distribution. With locations obtained for each transformer, a ratio cost is established between conductors of each voltage level and proceeds to calculate total cost which depends on the amount and type of transformer located based on their capacity and distance and type of electric conductor used to establish the complete connection.
The routing for the deployment of the total grid is achieved using Dijkstra's algorithm which explores all routes declared as feasible in selecting the best weight criteria which are defined by cost and distance. The process consists of finding the optimal solution by making a smart shift of the positions

assigned by minimizing candidates with relocation within the feasible path described by Dijkstra as shorter to deploy an air grid at different voltage levels. Later, a recalculation of the objective function cost and loss is achieved, keeping only the solution that meets the criterion of lowest total cost. Thus the optimal route and location of the processing centers is obtained in order to meet the criteria for the electrification in rural areas.

Nomenclature	Variables
C	Rural Houses
S	Sites
P	Transmission connection point
I_i	Current in the node i
R_i	Resistance in the node i
cap	Capacity Substation
CostXuC	Cost for a unit distribution cable
CostXuS	Cost for a unit transmission cable
$CostoT_i$	Total cost
CostoO	Optimal Total Cost
Subest	Optimal location Substation
PuntSal	Transmissions
slnop_i	Optimal placement of Transformers
Pcand	Candidate sites
Transf	Electric transformers
ΔP	Losses

Table 1. Shows the notation of the variables used in the algorithm in order to achieve a better understanding of it.

```
Step 1 Input: C; S; K; CostX uC;
Step 2 Output: costTc; Transf;
Step 3 Initialize: noCon = S; Ns(U_j) = ∅;
Step 4 Clusters = Kmeans(X_C; Y_C; K)
    for centroide ∈ K_C
        S´_k = argmin = {Dck, S};
        Costomincluster = inf;
        for Δx ∈ {−d, o, d}
            for Δy ∈ {−d, o, d}
    S´_k = S_i ∈ S | x_i = x_k + Δx_i; y_i = y_k + Δy;
                Costo = Σ^m_{i∈Ck} ds´_k, C_i ;
                if costo < Costomincluster
                    Costomincluster = Costo;
                Sm = S´_k ;
            endif ; endfor ; endfor ;
        S_e = S_e ∪ S_m;
        costoT = costoT + Costomincluster;
    endfor ;
Step 5: Return: costT c; Transf
```

Algorithm 1. Function *CostoMinDist*

```
Step 1 Input: C(xc, yc); M; K; P (xp, yp);
        CostXuC; CostXuS; cap; P; no; n; I_i; R_i;
Step 2 Output: Transf; Subest; PuntSal; CostoOP;
Step 3 Initialize: Costoi=inf; CostoT=inf;
        CostoC_T=inf; dis=inf; padre= null;
        CostoOP=inf; S(xs,ys); Transf = null;
        Pcand = [−n − 1, −n, −n + 1, − 1, +1, n − 1,
            n, n + 1];
Step 4  [Transf, costTc] ⇐ CostoMinDist (C, S, K);
Step 5  forall i ∈{1, . . . , |no|}
        MSP = Transf ∪P (i);
        ClustersT ⇐ Kmeans (X_msp; Y_msp; K);
        MSP = Transf\P (i); Sk = argmin {D_ck, S};
            for Δx_mps ∈{−d, o, d}
                for Δy_mps ∈{−d, o, d}
                S´_k = S_i ∈ S | x_i = x_k + Δx,
                    y_i = y_k + Δy;
                Costo = Σ^m_{i∈Ck} ds´_k, C_i
                Costo = Costo * CostXuS;
                Costoi = Costo + costTc;
                if Costoi < CostoCT
                    CSubest = ClustersT;
                    PuntSal = P (i);
                endif ; endfor ; endfor;
    endforall
                Actualizar (cola, < j, dis[j] >)
            endif ; endfor ;
        endwhile ;
    endforall

CostoSal = distpuntSal * CostXuS;
Step 7  for k ∈{1, . . . , |Subest|}
            CostoT = costTc + CostoSal;
            if CostoT < CostoO
                CostoO = CostoT ;
                Transf END = Transf;
            endif
        for i ∈{1, . . . , |xt|}
            dist=dist+sqrt((xc(j)−c(idx(j), 1))^2 + . . .
            ... +(yc(j)-c(idx(j),2))^2);
        endfor
            CCasa = Σ_{i=1}^{|dist|} dist (i,j)CostXuC ;
            ΔP = Σ_{i=1}^{m} I_i^2 * R_i * dist_i
Step 8  tray2ord=cola; CostoCT =costTc (CostXuC);
        CostoTS = dis (CostXuS);
        CostoT = CostoCT i + CostoTS ;
        if CostoT < CostoOP
            CostoOPk = CostoT ;
            ColaOp = colak ;
            Transfk = Transfk;
```

$V = S \cup MSP \cup Subest;$
Step 6 **forall** $i \in \{1, \ldots, |V|\}$
$dis[i] = inf; \quad padre[i] = null;$
$add\,(cola, < i, dis[i] >); \quad dis[s] = 0;$
while $cola \neq 0$
$i = cola \setminus min\,(cola)$
for j *adjacent* a i
if $(j \notin cola)$and $(dis[j] > peso\,(i, j))$
$padre[j]=i; dis[j] = peso\,(i, j);$

Algorithm 2. Optimal deployment of the grid

endif
if $sTransfk \neq Sub$
$cola_k\,(cola_k == transf_k) = [\ ];$
$Transf_k = tray2ord_k\,(:, b + 1);$
$dis_k = dis_k - 1;$
endif
endfor
Step 9 Return: *CostoOP* ; *Transf* ;*Subest; PuntSal*;ΔP

Algorithm 2 (Cont). Optimal deployment of the grid

The proposed algorithm describes a heuristic based on an intelligent search of feasible solutions, in which this case are defined candidate sites and that meet the characteristics for the location of a transformer center or a trace of the air line of either distribution or transmission. The algorithm also calls for other algorithms for calculating distances and location of reference clusters.

4. Case studies and results analysis

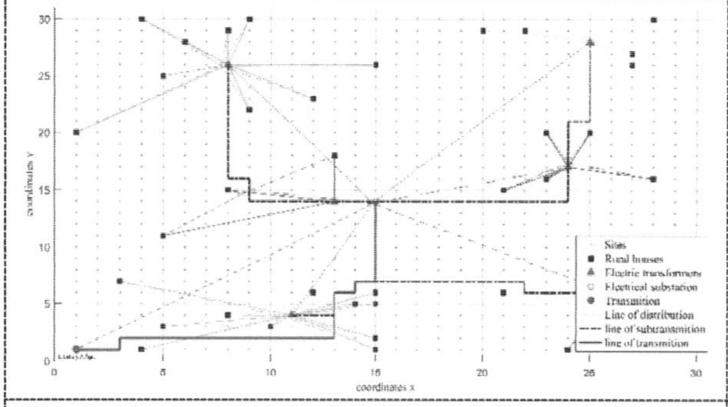

Figure 2. Optimal deployment of six transformers, substation and transmission. Rural areas with equal residential loads were considered, with average current values of *I* equal to 3.62 A, gauge power cables 2 P/u and distribution 1/u, for 40 houses located randomly between 900 sites that define the territory, and a capacity constraint that selects only 6 transformers for power supply. The cost between the distribution grid and transmission grid is 0.5/u to 1/u.

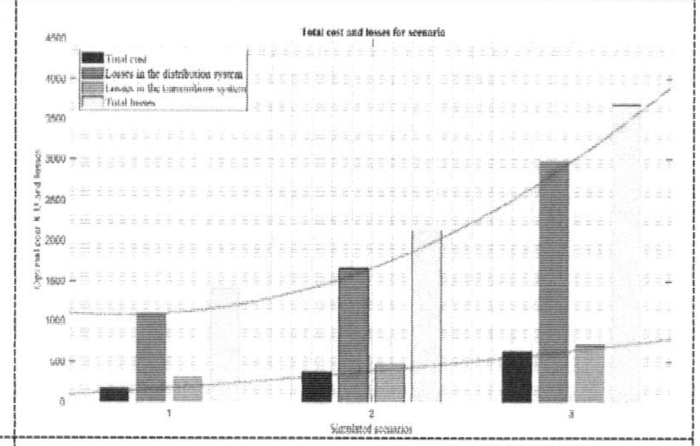

Figure 3. Comparison of cost metrics for different scenarios. The optimization model was evaluated under three different simulated scenarios, in which metrics are obtained and define cost function and loss for each one, as shown in figure.

The scenarios are described for different capacity allocations that define the amount of subtransmission transformers located at distribution, different cost ratios between electric conductor that describe the distribution and transmission, and losses in distribution circuits. It is necessary to point out that for each case study, the rural residential areas are considered with equal loads, with the average current *I* is 3.62A, gauge of transmission cables 2 P/u and distribution 1/u. By changing the amount of selected transformers in distribution according to the defined capacity, different costs are acquired that would be dependent on the nominal power to meet demands. The heuristic algorithm described in Matlab generates Figure 2 and figure 3, which shows the path of optimal deployment for grids and optimal location of the transformation centers for this scenario. Centers and high capacity lines show the results obtained from k-means.

5. Conclusions and future works

This strategy is justified by 'sacrificing' the accuracy of the solution by seeking lower computational costs of data processing and reducing the mathematical effort, under the criterion that, it is not always convenient to describe a problem with all its information assuming a high computational time. By establishing a trade-off between accuracy and effort, the uncertainty is managed more efficiently and vigorously.

For future developments concerning this study, in addition to the orientation and development of the algorithm towards urban or underground electrification, other algorithms and heuristic optimization would be developed in order to calculate costs for Microgrid designs with Distributed Generation. This algorithm would help implement a decision between electrification with the national system, implementing a network with only micro distributed generation, or determine the level of penetration of clean energy to the distribution system of the national network in certain areas.

In future studies, a georeferenced algorithm may be used that recognizes images by sectors that allow the identification and location of candidate sites.

6. References

[1] J. Triadó-Aymerich, L. Ferrer-Martí, A. García-Villoria, and R. Pastor 2016 *MILP-based heuristics for the design of rural community electrification projects*(Computers and Operations Research) vol. 71, pp. 90–99

[2] Y. Gonoso 2007 *Multi-Objective Optimization in Computer Networks Using Metaheuristics* (Auerbach Publications Boston, MA, USA) vol. 53

[3] G. Gigerenzer, R. Hertwig, and T. Pachur 2011 *Heuristics. The foundations of adaptive behavior* (Oxford University Press) pp. 872

[4] A. Aguila, D. Carrión, and L. Ortiz 2015 *Analysis of power losses in the asymmetric construction of electric distribution systems* (IEE Latin America Transactions) vol. 13, no. 7, pp. 2190–2194

[5] C. H. Papadimitriou and K. Steiglistz 1998 *Combinatorial Optimization, algorithms and Complexity* (New YYork: Englewood Cliffs, N.J) pp. 7-13,

[6] E. N. G. El, D. M. Ali, E. N. G. A. Hassan, and E. N. G. L. Mohy 2001 *The Project the Total Population* no. 482, pp. 18–21

[7] V. Orejuela, D. Arias, and A. Aguila 2015 *Response of Residential Electricity Demand Against Price Signals in Ecuador* (PROCEEDINGS of the 2015 IEEE Thirty Fifth Central American and Panama Convention, Concapan XXXV) pp. 373 – 378

[8] D. Torres Riascos, E. Inga Ortega, and A. Aguila Tellez 2015 *Advanced metering infrastructure for microgeneration in rural zones using TV white spaces based on spectrum opportunistic* (IEEE Latin America Transactions) vol. 13, no. 12, pp. 3813–3819

[9] D. Thirault, Y. Besanger, N. Hadjsaid, a. Almeida, and G. Huard 2002 *Cost estimation of different electric systems for rural areas in developing countries* (IEEE Power Engineering Society Summer Meeting) vol. 1, pp. 23–28

[10] N. R. Karki 2004 *Reducing the cost of rural electrification: A key to improve the quality of life in rural areas in developing countries* (2004 International Conference on Power System Technology, POWERCON 2004) vol. v 1, pp. 447 – 452

A Taxonomy on Accountability and Privacy Issues in Smart Grids

Ameya Naik and Hamid Shahnasser
Electrical & Computer Engineering Program, San Francisco State University, San Francisco, U.S.A

Email: ameyunaik@gmail.com;

Abstract- Cyber-Physical Systems (CPS) are combinations of computation, networking, and physical processes. Embedded computers and networks monitor control the physical processes, which affect computations and vice versa. Two applications of cyber physical systems include health-care and smart grid. In this paper, we have considered privacy aspects of cyber-physical system applicable to smart grid. Smart grid in collaboration with different stockholders can help in the improvement of power generation, communication, circulation and consumption. The proper management with monitoring feature by customers and utility of energy usage can be done through proper transmission and electricity flow; however cyber vulnerability could be increased due to an increased assimilation and linkage. This paper discusses various frameworks and architectures proposed for achieving accountability in smart grids by addressing privacy issues in Advance Metering Infrastructure (AMI). This paper also highlights additional work needed for accountability in more precise specifications such as uncertainty or ambiguity, indistinct, unmanageability, and undetectably.

1. Introduction

The swift developments in technology along with the improvement and up gradation in physical systems have emerged into integrative and multi-faceted engineering systems called Cyber-Physical System (CPS). This system relies on collecting data for making controlled decisions for the physical system and managing communication. Smart grid is a good example for the Cyber-Physical System. Smart grid is a promising power delivery infrastructure incorporated with communication network and information technologies. Its main objective being improvement in power system, checking the transmission, distribution and consumption and its various collaborations to aid the consumer sector. With an upgraded sensing and control technologies, CPS can help in analyzing and determination of power usage, delivery and generation in a real-time. CPS can also be used to enhance proper assimilation of energy resources being distributed and the use of intelligent appliance control for boosting the energy efficiency. [1]

Also, to protect and overcome problems arising on the aspects of privacy, honor and confidentiality, accountability is required. Jing Liu and Yang Xiao [2] describe accountability as a system that is recordable and noticeable, thus making it legally answerable to communication principles for its actions. If security issues arise, the in-built accountability mechanism will play an integral role in investigating who is responsible for it. Thereby, once the problem has been identified, a pre-defined program can be used to fix it automatically or it can be also resolved by releasing relevant information to experts for evaluation. [2]

2. Accountability and Authentication

"Accountability logic" is considered as most appropriate way to analyze the accountability of a secure system. Most of accountability logic structured for e- transactions. The usage accountability is critical to minimize and prevent power theft and in promotion of the billing. Thus, the user is responsible for demanding the amount of electricity and is billed with respect to his usage value. Moreover, the utility can help keeping a track of amount of electricity used by an individual compared to that of bulk energy.

Authentication plays a very important role in analysis of accountability. Since time-critical applications require proofs that guarantee the temporal activities of each principal, Kailar's accountability logic can be extended to analyze such applications with logic for e-commerce protocols such as payment and public key distribution. He defined accountability as a property wherein unique identification of original client can be associated with an object or can be verified with third party. [3]

Although Kailar's accountability logic allows some temporal context, which must be added to represent the validation period of security-related information, such as a time-critical aspect, M. Kudo [4] expands the Kailar's logic so that it could represent temporal accountability. He emphasized on different kind of attacks on temporal records of payment times and an error that could occur during closing time of electronic submission. For this type of attack, a third-party architecture can be introduced for security.

Based on Kudo's secure electronic submission protocol, accountability logic has been developed in home area network (HAN) of smart grid. Jing Liu, Yang Xiao and Jingcheng Gao [5, 6] jointly proposed a possible architectural framework for implementation of accountability logic and identification potential security problems in home area network (HAN) and Neighborhood Area Network (NAN). Based on communication capability, electrical appliances are divided into two categories: the smart appliances and the other regular appliances.

Jing Liu, Yang Xiao and Jingcheng Gao [5,6] proposed the protocol that addresses both power usage and real-time market price change issues. Focusing on the privacy concern of the client, an ideal meter should not disclose any information to unauthorized or unapproved requests. Adopting pseudonym mechanism in NAN, communication flow will become anonymous. Thus, systems are responsible in setting up of various witnesses to screen activities in observation process of an object. Once a variation in conduct is identified, those witnesses will give significant proof with a specific end goal to bolster their discoveries.

M. Stegelmann et al. [7] proposed a scheme, in which a smart meter transmits the valuable metering data to neighborhood aggregator. This aggregator applies anonymous protocol and resends the information to service providers. The information sent for billing is not anonymous but the same one is anonymous for service providers.

X. Lin [8] presents a routing protocol system with a secured protocol for ad-hoc network. This protocol enables secrecy of the source, destination and path. In this protocol, a source starts initializing the path request number and broadcasts it. This request consists of sequence number of the path and an encrypted destination address. The nodes in channel rebroadcasts the path requests after noting it. The destination responds back the path request and these nodes along the channel are reserved by checking and matching the information about several iterations like previous hops and next hops.

IOP Conf. Series: Earth and Environmental Science **73** (2017) 012021 doi:10.1088/1755-1315/73/1/012021

3. Privacy Analysis

Security and privacy are not the same. Security of data is critical with regards to smart metering. As the data is traded over open systems, it is defenseless to being seen or altered in its transmission by unintended individuals and this could have unfortunate outcomes for the people to whom the data belongs. Some of the privacy concerns resulting from smart metering systems, Attacker models, and potential security attacks on smart metering systems are:

3.1. Software vulnerabilities:

These Issues can occur at programming level that can leave a Frame work open to attack. This could be an Operating System (OS) on a modern controller, or attack on firmware in a security gadget, or a zero-day powerlessness on a server working framework.

3.2. Hardware vulnerabilities:

This refers to physical hardware installed in the network. The basic purpose of hardware will be in the product which runs it, i.e. the firmware. Attack on firmware can be serious as it can make the remote gadgets which run the firmware useless. These firmware usages can be redesigned remotely over a system. If this firmware updated is initiated deliberately with malevolent or incorrect code it can possibly "block" the gadget, making it useless. This could bring about extreme downtime for whatever framework it relates to.

3.3. Network vulnerabilities:

Security of network systems is very important. Basic frameworks need to be set up in disconnected systems with constrained user access. Any entrance or departure to these frameworks need to be secured by vigorously fortified firewalls with strict access control instruments. Lower level frameworks, for example, production network needs utmost care and should be secured with firewall, Intrusion Detection Systems (IDS)/Intrusion Prevention Systems (IPS) and conceivably some 'hacker trap' frameworks should be introduced. WAN access to production network need to be constrained and very secured. Remote destinations and remote clients also should be secured and strict IPSec VPNs, or stay on private systems should be implemented.

3.4. Infrastructure/Server vulnerabilities:

This refers to bad setup in infrastructure of Information Technology (IT) frameworks, for example, setting all clients up with the same passwords, or passwords considering their name, and so on are extremely risky. Unpatched server vulnerabilities, especially those in view of the edge of a system or in a 'Neutral territory' (DMZ), are more exposed for hacking. All remote access should to be secured over Virtual Private Network (VPN) with a two-element verification framework set up and in addition to this for more security in basic frameworks; two-element confirmation should be set up over all frameworks.

3.5. Physical vulnerabilities:

Physical vulnerabilities can be referred to as physical setup like bad substations, wind ranches, or server rooms all posture potential danger to a framework. If these are a piece of a protected or basic system then they should be vigorously secured. Access control with bookkeeping is important at these physical areas. There should be more security protocols for gadgets. Switch security can be utilized on systems to prevent potential assailants from putting in their own Ethernet link (by locking interchanges

to the first MAC address as it were). USB ports on PCs and Servers should be debilitated by equipment so that malware would be extremely hard to be installed physically. [9]
General outline of the attack can be defined using simple flowchart which includes various phases.

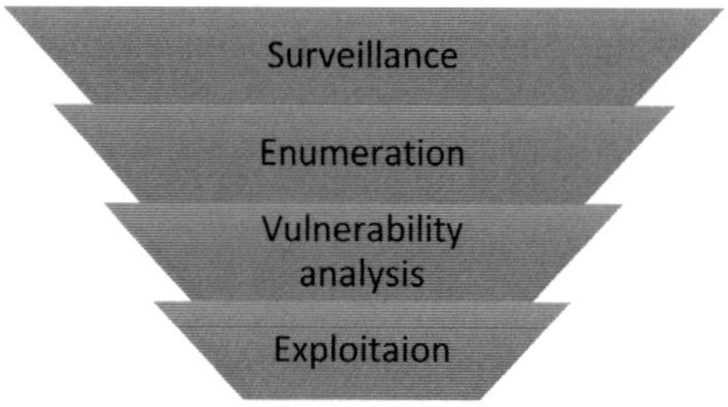

Figure 1. General Outline of the attack.

The various types of attacks are:

3.6. Eavesdropping:
Eavesdropping or spying is a serious issue and is considered as passive attack including the assailant latently listening to the data passed on by a smart meter or a smart metering passage to the administration suppliers. Such attack influences the protection of client and can be effectively performed over a remote correspondence channel or an electrical cable. The discovery of such attack is exceptionally troublesome. Such an attack can generally be dispatched through the WAN and by an inquisitive neighbor or a thief for different reasons.

3.7. Denial of Service Attacks:
Denial of Service attack focuses in general, smart metering foundation, the smart grid or part of the network. Such attack can be performed, for instance by an enemy by sending numerous more charges than anticipated to the smart metering gateways or on the flip side to the utility servers. This will make the framework at risk level which might not be ready to react to the authentic solicitations. It will basically close the matrix or piece of framework for vital administrations. The extent of this attack is the entire matrix or parts of a network. Such attacks can be dispatched through the WAN.

3.8. Packet Injection Attacks:
Packet Injection attack can be dispatched by infusing false packets e.g., false commands for a smart meter, into the system. This may be done to slice power, to parts of the smart metering framework. This attack can also be used to trade off the charging process by creating false bills costing the clients and utilities a considerable measure of cash. Such attacks will regularly be propelled from the WAN.

3.9. Malware Injection Attacks:
Malware Injection attack is mainly caused by infusing malware into the system which can propel this attack. This will influence the correspondence between gadgets in the matrix and bargain the charging and reporting forms. The interest/utilization status of the grid is disturbed in order to destabilize the heap on the matrix. The extent of this attack can possibly be up to the entire smart metering framework. Such attacks can be propelled from the WAN.

3.10. Man-in-the-center Attacks:
A man-in-the-center attack is a sort of attack in which an attacker "embeds" himself between communicating clients. An attacker makes an association with both the clients, catches their messages

and transfers them to the next end and makes each of them believe that they are talking specifically to each other. In this manner, the attacker has the opportunity of either inactively listening stealthily or adjusting the data being traded between the clients.
B Min [10] discusses further Malware attacks targeting cyber physical systems and proposes these types of attacks as the attacks which remain undetected as their impact is physical. This type of attack can be considered as deceptive attack as the attackers ensure that the operators do not notice their activity and whether system is under attack. Genuine malware attacks effectively infect and modifies several CPS and makes them malfunction. It also dishonestly reports general data and prevents it from recovering. Three basic requirements for deceptive attack: Modifying the behavior of target physical devices, hiding physical impact via software or data manipulation, preventing automatic and manual recovery attempt. These attacks are multistage attacks since most of CPS attacks include power plants and other smart grid subsystems and are set up using customized commercial off the shelf COTS hardware and software solutions.
B. Min [10] also discusses a very important attack which is a combination of characteristics of smart grid and above discussed CPS related malware attacks.

3.11. Smart Grid Blackout Attack:
This focusses on both nullification of built in security structures like self-healing and validation in smart grid and CPS related malware attacks. There are several phases during attack. First phase is initialization phase. In this target subsystems are attacked by penetrating wide range of attack vectors. After this the system installs cyber espionage malware on attacked targeted subsystems and starts gathering data on the systems and configures the smart grid systems and networks. As the architecture of smart grid is huge and complex it's not easy to get total control over the system, for that reason it targets small systems one by one and penetrates from one system to other. During propagation from one system to other malware can self-replicate to other systems on LAN to find target network systems and field devices. This process continues until the attack finds target field devices to modify at the phase of deceptive attack. Next phase involves gathering of information and organization. In this phase already installed malware carries out cyber surveillance tasks and collects required information from SCADA systems. The malware uses HTTP and RPC with encrypted protocol which makes security systems to detect it. Also collected data stored in the infected system is encrypted to make it difficult for analysis. [10]

4. The main Solutions and Architectures

4.1. Encrypted Communication:
The main goal of Encrypted Communication technique is to see that correspondence between the elements in the smart metering frameworks is encrypted. The encrypted correspondence is regularly accomplished by utilizing existing conventions, for example, TLS. TLS is not viewed as secure any longer and in this way different methodologies must be consolidated with TLS to make encryption stronger. These incorporate application layer encryption notwithstanding TLS to fortify the general encrypted correspondence. An illustration is the insurance profile (PP) based methodology created in Germany by the BSI, otherwise called the BSI PP. They propose double encryption, at the application layer to guarantee end-to-end encryption and at the vehicle layer.

4.2. The Privacy-saving Authentication Scheme for a Smart Grid Networks (PASS):

The Privacy-saving Authentication Scheme for a Smart Grid Networks (PASS) includes the utilization of smart apparatus (situated at clients' homes) connected with an alter safe gadget for producing pseudo personalities and marks on messages. A client is given this gadget when he/she opens a record or registers a recently obtained smart machine.

B.Min, V.Varadharajan [10] describes how the attacks can be detected and blocked by applying suitable defensive techniques.

4.3. Gateway based Approach:

Gateway is a latest way to deal with smart metering, proposed by European nations, like Germany and the UK. A gateway is an empowering innovation for future communications between the smart meters and the smart grid. A gateway (or smart metering gateway) goes about as a correspondence unit between the metering gadgets in the client premises and the utility. The portal not just serves as a correspondence and control unit between the meters and the utility, but on the other hand is responsible for the guarantee of the protection of client. One passage oversees the smart meters and other smart gadgets introduced in a house or a premise. The gateway intermittently speaks with the utility servers by means of WAN

4.4. Interruption Detection Systems:

Interruption Detection Systems (IDS) and Intrusion Prevention Systems (IPS) could be utilized as a part of the systems of a smart metering framework. This gives recognizable proof of interruptions, location of rebel hubs or wellspring of attacks and rejection of these hubs from further correspondence in the system.

F.Siddiqui, S.Zeadally, C. Alcaraz, S.Galvao [11] further proposes some more architecture for smart grid Privacy.

4.5. Blind signature:

Blind signature consists of following procedures:

First party signs a message created by second party with no knowledge of its original data. Third party receives the signed message. It verifies that the message is authorized by first party. The consumers prepare a set of credentials, each stating the amount of electricity requested. Customer requests the control center to sign them unseeingly so that they can submit any of these credentials to request of electricity. In this process Client 1 is unaware of the valid data sent by Client 2, this data is verified using a protocol which is used in e-cash system. Client 2 generates ‘a ‘data using various blinding platforms. It then blinds the ‘a’ data and transmits it to Client 1. Client 1 randomly selects ‘b’ data (b<a) and asks Client 2 to reveal the b blinding factor. If ‘b’ blinding factors are valid, Client 1 accepts the request for authentication and verifies the remaining (b-a) data. When a consumer presents a credential secretly, the control center cannot verify which customer is making the request but can verify the signature to confirm that it has been received from valid consumer. [11]

4.6. The 3rd Party Escrow Architecture

The 3rd Party Escrow Architecture helps in anonymizing high frequency energy measurement data using pseudonymous identity (ID)

- It has two separate IDs
- Two Ids are High Frequency IDs (HFID) and one Low Frequency Id (LFID)
- HFIDs are anonymous

- LFID can be assigned to a consumer or smart meter

The confidentiality is implemented by keeping HFID secret to the utility or the smart meter installer. The HFID is 'secret' inside the smart meter, or hard-coded to be used for all HFID related messages. For the utility to authenticate the legality of the HFID, a 3rd party Escrow mechanism is implemented. The 3rd party can build smart meter itself or can outsource the work to a valid third party. This third party will also have an authentication to access this information. The manufacturer can assign two unique IDs to each smart meter that is produced, only one of which (LFID) is visible to the utility, both during the procurement and distribution procedures. [11]

5. Conclusion

Cyber security and privacy issues in the smart grid are new areas in the fields of power industry and electrical engineering. Increased interconnection and integration introduces cyber vulnerabilities into this grid and should be provisioned in smart grid designs. This requires different designs of traffic routing to meet the required privacy properties. Current reviewed schemes of accountability don't address all the above issues. Thus, in conclusion any communication paradigm used in a smart grid should support all aspects of privacy such as ambiguity, uncertainty, indistinct, unmanageability and undetectably. In the coming years, the trend in energy management system of the home of intelligent home will become increasingly apparent. Application and development of Bluetooth and ZigBee are most appreciated by the industry in the recent years. Of the two types of technology, ZigBee supports low cost and more knots, and with its low power consumption, is set to become an important part of communication technology for the intelligent home. Because the ZigBee specification SEP2.0 support multiple functions of the network technology, a product can be compatible with a variety of communication methods to connect to the home network and home automation. [12].

References

[1] "Security Issues in Smart Metering Systems "Obaid Ur-Rehman, Natasa Zivic, Christoph Ruland
[2] "Cyber Security and Privacy Issues in Smart Grids" J liung, Y. Xiao, S Lui, C.L. Phillip Chen.
[3] R. Kailar, "Accountability in electronic commerce protocols," IEEE Trans. Softw, Eng., vol. 22, no. 5, pp. 313–328, May 1996.
[4] M. Kudo, "Electronic submission protocol based on temporal accountability," in Proc. 14th Annu. Comput. Secur. Appl. Conf., 1998, pp. 353–363
[5] "Accountability in Smart Grids" Jing Liu, Yang Xiao, Jingcheng Gao.
[6] Hastings, John, David M. Laverty, and D. John Morrow "Securing the smart grid", 2014 49th International Universities Power Engineering Conference (UPEC), 2014.
[7] M. Stegelmann and D. Kesdogan, ``GridPriv: A Smart metering architecture offering k-anonymity," in Proc. IEEE 11th Int. Conf. Trust, Security Privacy Comput. Commun., Jun. 2012, pp. 419_426.
[8] X. Lin, R. Lu, H. Zhu, P.-H. Ho, X. Shen, and Z. Cao, ``ASRPAKE: An anonymous secure routing protocol with authenticated key exchange for wireless Ad Hoc networks," in Proc. IEEE ICC, Jun. 2007, pp. 1247_1253.
[9] Hastings, John, David M. Laverty, and D. John Morrow, "Securing the smart grid", 2014 49th International Universities Power Engineering Conference (UPEC), 2014.
[10] "Design and Analysis of Security Attacks against Critical Smart Grid Infrastructures "B. Min, V. Varadharajan.

[11] "Smart Grid Privacy: Issues and Solutions "F. Siddiqui, S. Zeadally, C. Alcaraz, S. Galvao.
[12] "Application and Development of ZigBee Technology for Smart Grid Environment" Sheng-Fa Chang1, Chi-Feng Chen1, Jyh-Horng Wen1, Jih-Hsin Liu1, Jun-Hong Weng1, Jian-Li Dong2.

IOP Conf. Series: Earth and Environmental Science **73** (2017) 012022 doi:10.1088/1755-1315/73/1/012022

Distribution Network Reconfiguration based on Improved Gravitational Search Algorithm

Yanshuo Wang, Bo Zhang
School of Electrical Engineering, Shandong University, Jinan 250061 China.

E-mail: wangyanshuo6868@qq.com

Abstract. Network reconfiguration for loss reduction in the distribution system is a significant way to save resources. It's an extremely hard optimization problem because of a plenty of switches in a distribution network. This paper presents an efficient Gravitational Search Algorithm (GSA) and its improvement based on elite strategy and adaptive position (ES-APGSA) to reduce grid loss using reconfiguration. The new algorithm can improve the speed of iterations and avoid the local convergence. In addition, the encoding method based on ring network is proposed which makes the dimensions of the particles much lower leading to an acceleration of the process. The correctness and effectiveness of the methods suggested in this paper is verified through the simulation results of IEEE33-node system.

1. Introduction

With the development of society, the requirements of energy economy and quality are increasing gradually. Distribution network as a part which is directly connected to power consumers has a profound impact on the users' feeling and evaluation. How to make the distribution network operation better and securer is becoming a research hot spot today.

Serving as one of the important measures to facilitate optimal operation of distribution, distribution network reconfiguration is lucrative in both economic and social benefit. It is a complex combinatorial optimization process aimed at finding a radial operating structure that minimizes the system power loss, balances the overload of the network, and improves the reliability of power supply [1] while satisfying operating constraints. Because of a large amount of section switches and contact switches in network, the optimization is a multi-objective nonlinear mixed integer problem. The traditional solutions are mathematical optimization theory, optimal power flow and switch exchange method. In recent years, intelligent optimization algorithms have been booming. Genetic algorithm (GA) is applied for reconfiguration in [2], but the coding isn't improved. Ant colony search algorithm (ACSA) improves the speed of iterations with a better performance than GA in [3]. With the development of computers, artificial neural network (ANN) provides a new method to this problem but can't explain the essence of reconfiguration in [4].

This paper mainly focuses on a new method based on improved GSA including elite strategy and adaptive position to solve network reconfiguration of distribution system for power loss reduction, and encoding strategy of GSA is also studied.

2. Distribution network reconfiguration model

2.1. Objective function

The objective function of network reconfiguration is to minimize the total power losses in the condition of operating constraints in the distribution system. With the assumption that the loss due to line reactance is negligible, the objective function can be expressed by equation (1).

$$\min f = \sum_{j=1}^{N} R_j \frac{P_j^2 + Q_j^2}{U_j^2} \tag{1}$$

where f is line losses in the whole system, N is the number of branches, R_j is the resistance in the j-th branch, P_j and Q_j are the power flow, U_j is the voltage at the terminal of j-th branch.

2.2. Constraints

Reconfiguration must meet the constraints or else, the objective function value is set to a penalty value for avoiding this computation.

2.2.1. Power flow constraint.

Power flow must be met in the process of every reconfiguration which makes distribution network topology change.It is sure that the result is correct.The forward-backward sweep method is generally adopted in power flow calculation of distribution network while its radial structure.

2.2.2. Inequality constraints.

The results of line flows and node voltages can't be exceeded as in equation (2).

$$\begin{cases} S_j \le S_{j\max} \\ U_{i\min} \le U_i \le U_{i\max} \end{cases} \tag{2}$$

where S_j and $S_{j\max}$ are power flow and maximum capacity in j-th branch, Ui is i-th node voltage, $U_{i\min}$ and $U_{i\max}$ are minimum voltage and maximum voltage.

2.2.3. Radial structure constraint.

A distribution network is designed in a closed loop and operated in an open loop. The topology after reconfiguration must be radial while there are no islands or acnodes.

2.3. Forward-backward sweep method

The distribution network power flow calculation based on forward-backward sweep method is faster than Newton-Raphson method, and the former requires initialization less. To make the result correct, hierarchical node optimization is needed because of data about head-foot nodes in the branches changing with each reconfiguration.The each node is picked out in layers in accordance with the current flow. The power supply node is the first layer, all the nodes connected with supply node are belong to the second layer, and so on. The nodes are renumbered according to the layered order which is used for the next power flow calculation.

3. Coding and feasibility analysis

A distribution network is designed in closed loop and operated in open loop. Generally, the reconfiguration is encoded in each switch state and it leads to too long codes and complicated calculation. Therefore, ring-based coding is necessary to solve this problem.

3.1. Ring-based coding principles

According to the number of contact switches, the system is divided into equivalent minimum loops which require a ring with the shortest distance and no common branch with each other.In the process of reconfiguration, only the switches in loops are open.

3.2. Radial structure judgment

In order to avoid rings in the system, the switches in noncommon branches must be open. As multi-loops, judging the radial network is the same as judging the open switch in common branch or for two loops. There are two undesirable cases, one is that the open switch is same and the other is that the open switch is in common branch in two loops.

3.3. Connectivity judgment

The judgment of connectivity is mainly generated connectivity matrix which is defined as:

- The rows and columns of the matrix are different nodes.Each element t_{ij} is stated for the branch existence that '1' represents a branch from *i*-th node to *j*-th node and '0' stands for no branch.
- The diagonal of this matrix represents the total number of branches in connection with this node. Its value equals to the sum of '1' in this row and column.

If a diagonal element is '0', the node corresponding to it is solitary. The island is no power flow through it and can be found in the process of hierarchical node optimization.

4. Gravitational search algorithm and improvements

4.1. Gravitational search algorithm

Gravitational search algorithm is a heuristic optimization method based on the law of universal gravitation and Newton's second law, referred to as GSA[5].The GSA seeks the optimal solution by the masses of the objective space moving through the medium of information communication, gravitational force. The optimal solution will be found when the masses move to the optimal position.

The GSA initializes the position and velocity in the solution space and velocity space firstly, where the position represents the solution. The position and velocity of the *i*-th search particle in the D-dimensional space are expressed as in equation (3).

$$\begin{cases} X_i = (x_i^1, \cdots, x_i^d, \cdots, x_i^D) \\ V_i = (v_i^1, \cdots, v_i^d, \cdots, v_i^D) \end{cases} \tag{3}$$

where x_i^d and v_i^d represent position and velocity.

The mass of ith particle is defined as in equation (4) and equation (5).

$$q_i(t) = \frac{fit_i(t) - worst(t)}{best(t) - worst(t)} \tag{4}$$

$$M_i(t) = \frac{q_i(t)}{\sum_{j=1}^{N} q_j(t)} \tag{5}$$

where $fit_i(t)$ and $M_i(t)$ represent the fitness value and the mass of *i*-th particle at the *t*-th iteration; $best(t)$ and $worst(t)$ represent the best and worst fitness value of all the particles at the *t*-th iteration.

The algorithm originates from the simulation of the law of universal gravitation but does not rigidly adhere to the exact formula of universal gravitation. According to the experimental results, on the *d*-dimension, the gravitational force of *i*-th particle is defined as equation (6).

$$F_{ij}^d(t) = G(t)\frac{M_i(t)M_j(t)}{R_{ij}(t) + \varepsilon}(x_j^d(t) - x_i^d(t)) \tag{6}$$

where $R_{ij}(t)$ is the Euclidean distance from *i*-th to *j*-th particle. ε is a constant for denominator not zero. $G(t)$ represents the value of the gravitational constant at the *t*-th iteration as equation (7).

$$G(t) = G_0 e^{-\alpha t/T} \tag{7}$$

where G_0 and α are constants, T is the maximum number of iterations.

The resultant force of *i*-th particle on the *d*- dimension is represented as equation (8).

$$F_i^d(t) = \sum_{j \neq i}^{N} F_{ij}^d(t) \tag{8}$$

According to the Newton's Second Law, the acceleration of *i*-th particle on *d*-dimension is defined as equation (9).

$$a_i^d(t) = \frac{F_i^d(t)}{M_i(t)} \tag{9}$$

Then velocity and position are updated as in equation (10).

$$\begin{cases} v_i^d(t+1) = r \times v_i^d(t) + a_i^d(t) \\ x_i^d(t+1) = x_i^d(t) + v_i^d(t+1) \end{cases} \tag{10}$$

The basic GSA is sufficient to optimize some problems but there are many challenges such as a waste of time and difficulty of convergence. Therefore, the GSA is supposed to be improved.

4.2. Improvements of GSA

Elite strategy and adaptive position are both applied for GSA. Improved GSA is called ES-APGSA.

4.2.1. Elite strategy.

When calculating the resultant force, some particles with pretty low mass can be neglected due to slight influence on others. It is called "Elite Strategy". It is wise to take the particles with high mass into consideration which have strong gravitational force. A larger particle occupies a better position, representing a better solution. Doing so can eliminate the influence on a large mass caused by a short distance. So all the particles move closer to the one with the largest mass gradually and ultimately reach the position representing the optimal solution. Therefore, it's necessary to make changes to resultant force in equation (8) as equation (11).

$$F_i^d(t) = \sum_{j \in kbest, j \neq i}^{N} rand_j \cdot F_{ij}^d(t) \tag{11}$$

where $rand_j$ represents a random variable which is uniformly distributed between [0,1]. *kbest* indicates particles' mass is in descending order of the first *k* individuals and the *k*-value decreases linearly from *N* to 1 with the iterations changing.

4.2.2. Adaptive position.

There are essential differences between the particle swarm optimization algorithm (PSO) and GSA, but their models are both based on updates of the velocity and position.

Inspired by improvement of PSO[6], the paper introduces a new coefficient *s* in the position updating equation of GSA as equation (10). This coefficient is a kind of constriction factors and makes rangeability of position accommodate iterations. Constriction factor 's' is better to take the form of the variable and its expression is equation (12).

$$s = s_{\max} - (s_{\max} - s_{\min})\frac{t}{T} \tag{12}$$

where *t* represents the iterations at present, *T* represents the maximum number of iterations. s_{max} and s_{min} represent the upper and lower bounds of *s* which are 1 and 0.

Taking *s* into consideration, the position updating changes into equation (13) .

$$x_i^d(t+1) = s \cdot x_i^d(t) + v_i^d(t+1) \tag{13}$$

5. Steps of reconfiguration

1) Read the raw data including the node powers, the branch data, the reference voltages and powers.
2) Calculate the voltage of each node and the total loss.
3) Do ES-APGSA.
   a. Initialize velocity and position.
   b. Calculate the fitness values of all the particles. The best is global optimal value expressed as F_{best} and the position of the particle is global optimal solution expressed as X_{gbest} while the position is X_{pbest} in every iteration。
   c. Update mass and gravitational constant.
   d. Calculate the resultant force with ES.
   e. Calculate the acceleration and update velocity and position with constriction factor. Note that the positions should be rounded.
   f. Repeat steps b~e until maximum iteration.
4) Optimal computation more than once and analyze results.
5) Choose the best in step(4) as final result and compare it to result in step(2).

IOP Conf. Series: Earth and Environmental Science **73** (2017) 012022 doi:10.1088/1755-1315/73/1/012022

6. Case study

A typical distribution system IEEE33-node, as shown in figure 1, which was studied in [1] is taken as a case study to test the performance of the ES-APGSA.

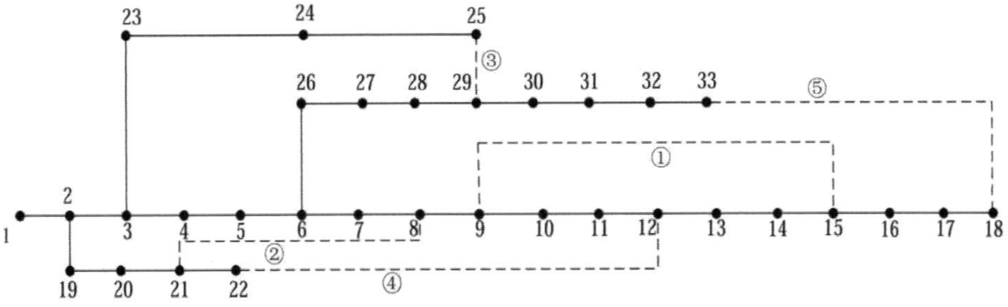

Figure 1. IEEE33-node distribution system

In this system, the total load power is 3715kW+j2300kvar, the base voltage is 12.66kV and the base capacity is 10MVA. There are 32 section switches and 5 contact switches. 1st node is the power supply. Before reconfiguration, the line loss is 202.67kW and the lowest voltage is 0.9131p.u. for 18th node in the context of the contact switches open. According to coding principle, the dimension of each particle in GSA is 5 and the number of particles in a group is set to 10. It iterates 100 times every GSA and the reconfiguration is repeated 30 times.

There are convergence curves using GSA and ES-APGSA for 30 times as figure 2 and figure 3.

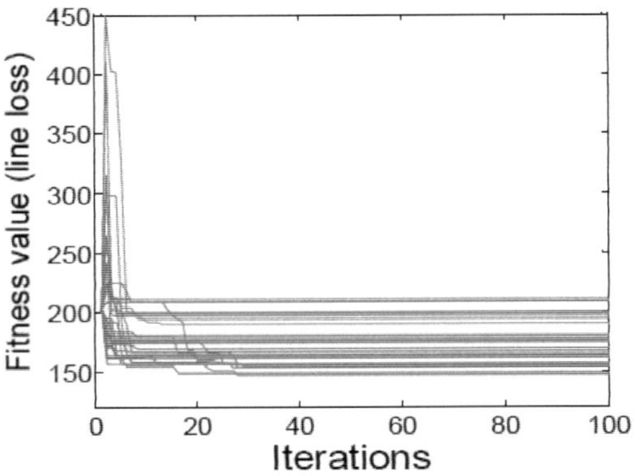

Figure 2. The convergence curves in GSA

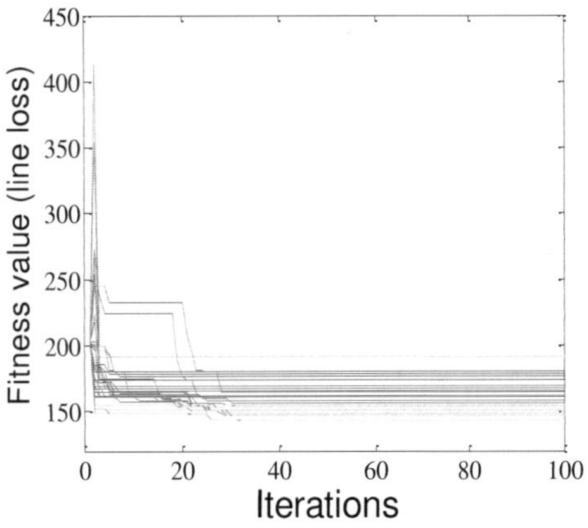

Figure 3. The convergence curves in ES-APGSA

Figure 2 shows that in GSA, some convergence results is unsatisfactory because they are worse than before, and a few results can reach around best. ES-APGSA has a better performance that all the results are better in some degree and most of them are near the best. The detailed results are shown in table 1 and compared to PSO-w studied in [6].

After reconfiguration using ES-APGSA, the total line loss decreases from 202.6762kW to 139.9767kW by 30.94% and the voltage levels are improved in this system. The performance of ES-APGSA is much better than others.

Table 1. Statistical results of the different methods.

	Before Optimization	PSO-w	GSA	ES-APGSA
Minimum value	202.6762	143.2957	144.0234	139.9767
Mean value	--	165.3088	167.8482	158.3088
Variance	--	385.8687	324.2321	104.0764
Optimal open branch	9-15, 8-21, 25-29, 12-22, 18-33	14-15, 7-8, 27-28, 9-10, 32-33	14-15, 7-8, 27-28, 10-11, 32-33	14-15, 7-8, 28-29, 9-10, 32-33
Lowest voltage	0.9131(18th node)	0.9401(32nd node)	0.9398(32nd node)	0.9413(32nd node)

7. Conclusions

An improved GSA has been proposed for distribution loss minimum reconfiguration. The gravitational force is a way of transferring information between different particles and leads the particles to the optimal position representing the solution. ES-APGSA is improved in elite strategy and adaptive position to accelerate the calculations. Also, the encoding method based on ring network is successfully used for dimensionality reduction. The effectiveness of the ring-based encoding and ES-APGSA has been investigated on IEEE33-node system showing promising results when compared with PSO-w and others. This gives an advice to the decision makers to optimize system operation.

References

[1] Baran M E and Wu F F 1989 Network reconfiguration in distribution systems for loss reduction and load balancing *IEEE Trans. Power Del.* **4**(2) 1401-1407

[2] Zhu J Z 2002 Optimal reconfiguration of electrical distribution network using the refined genetic algorithm *Electr. Power Syst. Res.* **62**(1) 37-42

[3] Su C T, Chang C F and Chiou J P 2005 Distribution network reconfiguration for loss reduction by ant colony search algorithm *Electr. Power Syst. Res.* **75**(2–3) 190-199

[4] Kayal P, Chanda S and Chanda C K 2011 An ANN based network reconfiguration approach for voltage stability improvement of distribution network *IEEE Int. Conf. on Power and Energy* vol 15 p 1-7

[5] Rashedi E, Nezamabadi-Pour H and Saryazdi S 2009 Gsa: a gravitational search algorithm *Inform. Sciences* **179** 2232-2248

[6] Dahalan W M and Mokhlis H 2012 Network reconfiguration for loss reduction with distributed generations using PSO. *IEEE Int. Conf. on Power and Energy* vol 48 p 823-828

A Method of Fault Line Selection in Small Current Grounding System Based on VMD Energy Entropy and Optimized K-means Clustering

H Li[1], Y Sun[1], K Li[1], J Liu[1], S Li[1] and M Xu[2]
[1]School of Electrical Engineering, Shandong University, Jinan 250061, China
[2]Laiyang Power Supply Company, State Grid Shandong Electric Power Company, Yantai, China

Email: 969012980@qq.com; sy@sdu.edu.cn; lkjun@sdu.edu.cn; 249716289@qq.com; dereklsy@163.com; guoanqi630@163.com;

Abstract. When single-phase to ground fault occurs in small current grounding system, the zero-sequence transient current contains a lot of complex fault information. In order to extract characteristic from the zero-sequence current accurately as the criterion of fault line selection, this paper puts forward a method of fault line selection based on VMD energy entropy and optimized K-means clustering. Firstly, the zero-sequence current of each line is decomposed by VMD (variational mode decomposition, VMD), and each sub mode of the original current is obtained. Secondly, the energy entropy of each sub mode is calculated as the characteristic quantity of each line state. Finally, the optimized K-means clustering algorithm is used to cluster the fault feature to realize fault line selection. In this paper, Simulink is used to simulate the single-phase to ground fault in resonant grounded system and the results show that the method that has high accuracy can provide reference for practical engineering application.

1. Introduction

In recent years, there have been a lot of fault line selection methods, mainly divided into the line selection method applying the steady-state signal and the transient signal [1]. The transient current signal has large intensity, and is not affected by the arc suppression coil, so the method of line selection applying transient signal is more suitable for the identification of faults [2-3].

A method of Hilbert transform and probabilistic neural network routing is proposed in [4]. In the proposed method, it is shown that the zero-sequence transient current of each line is decomposed into a series of intrinsic mode functions by the empirical mode decomposition (EMD) and the intrinsic mode function energy is calculated as the input of PNN network to select the fault line. However, the empirical mode decomposition is prone to the phenomenon of modal aliasing, which cannot accurately extract the fault information[5].In view of the problem of mode mixing in EMD, the Hilbert transform is improved by ensemble empirical mode decomposition (EEMD), and the fault feature components are extracted to identify the fault line[6]. However, EEMD is affected by the addition of white noise amplitude and integration times, and the computational complexity is large.

Variational modal decomposition [7] is a new signal processing method that decomposes the original signal into k finite bandwidths with different center frequencies. This method not only effectively avoids the modal aliasing problem of EMD, but also greatly reduces the computational complexity and improves the computational efficiency relative to EEMD. K-means algorithm is a widely used clustering method [8], usually in European distance as a similarity measure. The algorithm is simple and has a fast convergence rate.

Based on the above analysis, this paper presents a fault line selection method based on variational mode decomposition energy entropy and optimized K-means clustering. Firstly, the zero-sequence current signal of each line is decomposed into sub-modal with the main frequency component by using VMD. Secondly, calculate the energy entropy of each sub modal, which is used as the fault characteristic. Finally, the optimized K-means clustering algorithm is used to cluster the energy entropy, so as to select the fault line more effectively. The simulation results show that the method proposed in this paper can accurately identify the fault line, which has good effectiveness and feasibility.

2. VMD and energy entropy

2.1. Variational modal decomposition principle

The core of VMD is to decompose the input signal into k finite bandwidths with central frequencies ,minimizing the sum of the bandwidth estimates for each modal, which is called the mode u_k. The signal decomposition process is the solution of the variational problem [9]. The variational problem model with constraints is shown in (1).

$$\begin{cases} \min\limits_{\{u_k\},\{\omega_k\}} \{\sum_k \left\| \partial_t[(\delta(t)+\frac{\mathrm{j}}{\pi t})\cdot u_k(t)]\mathrm{e}^{-\mathrm{j}\omega_k t} \right\|_2^2 \} \\ s.t. \quad \sum_k u_k = f \end{cases} \quad (1)$$

Where $\{u_k\}:=\{u_1,\ldots,u_k\}$ is each modal component. $\{\omega_k\}:=\{\omega_1,\ldots,\omega_k\}$ is the center frequency of modal component. $\delta(t)$ is the impulse function. ∂_t is the partial derivative of t. f is the original signal. In order to obtain the optimal solution of the constrained variational problem, Lagrange multiplication operator $\lambda(t)$ is introduced and the constrained variational problem is transformed into a non-constrained variational problem which is given by (2).

$$L(\{u_k\},\{\omega_k\},\lambda):=\alpha\sum_k\left\|\partial_t[(\delta(t)+\frac{\mathrm{j}}{\pi t})\cdot u_k(t)]\mathrm{e}^{-\mathrm{j}\omega_k t}\right\|_2^2+\left\|f(t)-\sum_k u_k(t)\right\|_2^2+\left\langle\lambda(t),f(t)-\sum_k u_k(t)\right\rangle \quad (2)$$

Where α is the second penalty factor. The saddle point of the above Lagrange function is obtained by alternate direction method of multipliers (ADMM), which is optimal solution. The principle of variational modal decomposition is described detailedly in [10], so this article will not repeat.

2.2. Energy entropy

In order to further extract the fault information contained in the zero sequence current, on the basis of the variational modal decomposition, this paper calculates the sub-mode energy entropy that is used as an important criterion for fault line selection. The principle of energy entropy and the solving process are described as follows [11].

1) Sub modal energy E_k. $E_k=\int_{-\infty}^{\infty}|u_k|^2\,dt \qquad k=1,2,\ldots K$ (3)

2) Energy normalization of K sub modal. $E=\sum_{k=1}^{K}E_k$ (4) $p(u_k)=E_k/E$ (5)

3) VMD energy entropy. $H(u)=-\sum_{k=1}^{K}p(u_k)\log p(u_k)$ (6)

3. The optimized K-means clustering algorithm

3.1. Linear decreasing weight PSO

PSO is an intelligent optimization algorithm, which is widely used in the optimization of dynamic and multi-objective problems. Compared with the traditional optimization algorithm, it has faster computing speed and better global search ability [12]. Therefore, PSO is applied to the optimization of K-means clustering algorithm to solve the problem that K-means algorithm is easy to fall into local optimal problem improving the final clustering accuracy. Because PSO is easy to be premature and prone to oscillation in the vicinity of the global optimal solution, this paper improves the problem by the method of linear decreasing weight.

3.2. K-means clustering algorithm

The Euclidean distance is used as similarity measure in K-Means algorithm, and the squared error criterion function is used as the clustering criterion function [13]. For the problem of fault line selection, there is no need to consider the selection of the number of cluster centers and we only need to divide the number into two categories. However, the algorithm is sensitive to the initial cluster center selection, and it is prone to local optimum. So the algorithm needs to be optimized properly.

3.3. Linear decreasing weight PSO- K-means clustering algorithm

Based on the above analysis, this paper proposes a method of linear decreasing weight PSO-K-means clustering, which uses PSO algorithm to solve the problem of selecting the initial cluster centers of K-means. In this paper, we use the method of cluster center coding, and design the Euclidean distance function of the sample data points and the center of mass as the fitness function of linear decreasing weight PSO. The whole clustering process is described in Figure 1.

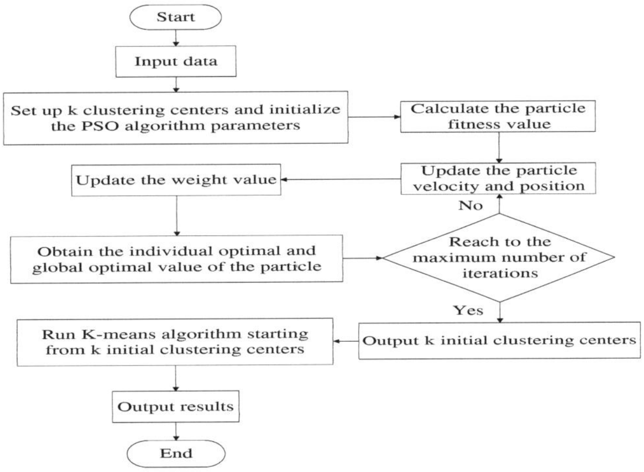

Figure 1. The algorithm flow chart of linear decreasing weight PSO-K-means clustering

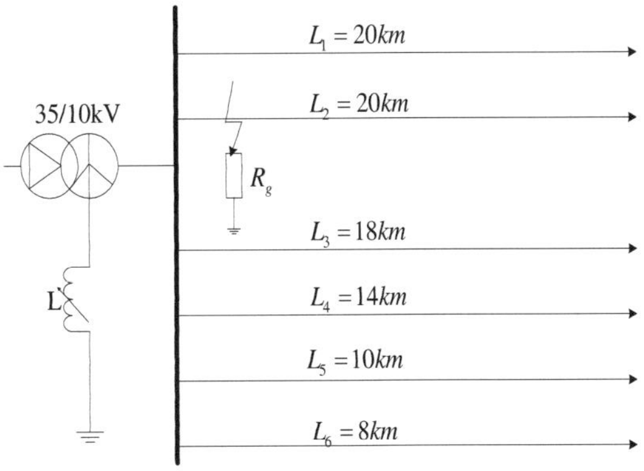

Figure 2. System simulation model

4. Simulation analysis

In this paper, a 10kV system is simulated by Simpower System module in Simulink, which is shown in Figure 2. Some of the parameters in the circuit are as follows: In the model there are 6 return lines in

IOP Conf. Series: Earth and Environmental Science **73** (2017) 012023 doi:10.1088/1755-1315/73/1/012023

total, and the cable length is shown in the figure. The positive sequence resistance of the line is 0.17Ω/km. The zero sequence resistance is 0.23Ω/km. The positive sequence inductance is 1.2mH/km. The zero sequence inductance is 5.48mH/km. The positive sequence capacitance is 12.74×10^{-9}F/km. The zero sequence capacitance is 7.75×10^{-9}F/km. The transformer rated voltage ratio is U_{N1}/U_{N2}=35kV/10kV. Single phase to ground fault occurs in line 2 and the fault location is 4 km away from the bus-bar. The operating mode of the system is over compensation. The compensating inductance is 3.3H and the compensation degree is 10%. The sampling frequency is 6 kHz during the simulation.

VMD is used to decompose the zero sequence current signal of line 3 when the fault angle is 90°. The time frequency of zero sequence current for the line 3 is shown in Figure 3. The VMD results of original signal are shown in Figure 4, Figure 5 and Figure 6.The original signal is decomposed into three sub-modal.

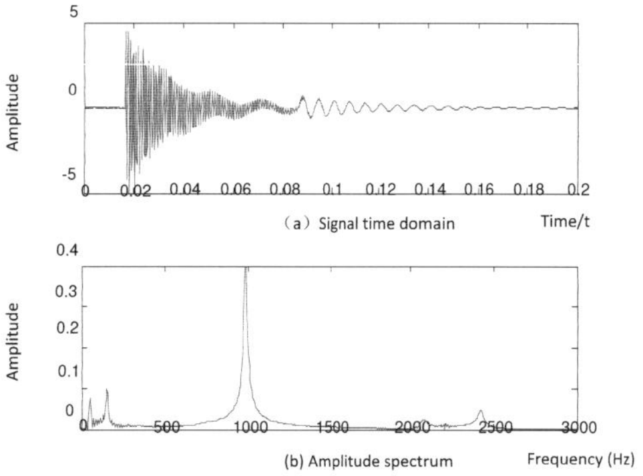

Figure 3. Time-frequency of zero sequence current signal

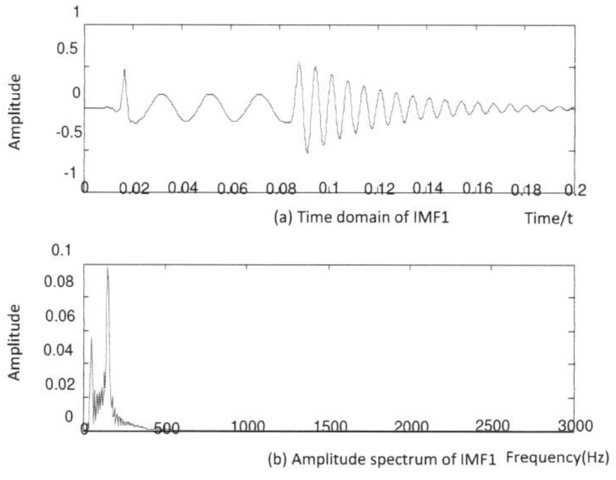

Figure 4. Time-frequency of IMF1

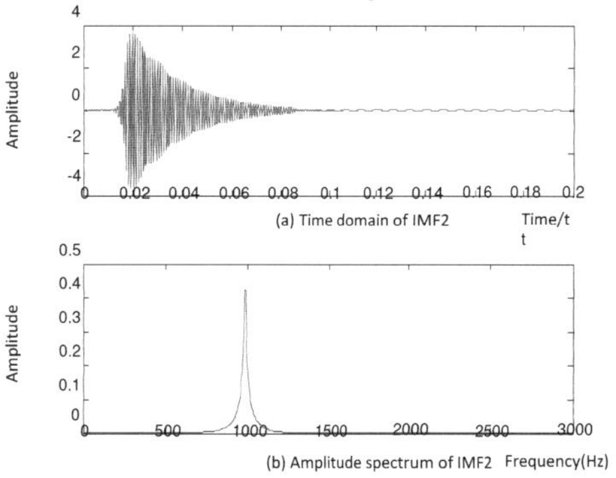

Figure 5. Time-frequency of IMF2

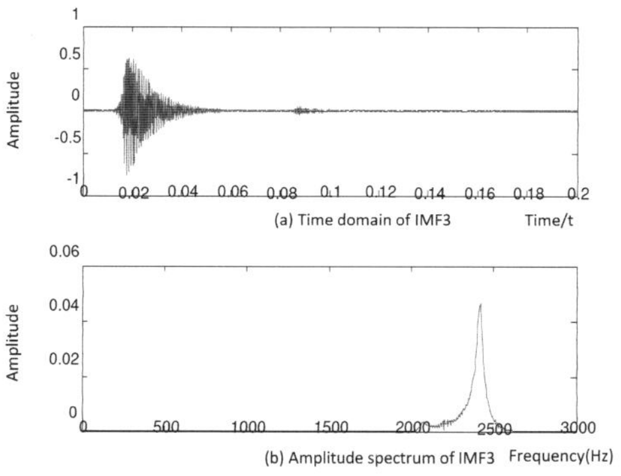

Figure 6. Time-frequency of IMF3

The results show that VMD can effectively decompose the zero sequence current signal. and the frequency component of each IMF component is relatively concentrated.

When the fault angle is 30° and 60°, and the grounding resistance is 20Ω and 200Ω, the system is simulated. The VMD energy entropy of each line is calculated and then clustered by the linear decreasing weight PSO-K-means clustering algorithm. The clustering results are described in Table 1 and Table 2.

The clustering results of Table 1 and Table 2 show that whether it is a small fault angle or high resistance fault, the proposed method can accurately determine the fault line.

Table 1. Clustering results of different fault angles

Fault setting R_f=20 Ω	Line	Energy entropy	Clustering results
θ=30°	L_1	32.32 24.32	Normal
	L_2	28.73 17.06	Fault
	L_3	29.51 23.27	Normal
	L_4	30.81 24.65	Normal
	L_5	30.85 20.87	Normal
	L_6	28.93 20.71	Normal
θ=60°	L_1	37.52 62.40 24.51	Normal
	L_2	36.82 60.51 72.14	Fault
	L_3	28.58 81.45 25.06	Normal
	L_4	29.49 12.11 26.64	Normal
	L_5	27.74 18.27 24.71	Normal
	L_6	21.15 11.64 24.96	Normal

Table 2 Clustering results of different fault resistances

Fault setting θ=0°	Line	Energy entropy	Clustering results
R_f=20Ω	L_1	29.22 17.34	Normal
	L_2	22.25 16.94	Fault
	L_3	24.06 20.71	Normal
	L_4	25.69 19.14	Normal
	L_5	24.42 20.49	Normal
	L_6	25.55 21.68	Normal
R_f=200 Ω	L_1	30.00 23.74	Normal
	L_2	26.52 20.99	Fault
	L_3	28.01 21.50	Normal
	L_4	30.04 22.63	Normal
	L_5	27.01 21.86	Normal
	L_6	27.51 22.00	Normal

5. Conclusion

This paper presents a method of fault line selection based on VMD energy entropy and optimized K-means clustering. This method not only avoids the modal aliasing problem of EMD, but greatly reduces the computational complexity relative to EEMD. K-means algorithm is optimized by using linear decreasing weight PSO so that the problem has been solved that K-means algorithm is easy to fall into local optimal problem. It is shown that the proposed method in this paper can effectively identify the fault line, which provides a new idea for fault line detection in small current grounding system. The method of feature extracting based on VMD has its own drawbacks. For example, the value of sub modal K should be given in advance, so the method needs to be further improved.

Acknowledgments
The authors gratefully acknowledge the assistance of Professor Ying Sun and Ke-Jun Li in research. And at the same time the authors gratefully acknowledge to Shandong University.

References
[1] Jiang B, Dong X and Shi S 2014 A method of single phase to ground fault feeder selection based on single phase current traveling wave for distribution lines *Proceedings of the Chinese Society of Electrical Engineering* **34** 6216-6227
[2] Zhang S, Yue M and Pan L 2015 Application of Improved Generalized Harmonic Wavelet Packet Decomposition and Chaos Oscillator to Fault Line Detection in Small Current Grounding System Transactions of China Electro-technical Society 30 13-20
[3] Lai P, Zhou X and Qiu D 2015 Research on transient-current frequency analysis and faulty line detecting method in indirectly grounding power system *Power System Protection & Control* **43** 51-57
[4] Li Z, Wang S and Wu L 2014 A new method and simulation for arcing high-impedance-grounding fault line selection in resonant grounded system *Power System Protection & Control* **42** 44-49
[5] Yang Y, Cheng J and Zhang K 2012 An ensemble local means decomposition method and its application to local rub-impact fault diagnosis of the rotor systems *Measurement* 561-570
[6] Ren Z and Zhang Y 2015 Fault line selection of distribution network based on improved Hilbert-Huang Transform and identification confidence degree *Power System Protection & Control* **43** 8-13
[7] Konstantin D and Dominique Z 2014 Variational mode decomposition *IEEE Transactions on Signal Processing* **62** 531-544
[8] Renato C and Mirkin B 2012 Minkowski metric, feature weighting and anomalous cluster initializing in k-means clustering *Pattern Recognition* **45** 1061-1075
[9] Liu C, Wu Y and Zhen C 2015 Rolling bearing fault diagnosis based on variational mode decomposition and fuzzy C means clustering *Proceedings of the Chinese Society of Electrical Engineering* **35** 3358-3365
[10] Zhang N, Zhu Y and Gao Y 2016 An on-line detection method of transformer winding deformation based on variational mode decomposition and probability density estimation *Power System Technology* **40** 297-302
[11] Xie B, Chen W and Zhou Q 2016 Partition of the development stage of air-gap discharge in oil-paper insulation based on wavelet packet energy entropy *IEEE Transactions on Dielectrics & Electrical Insulation* **23** 866-872
[12] Benidris M, Elsaiah S and Mitra J 2015 Power system reliability evaluation using a state space classification technique and particle swarm optimisation search method *IET Generation, Transmission & Distribution* **9** 1865-1873
[13] Boutsidis C and Magdon-Ismail M 2013 Deterministic Feature Selection for K-Means Clustering *IEEE Transactions on Information Theory* **59** 6099-6110

IOP Conf. Series: Earth and Environmental Science **73** (2017) 012024 doi:10.1088/1755-1315/73/1/012024

Parameter Identification of Equivalent Circuit Models for Li-ion Batteries Based on Tree Seeds Algorithm

W J Chen, X J Tan, and M Cai

School of Engineering, Sun Yat-sen University, No. 135, Xingang Xi Road, Guangzhou, 510275, China

Email: 284697081@qq.com; zsutxj@163.com; caiming@mail.sysu.edu.cn;

Abstract. Parameter identification method of equivalent circuit models for Li-ion batteries using the advanced tree seeds algorithm is proposed. On one hand, since the electrochemical models are not suitable for the design of battery management system, the equivalent circuit models are commonly adopted for on-board applications. On the other hand, by building up the objective function for optimization, the tree seeds algorithm can be used to identify the parameters of equivalent circuit models. Experimental verifications under different profiles demonstrate the suggested method can achieve a better result with lower complexity, more accuracy and robustness, which make it a reasonable alternative for other identification algorithms.

1. Introduction

Usage of rechargeable Li-ion batteries has aroused considerable attentions in a variety of industrial areas. Hybrid electric vehicles (HEVs), plug-in hybrid electric vehicles (PHEVs) and battery electric vehicles (BEVs) are the significant areas of applications of battery technology. Accurate monitoring of battery statuses and efficient management of battery power are presently the technique bottlenecks of these electrified vehicles. Generally speaking, to enable safe, reliable and efficient operations of the power batteries under the most demanding and grueling driving conditions, an effective battery management system (BMS) must be used [1]. This system is used to inspect the conditions and states of the power battery, such as state of charge (SOC) and state of health (SOH). However, since these states are usually immeasurable by any sensors, many model-based estimation methods are used to solve the problem [2-6]. It is pretty necessary to guarantee the accuracy of battery models in order to acquire a satisfied estimated result. Besides, it is also important to balance the model complexity and accuracy in purpose to ensure acceptable estimation result while reducing calculation in real-time [7]. In brief, these models should be accurate enough and has reasonable computational burden.

The models can be categorized into two main types including electrochemical models and equivalent circuit battery models [8]. As with the former type, it is usually suitable for understanding the distributed electrochemistry reactions in the electrodes and electrolyte [9]. However, they typically deploy partial differential equations with a large number of unknown parameters, which remarkably increases the computation load. Efforts are made to simplify the structure of electrochemical models in some researches, but meanwhile several effects are ignored. Therefore, these electrochemical models are not desirable for real time battery management in electric vehicles.

For the equivalent circuit battery models, they are often lumped models with relatively fewer numbers of parameters, thus being widely applied in impedance analysis [10], SOC estimation [11], charging control [12] and SOH prediction [13]. For example, Plett [14] proposed some lumped models, such as the simple, zero-hysteresis, one-state hysteresis, combined, and enhanced self-correcting (ESC) models and the battery SOC is estimated by extended Kalman filter (EKF) through the proposed models. Among the different kinds of equivalent circuit models, the resistance-capacitance (RC) network based models are widely researched, including the first-order RC [15-17], the second-order RC [18] and the third-order RC [19] etc. Seaman et al. [20] have conducted a comprehensive review for different types of equivalent circuit battery models. Considering the balance of model complexity and accuracy, the second-order model is regarded to be suitable for most of the applications.

However, the accuracy of circuit-based models is substantially sensitive to parameters, especially when dealing with some control-oriented problems which are essential in electric vehicle design, and thus making the identification of parameters very challenging. To be specific, the difficulty of parameter identification may due to: (1) The inconvenience of conducting EIS for frequency-domain identification; and (2) the high demand of accuracy and risk of trapping into local optimum of time-domain identification. For the latter, the unknown parameters in the governing dynamic equations can be identified from time-domain experimental data [21]. The estimation result can be improved by applying a proper global optimum algorithm. There have been some researches in the literature for parameter identification of batteries. For instance, Ouyang et al. [8] applied the genetic algorithm (GA) and Ramadesigan et al. [22] used Gauss-Newton method, a Jacobian-based scheme, for the process of nonlinear optimization in their parameter identification efforts, by minimizing an objective function that presents the discrepancy between the model outputs and the experimental one. However, those classical optimization methods are hard to achieve a satisfied result with relative lower time consume. Thus a more efficient algorithm should be applied.

In this paper, we focus on the parameter identification problem for the equivalent circuit models by using the tree seeds algorithm (TSA). Compared with other methods like Gauss-Newton method [22] and genetic algorithm [8], the tree seed algorithm (TSA) [23] has more advantages: (1) it does not require good initial values and gradient information and (2) it is proven to be more powerful to tackle with multiple mode functions. In order to achieve higher accuracy and lower time consume in parameter identification problem, the TSA is introduced. And then, the objective function based on the difference between the measured voltage and the estimated one is developed. Afterward, the improved algorithm is used to optimize the objective function and finally acquires the system parameters. The remainder of this paper is organized as follows. In section 2, we introduce the equivalent circuit battery models. In section 3, we elaborately illustrate the TSA. In section 4, the TSA is applied to parameter identification, and the comparison with genetic algorithm is presented. In section 5, some conclusions can be drawn.

2. Battery modeling and parameter identification

2.1. Description of equivalent circuit model

The equivalent circuit model used in this paper is shown in figure 1 [19], with a voltage source, an ohmic resistance and two pairs of RC networks in series. The battery dynamic and static performance can be simulated by the combination of these components. In this system, the voltage source is parameterized as a kind of nonlinear function of battery SOC and to express the open circuit voltage characteristic at different SOC. The two RC networks denote the time-dependent polarization and diffusion effects of the cell. The ohmic resistance is used to describe the instant voltage drop after an excitation current in the battery.

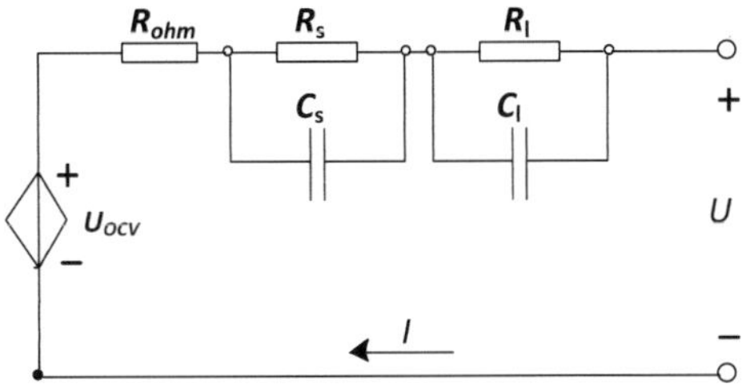

Figure 1. Battery equivalent circuit model.

2.2. *Parameter identification*

In the proposed method, we use the TSA to identify the ohmic resistance R_{ohm} and polarization resistances R_s, R_l and polarization capacitances C_s, C_l.
By observing figure 1, the following equations can be acquired

$$u_{ohm} = IR_{ohm} \tag{1}$$

$$C_s\frac{du_s}{dt} + \frac{u_s}{R_s} = I \tag{2}$$

$$C_l\frac{du_l}{dt} + \frac{u_l}{R_l} = I \tag{3}$$

$$u = u_{ocv} - u_{ohm} - u_s - u_l \tag{4}$$

where u is the battery terminal voltage, I is the battery current, and u_{ocv} is the open circuit voltage. R_{ohm}, R_s, R_l, C_s, C_l are the battery parameters which reflect the battery dynamic response and are needed to be identified. u_{ohm} is the voltage drop on R_{ohm} and u_s, u_l are the voltages across the two RC networks.
For further analysis, we apply discretization to the system and acquire:

$$u_k^{ohm} = I_k R_{ohm} \tag{5}$$

$$u_k^s = I_{k-1}\frac{R_s}{1+R_sC_s} + \frac{R_sC_s}{1+R_sC_s}u_{k-1}^s \tag{6}$$

$$u_k^l = I_{k-1}\frac{R_l}{1+R_lC_l} + \frac{R_lC_l}{1+R_lC_l}u_{k-1}^l \tag{7}$$

$$u_k = u_k^{ocv} - u_k^{ohm} - u_k^s - u_k^l \tag{8}$$

where k represents the time step and u^{ocv} is a nonlinear function of SOC. The function can be linearized within a short time interval:

$$u_k^{ocv} = a_k SOC_k + b_k \tag{9}$$

According to the definition of SOC:

$$SOC_k = SOC_{k-1} - \mathrm{I}\frac{1}{C_N} \tag{10}$$

where C_N is the nominal capacity of battery and the time interval is chosen as 1s here.

Define $x_k = [u_k^{ohm} \quad u_k^s \quad u_k^l \quad SOC_k]^T$, $y_k = u_k$, $\mathrm{A} = \begin{bmatrix} 0 & 0 & 0 & 0 \\ 0 & \frac{R_sC_s}{1+R_sC_s} & 0 & 0 \\ 0 & 0 & \frac{R_lC_l}{1+R_lC_l} & 0 \\ 0 & 0 & 0 & 1 \end{bmatrix}$,

$\mathrm{B} = \left[R_{ohm} \quad \frac{R_s}{1+R_sC_s} \quad \frac{R_l}{1+R_lC_l} \quad -\frac{1}{C_N}\right]^T$, $\mathrm{C} = [-1 \quad -1 \quad -1 \quad a_k]$, $\mathrm{D} = [0 \quad 0 \quad 0 \quad b_k]$,

then these equations can be rewritten in a standard form:

$$x_k = Ax_{k-1} + BI_k \tag{11}$$

$$y_k = Cx_k + D \tag{12}$$

Based on equation (11)-(12), the output of the system can be acquired.
Here the current I is the input of system, and terminal voltage y is the output. The parameters in the system may intensively influence the input-output relation. Thus an efficient algorithm for high fidelity estimation is needed. For that, an objective function should be defined as below

$$Z = \sum_{k=1}^{N} e_k^2 \tag{13}$$

IOP Conf. Series: Earth and Environmental Science **73** (2017) 012024 doi:10.1088/1755-1315/73/1/012024

where e_k is the difference between the predicted output and the measured output of the model. And in this paper, the discrepancy of the voltage is used to build up the objective function, so it can be calculated as $e_k = y_k - v_k$, in which y_k is the predicted voltage at time step k and v_k is the measured voltage. Then the heuristic algorithm, such as the tree seeds algorithm can be used to minimize the error and obtain the systematic parameters.

3. Tree seeds algorithm

This heuristic algorithm is actually proposed based on the natural phenomena of trees and their seeds [23]. In reality, trees usually spread to other places through their seeds. If assuming the place for these trees and seeds as a search space for the optimization problem, the location of trees and seeds can be regarded as feasible solutions for the problem. To acquire a location of one seed that would be produced from a tree is important for the optimization problem since this procedure contains the core of search. The implement of tree seeds algorithm is shown in figure 2.

In this algorithm, the two update rules can be given as below,

$$S_{i,j} = T_{i,j} + \alpha \times (B_j - T_{r,j}) \tag{14}$$

$$S_{i,j} = T_{i,j} + \alpha_{i,j} \times (T_{i,j} - T_{r,j}) \tag{15}$$

where $S_{i,j}$ is jth dimension of ith seed that would be generated ith tree. Meanwhile $T_{i,j}$ is the jth dimension of ith tree, B_j is the jth dimension of best-so-far tree location, and $T_{r,j}$ is the jth dimension of rth tree randomly picked from the colony size, α is the scaling factor arbitrarily produced in range of [-1,1], besides i and r are different indices.

To balance the search capacity of the proposed search modes, it can be controlled by a parameter called as search tendency (ST) in range of [0,1]. The higher value of ST offers a powerful local search and speed convergence while the lower value of ST renders slow convergence but strong global search. In brief, the exploitation and exploration abilities of the TSA are controlled by this parameter.

In the beginning of operation with TSA, the tree locations (possible solutions for the optimization problem) are produced by utilizing equation (16)

$$T_{i,j} = L_{j,min} + r_{i,j} \times (H_{j,max} - L_{j,min}) \tag{16}$$

where $L_{j,min}$ is the lower bound of the search space and $H_{j,max}$ is the higher bound, $r_{i,j}$ is a random number produced for every dimension and location, in range of [0,1]. As with the seeds number it can be determined by the colony size (approximately between 10% and 25% of the colony size). The exact number of seed generation is utterly random in TSA.

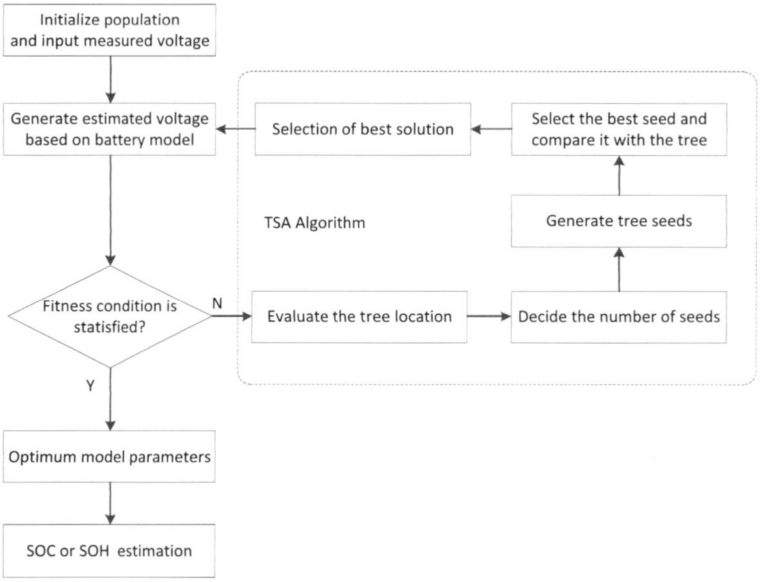

Figure 2. Flow chart of tree seeds algorithm for parameter identification.

IOP Conf. Series: Earth and Environmental Science **73** (2017) 012024 doi:10.1088/1755-1315/73/1/012024

4. Verification and comparison

In order to verify the identification result of the proposed method, a lithium nickel-manganese-cobalt oxide (LiNMC) cell with nominal capacity of 1600mAh is tested under different profiles. Here, the commonly used dynamic profiles DST and FUDS are chosen. The performance of the proposed method is evaluated by comparing the estimated voltage with the measured voltage. For further analysis, the statistic of voltage discrepancy such as MAE (the maximum absolute error) and RMSE (the root mean square error) are listed. Moreover, the identification result of genetic algorithm is attached for comparison.

The current curve of DST test is shown in figure 3. The identification results of TSA and GA under DST are shown in figure 4. For clearer comparison, the error between estimated voltage and measured voltage of the two algorithms are presented in figure 5.

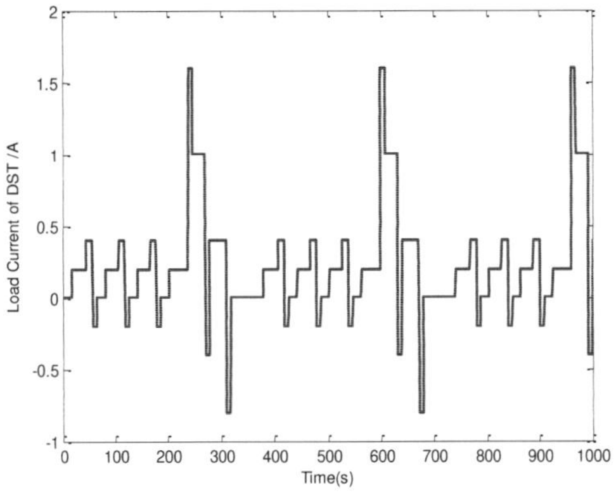

Figure 3. Current profile of DST test.

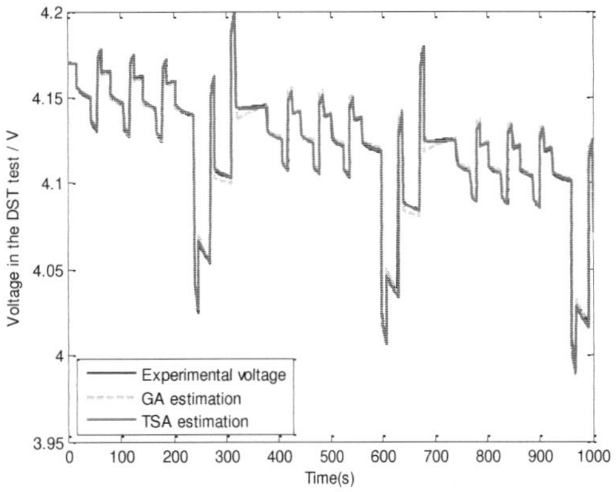

Figure 4. The measured voltage of DST test and estimation results of GA and TSA.

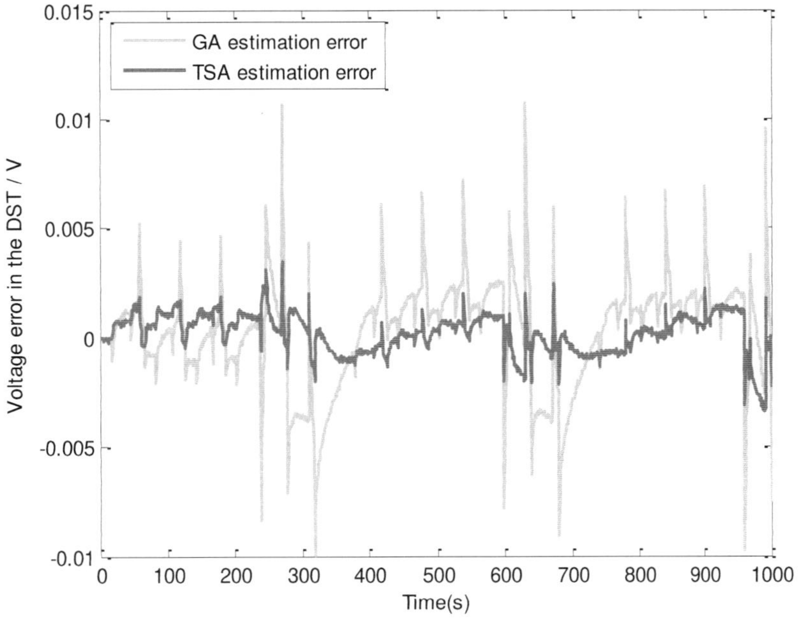

Figure 5. Voltage errors of GA and TSA in the DST test.

Similarly, the current curve of FUDS test is shown in figure 6. The identification results under FUDS are shown in figure 7 and figure 8. Table 1 lists the statistic results of identification under the two profiles.

The verification results illustrate that the proposed method has better performance in tracking the real voltage, compared with the classical genetic algorithm, thus the model accuracy can be improved.

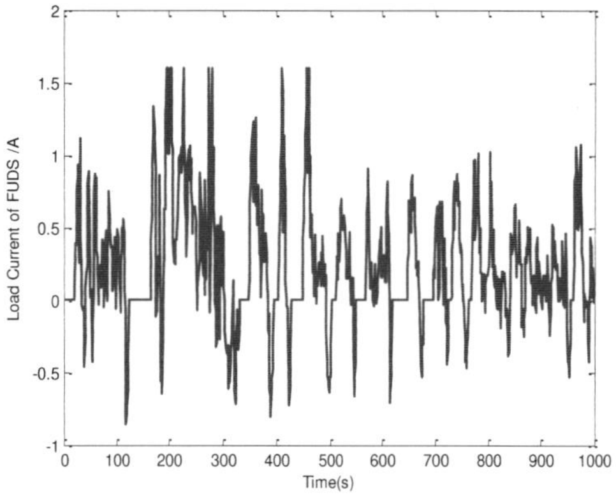

Figure 6. Current profile of FUDS test.

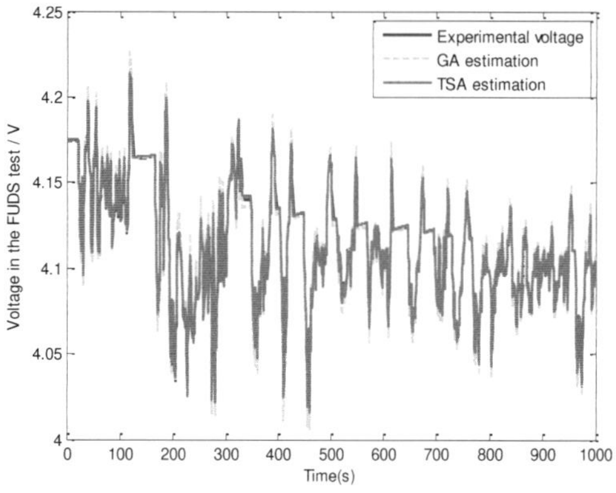

Figure 7. The measured voltage of FUDS test and estimation results of GA and TSA.

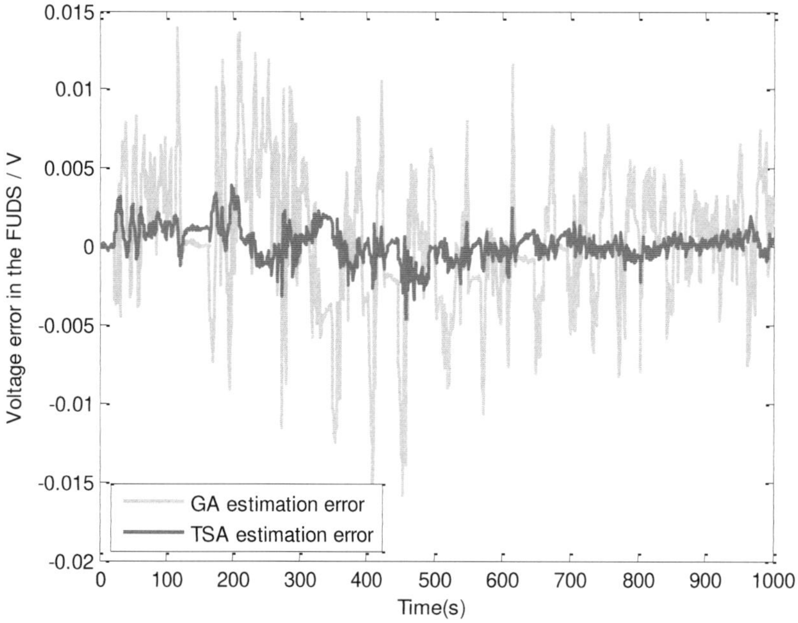

Figure 8. Voltage errors of GA and TSA in the FUDS test.

Table 1. Comparison of MAE and RMSE in the different tests.

Algorithm	DST test		FUDS test	
	MAE(mV)	RMSE(mV)	MAE(mV)	RMSE(mV)
GA	10.73	2.47	16.58	4.35
TSA	3.46	0.96	4.56	1.05

5. Conclusion

A parameter identification method based on tree seeds algorithm is presented. The sum of square of discrepancy between the measured and the estimated voltage is defined as the object function and the TSA is employed to find the optimum parameters by minimizing the function value. Experiment results show the superior performance of the proposed method over the classical genetic algorithm. The RMSE of the model is reduced by 61.1% under DST test and 75.9% under FUDS test. The proposed method can not only improve the accuracy of the model, but also can be used for online state estimation. State variables like SOC and SOH can be estimated as one of the unknown parameters. To verify the practicability of the proposed method, the algorithm will be implemented to the real BMS in the future.

References

[1] Lu L, Han X, Li J, Hua J and Ouyang M A review on the key issues for lithium-ion battery management in electric vehicles 2013 *J. power sources* **226** 272-288.
[2] Moura S J, Chaturvedi N A, and Krstić M Adaptive partial differential equation observer for battery state-of-charge / state-of-health estimation via an electrochemical model 2014 *J. Dyn. Syst. Meas. Control* **136** 11-15.
[3] He H, Xiong R, Guo H, and Li S Comparison study on the battery models used for the energy management of batteries in electric vehicles 2012 *Energy Convers. Manage* **64** 113-121.
[4] Plett G L Extended Kalman filtering for battery management systems of LiPB-based HEV battery packs: Part 3. State and parameter estimation 2004 *J. power sources* **134** 277-292.
[5] Junping W, Jingang G, and Lei D An adaptive Kalman filtering based State of Charge combined estimator for electric vehicle battery pack 2009 *Energy Convers. Manage* **50** 3182-86.
[6] He H W, Xiong R and Chang Y H Dynamic modeling and simulation on a hybrid power system for electric vehicle applications 2010 *Energies* **3** 1821-30.
[7] Zhang J, Ci S, Sharif H, and Alahmad M An enhanced circuit-based model for single-cell battery 2010 *Applied Power Electronics Conference and Exposition* 672-675.
[8] Ouyang M, Liu G, Lu L, Li J and Han X Enhancing the estimation accuracy in low state-of-charge area: A novel onboard battery model through surface state of charge determination 2014 *J. power sources* **270** 221-237.
[9] Subramanian V R, Boovaragavan V, Ramadesigan V and Arabandi M Mathematical model reformulation for lithium-ion battery simulations: Galvanostatic boundary conditions 2009 *J. Electrochem. Soc* **156** A260-A271.
[10] Schmidt A P, Bitzer M, Imre Á W and Guzzella L Experiment-driven electrochemical modeling and systematic parameterization for a lithium-ion battery cell 2010 *J. power sources* **195** 5071-80.
[11] He H, Xiong R and Fan J Evaluation of lithium-ion battery equivalent circuit models for state of charge estimation by an experimental approach 2011 *Energies* **4** 582-598.
[12] Hu X, Li S, Peng H and Sun F Charging time and loss optimization for LiNMC and LiFePO 4 batteries based on equivalent circuit models 2013 *J. power sources* **239** 449-457.
[13] Li S E, Wang B, Peng H and Hu X An electrochemistry-based impedance model for lithium-ion batteries 2014 *J. power sources* **258** 9-18.
[14] Plett G L Extended Kalman filtering for battery management systems of LiPB-based HEV battery packs: Part 3. State and parameter estimation 2004 *J. power sources* **134** 277-292.
[15] Liaw B Y, Nagasubramanian G, Jungst R G and Doughty D H Modeling of lithium ion cells—A simple equivalent-circuit model approach 2004 *Solid state ionics* **175** 835-839.
[16] Dubarry M, Vuillaume N and Liaw B Y From single cell model to battery pack simulation for Li-ion batteries 2009 *J. power sources* **186** 500-507.
[17] Chiang Y H, Sean W Y and Ke J C Online estimation of internal resistance and open-circuit voltage of lithium-ion batteries in electric vehicles 2011 *J. power sources* **196** 3921-32.
[18] Hu X S, Sun F C and Cheng X M Recursive calibration for a lithium iron phosphate battery for electric vehicles using extended Kalman filtering 2011 *J. Zhejiang Univ. – Sci. A* **12** 818-825.
[19] Hu X, Li S and Peng H A comparative study of equivalent circuit models for Li-ion batteries 2012 *J. power sources* **198** 359-367.
[20] Seaman A, Dao T S and McPhee J A survey of mathematics-based equivalent-circuit and electrochemical battery models for hybrid and electric vehicle simulation 2014 *J. power sources* **256** 410-423.
[21] Subramanian V R, Diwakar V D and Tapriyal D Efficient macro-micro scale coupled modeling of batteries 2005 *J. Electrochem. Soc* **152** A2002-08.

[22] Ramadesigan V, Boovaragavan V, Arabandi M, Chen K, Tsukamoto H, Braatz R and Subramanian V Parameter estimation and capacity fade analysis of lithium-ion batteries using first-principles-based efficient reformulated models 2009 *ECS Transactions* **19** 11-19.

[23] Kiran M S TSA: Tree-seed algorithm for continuous optimization 2015 *Expert Syst Appl* **42** 6686-98.

IOP Conf. Series: Earth and Environmental Science **73** (2017) 012025 doi:10.1088/1755-1315/73/1/012025

Impacts on the Voltage Profile of DC Distribution Network with DG Access

J J Tu[1, 2] and Z D Yin[2]

[1]China Power Engineering Consulting Investment Group co., LTD, Beijing 100011, China

[2]North China Electric Power University, Beijing 102206, China

Email: shell2005@126.com;

Abstract. With the development of electronic, more and more distributed generations (DGs) access into grid and cause the research fever of direct current (DC) distribution network. Considering distributed generation (DG) location and capacity have great impacts on voltage profile, so use IEEE9 and IEEE33 typical circuit as examples, with DGs access in centralized and decentralized mode, to compare voltage profile in alternating and direct current (AC/DC) distribution network. Introducing the voltage change ratio as an evaluation index, so gets the general results on voltage profile of DC distributed network with DG access. Simulation shows that, in the premise of reasonable location and capacity, DC distribution network is more suitable for DG access.

1. Introduction

With the development of electronic technology, more and more distributed generations (DGs) are introduced into grid. However, when the position and capacity are inappropriate, it will lead to the increase of the power loss and the irregular changes of voltage.

In order to solve the problem above, DC distribution network has been widely concerned. It is proposed in recent years, the output of DG are mostly DC, so compared with the AC distribution network, DC distribution network maybe more suitable for the access of DG [1].

At present, domestic and foreign research institutes and colleges, has carried out the relevant research for DC power distribution technology, some of which have begun to put into practice. Domestic DC power distribution technology is mainly used in ship and subway power supply system, as well as boost substation, communication room and such industries systems, but its large-scale commercial applications are still in further study [2, 3].

In this paper, mainly do the research of the voltage profile of the DC distribution network with DG access, and introducing the voltage change ratio as an evaluation index, and do simulation with the typical IEEE circuit [4], to compare the results of the AC and DC networks.

2. DC distribution network and voltage change ratio

2.1. DC distribution network

DC distribution network topology is mainly radial, ring and grid. At present, the research about DC is mainly focused on devices, but the research about voltage profile after DG access, the interaction between DC distribution network and large power grid is very lacking. In paper [3] and [5], the current research of DC grid efficiency, topology, system stability, power flow calculation and control, relay protection and so on are introduced in detail.

2.2. Voltage change ratio

This paper adopts radiation structure, shown in Figure 1. Set the system bus as the reference node, line voltage is $U_0 = U_0 e^{j0}$. After DG access, to keep power flow in the same direction, need to constraint the DG capacity, so the capacity of the installed DG does not exceed the load size of the node, i.e. $P_{DG} < P_K$.

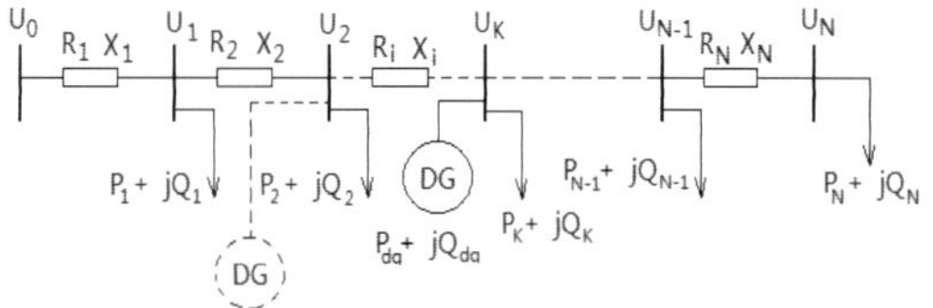

Figure 1. Radial distribution network

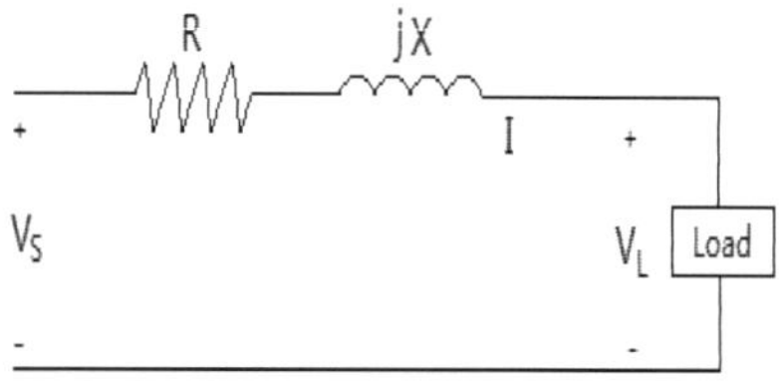

Figure 2. Single-phase equivalent circuit

At the same time, assuming all the loads are three-phase balance. The single-phase equivalent circuit of the feeder is shown in Fig. 2. According to Kirchhoff's law of voltage, there is:

$$V_S = V_L + (R + jX) \cdot I = V_L + R \cdot I + jX \cdot I \qquad (1)$$

Considering the phase between the power supply side and the load side is usually small. Therefore, the voltage drop across the line is approximately equal to:

$$V_{drop} \cong \mathrm{Re}(Z \cdot I) \qquad (2)$$

After DG access, the node voltage:

$$U_n' = U_0 - \sum_{l=1}^{n} I_l \sum_{i=1}^{l} Z_i - \sum_{l=n+1}^{N} I_l \sum_{i=1}^{n} Z_i + I_{dg} \sum_{i=1}^{n} Z_i \qquad (1 \le n < K)(3)$$

$$U_n' = U_0 - \sum_{l=1}^{n} I_l \sum_{i=1}^{l} Z_i - \sum_{l=n+1}^{N} I_l \sum_{i=1}^{n} Z_i + I_{dg} \sum_{i=1}^{k} Z_i \qquad (K \le n < N)(4)$$

$$U_n' = U_0 - \sum_{l=1}^{N} I_l \sum_{i=1}^{l} Z_i + I_{dg} \sum_{i=1}^{k} Z_i \qquad (n = N)(5)$$

According to the above derivation process, it can get a series approximate value of the node voltage of DC grid. The voltage change rate ε is:

$$\varepsilon = \frac{U_n' - U_n}{U_n} \cdot 100\% \tag{6}$$

The more the ε value changes, the more obvious the influence of DG on the node voltage.

3. Impacts Analysis on Voltage Profile after DG Access

3.1. Basic experiment

The basic experiment is to simulate the system before and after DG access, and compare the voltage profile of AC / DC distribution network. In the calculation of DC power flow, the line admittance is neglected, and the reactive power is not taken into account, and the Newtonian Loughson method is used for calculation [6-8].

Table 1. Active power generation and loss after DG accessing (9 notes) unit: pu.

DG access ratio	10%~15%	30~40%	60~70%
Active power loss (DC)	0.048	0.084	0.096
Active power loss (AC)	0.049	0.06	0.097
Active power supply (DC)	2.883	1.955	1.046
Active power supply (AC)	2.884	2.17	1.047

From Table 1, after DG access, the output of active power reduced, but active losses increased with the increase of DG access capacity.

Table 2. Active power loss after DG accessing (33 notes) unit: kW

DG access mode	centralized	decentralized	Different location
Active power loss (AC)	196.59	160.44	163.68
Active power loss (DC)	191.48	156.6	159.58

From Table 2, if DG access in different ways, when adopts the decentralized mode, the active power loss is minimum, and DC is less than the AC system.

3.2. Centralized and decentralized access

First take 9-node system as an example, DG access information is shown in Table 3.

Table 3. Location information of DGs (9 notes)

Node No. with DG	5	6	8	No DG access	Decentralized access DG	Centralized i
Location No.	W5	W6	W8	W0	FS	Wi

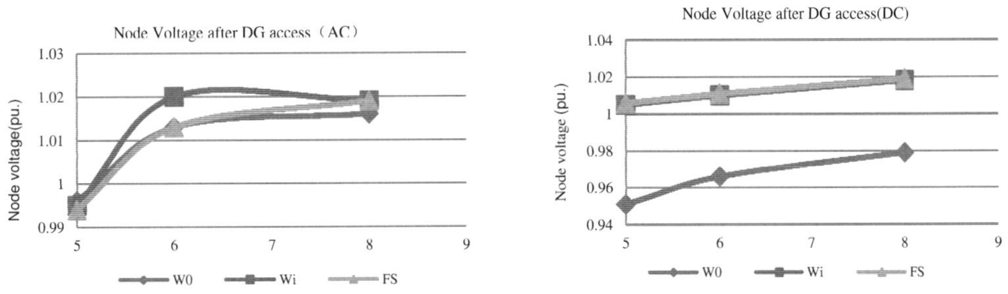

(a) Voltage curves of AC (b) Voltage curves of DC

Figure 3. Voltage curves of DG accessing in centralized and decentralized mode(9 notes)

From Fig.3, we can see in the AC system, when the DGs access centralized, the voltage of access point improves obviously, but the other nodes change small. In the DC system, no matter what access mode, after DG access, the voltage of each node is increased obviously.

Second, take 33-node system as an example, DG access in a centralized and decentralized mode, and access to the front, middle and end of the system, the results are shown in Fig.4.

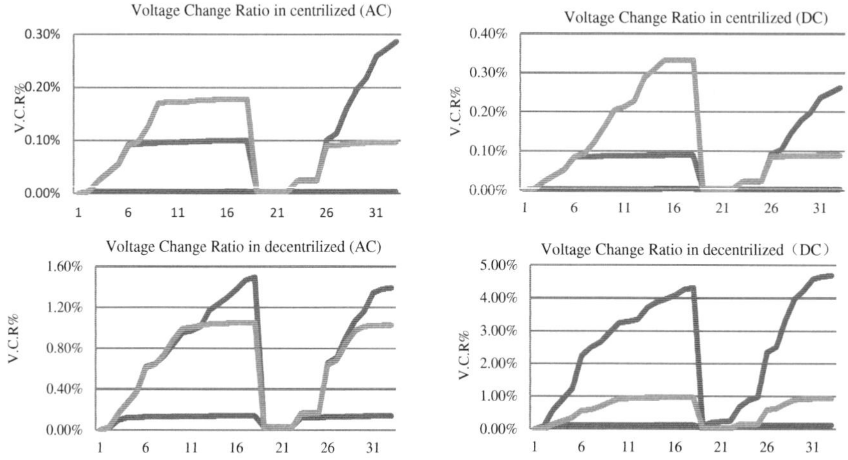

Legend of DG access location： red line denotes front; green line denotes middle; blue line denotes end.

(a) Voltage change ratio of AC (b) Voltage change ratio of DC

Figure 4. Voltage ratios of DG accessing in centralized and decentralized mode (33 notes)

From Fig.4, we can see in the AC system, when DGs connect to the end in a decentralized method, the voltage change ratio of each node is the most obvious, when DGs connect in a centralized method, the voltage change ratio of each node is small. So with DG access, the access position has a significant impact on the node voltage.

In the DC systems, the voltage variation trend of each node is similar to that of AC system. The difference is that when DGs connect in decentralized mode, the voltage variation of each node is larger than that in AC system.

In general, when DGs access to the central and end of the system, the impact on the node voltage is more obvious, and the DC system is more sensitive to the DG access mode.

3.3. Constant capacity access in different positions

In the premise of keeping the capacity unchanged, DGs connect in different position, and do the simulation to analyze the voltage profile. First, use 9-node system as an example, DG access information is shown in Table 4.

Table 4. Location information of DGs (9 notes)

Node No. with DG	5&6	5&8	6&8	No DG access
Location No.	W1	W2	W3	W0

Set the access capacity of DG up to 60% of the total load. From Fig.5, shown the results of voltage profile with DG access in the same capacity but different location in 9-node system. In the AC system, there is a decrease and increase in the voltage of each node with different DG access location. In the DC system, after DG access, the node voltage improve obviously; and with different location, node voltage change small, so it can be think that it is not affected much by the DG access location comparing capacity.

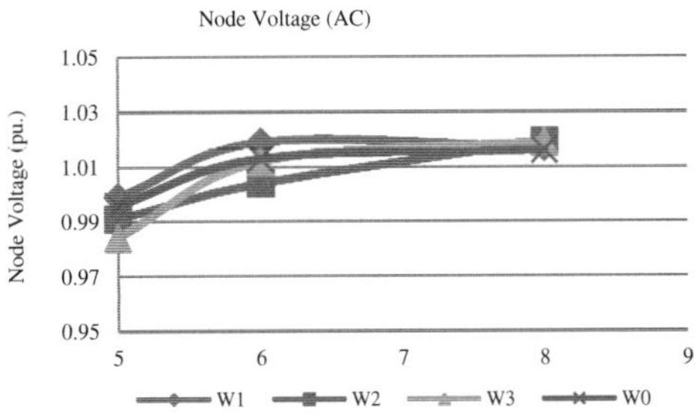

(a) Voltage curves of AC

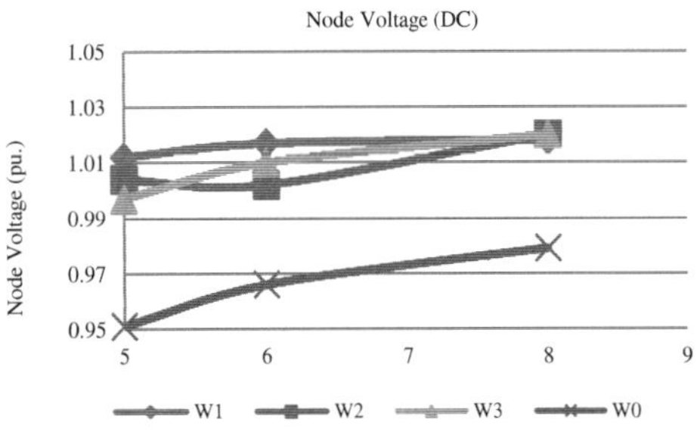

(b) Voltage curves of DC

Figure 5. Voltage curves of different locations with the same capacity (9 notes)

Second, take the 33-node system as an example, DG access also has a certain impact on the active loss and output. The DGs with certain capacity connect to the front, middle and end of the system respectively. The simulation results are shown in Table 5.

Table 5. Active power loss after DG accessing (33 notes) unit: kW

Access location	end	front	middle
Active power loss(AC)	160.44	195.7	165.31
Active power loss (DC)	156.6	190.43	161.17

From Table 5, when DG access to the end of the system, the system active loss is minimal and the DC system active loss is significantly smaller than that of the AC system.

After DG access the system, the node voltage significantly improved, especially when DGs access in the middle and end of the system, the system voltage increases the most obvious; and comparing AC system, the voltage of DC system increase more.

4. Results

Through the theoretical analysis and simulation with IEEE9 and IEEE33 system, the experimental data of the AC / DC distribution network are compared and analyzed in detail. First do basic examination; second, DG access in centralized and decentralized method; third, DG access in the front/middle/end part of the grid with the constant capacity. After DGs access in different location and capacity, the voltage profile of AC and DC grid are compared, and the active and reactive power loss are listing clearly.

The general conclusion is drawn. In the premise of proper access capacity and location, the main influences on voltage profile with DG access to DC distribution network are as follows: (1) the DG access position is less important to the DC system than the AC system; (2) After DG access, the voltage level of the DC system can be improved, and the grid voltage profile is more balance than the AC system; (3) After DG access, the active output is significantly reduced in DC system, and the power loss is smaller than that of AC system. In summary, the DC system is more suitable for DG access than the AC system.

While, there is still some challenges. In order to simplify the study, this paper does some equivalent, but the actual distribution network is more complicated. At the same time, this paper mainly considers DG access position, but the DG access location and capacity is a pair of inseparable factors to voltage profile of distribution network. So, under the consideration of the system output and power loss and etc., how to get the best DG access location and capacity in distribution network still needs further studied.

References

[1] JIANG Daozhuo, ZHENG Huan. Research Status and Developing Prospect of DC Distribution Network [J], Automation of Electric Power Systems, 2012, 36(8):98-104.

[2] XU Tong, WANG Yu-fei, ZHANG Yu, etc., Development Status and Application Future of DC Distribution Network [J], East China Electric Power, 2014, 42(6):1069-1074.

[3] ZHANG Zuping. Development prospect of DC power distribution technology [J], Distribution & Utilization, 2014, 7:16-19.

[4] GAO Kai, LIU Bowen, etc., Review on power flow calculation of AC / DC power systems [J], Electrical applications,2015,34(17):122-127.

[5] CUI Fubo, GUO Jianbo, JING Ping, etc., A Review of DC Power Distribution Technology, Power System Technology [J], 2014, 38(3):556-564.

[6] WANG Cheng-shan, ZHENG Hai-feng, XIE Ying-hua, CHEN Kai. Probabilistic Power Flow Containing Distributed Generation in Distribution System [J], Automation of Electric Power Systems, 2005, 29(24):39-44.

[7] WEN Jianchun, XIANG Zhanfeng. Loop power flow calculation method based on DC power distribution network and Thevenin equivalent in Distribution System [J], 2012, 9(27):27-28.

[8] WANG Shouxiang, WANG Chenshan. Modern distribution system analysis [M], Beijing: Higher Education Press, 2007.

Effects of Distributed Generation on Overcurrent Relay Coordination and an Adaptive Protection Scheme

Semih C ILIK, Aysen B ARSOY
Faculty of Engineering (B), Department of Electrical Engineering, Kocaeli University, Umuttepe Campus, Kocaeli 41380 Turkey

E-mail: semihcemilik@gmail.com; aba@kocaeli.edu.tr;

Abstract. Integration of distributed generation (DG) such as renewable energy sources to electrical network becomes more prevalent in recent years. Grid connection of DG has effects on load flow directions, voltage profile, short circuit power and especially protection selectivity. Applying traditional overcurrent protection scheme is inconvenient when system reliability and sustainability are considered. If a fault happens in DG connected network, short circuit contribution of DG, creates additional branch element feeding the fault current; compels to consider directional overcurrent (OC) protection scheme. Protection coordination might get lost for changing working conditions when DG sources are connected. Directional overcurrent relay parameters are determined for downstream and upstream relays when different combinations of DG connected singular or plural, on radial test system. With the help of proposed flow chart, relay parameters are updated and coordination between relays kept sustained for different working conditions in DigSILENT PowerFactory program.

1. Introduction

Distributed generation includes the power plants connected to low or medium voltage network and their output power varies between a few watts to 100 MW. Per module production ranges of commonly used DG sources rates are given as for micro turbines 35 kW-1 MW, wind turbines 200 W-3 MW, photovoltaic panels 20 W-100 kW, biomass 100 kW-20 MW and geothermal 5MW-100 MW [1]. According to power capacity, generator type and connection shape, DG has influences on electrical network. Some advantages appear due to the influences followed as; decrease in transmission and distribution line losses, lowering the pick power demands, and better voltage profile. However some disadvantages arise as; changing power flows directions, voltage rises in condition of low power demand, harmonic-flicker issues and protection problems such as selectivity loss [2]. Grid connected DG sources creates an impact which increases the short circuit level of network depend on its output power, power converter interface and generator electrical parameters. If a fault happens in network, short circuit contribution of DG creates a new branch element feeding the fault current. Radially constructed medium voltage network protection scheme is able to isolate the effect of one source feeding the fault which is actually the network itself. In another words, radial protection scheme cannot isolate the impact of multiple sources fault contribution when fault currents are bidirectional. Therefore existing protection scheme has to be made convenient for directional over current protection. Main and backup protection device settings may require revise for each fault current directions with considering the short circuit contribution of DG sources, in order to reachieve protection selectivity with coordination time interval between the protection devices.

IEEE Standard 1547 is pointed out; if a fault happen in DG connected network, DG sources should be disconnected. In addition to prevent unintentional islanding, DG sources have to disconnect within 2 seconds. For abnormal voltage drop situations, it is defined for the voltage amplitude of 0.5 p.u. (per unit), DG sources have to disconnect within 160 milliseconds [3].

Literature review shows that the main topic is utilization of adaptive relay. Typical relay protection will response due to predetermined protection curve characteristic but adaptive relaying is capable of determining relay response character for changing working conditions prevailing on network. This approach consist of a software determines the relay functions, communication featured relays and a control centre which commands remotely [4]. Adaptive relay concept is used in [5], [6] and [7]. In some studies; adaptive relay protection focus on grid connected DG or islanded working conditions scenarios which are generally uses communication between relays or breakers. Utilization of communication systems causes the high costs so in some studies authors interested in communication less or partially communication used solutions that are generally uses local measurement devices or switching positions of breakers [8]. Considering the working conditions prevailing on system, optimal adaptive relay settings are determined by using the algorithms as fuzzy inference module, neural network learning module [9] or symbiotic organism search optimization technique [10].

In addition to adaptive relaying, short circuit limiter is applied to limit growing fault currents due to presence of DG sources [11-12]. It consists of superconductive element and resistive shunt element which are installed at beginning or end of the line. When the measured current amplitude is over the determined critical value, it changes the operation mode to high resistive shunt element with the help of growing magnetic field or an external source and let the fault current flow over the resistance. In this study, negative effects of DG connections on OC protection is mentioned in Section 2, multiple source radial network protection scheme is given in Section 3, analysis and simulation studies are carried out in Section 4, conclusions and further work are given in Section 5 and Section 6. In a DG connected radial test system directional overcurrent relay parameters are determined for downstream and upstream relays when different combinations of DG connected as singular or plural, proper relay parameters may require update by means of adaptive relay protection.

2. Effects of distributed generation on overcurrent protection

2.1. Changing short circuit power and fault level

Generally large amount of the DG sources comprise rotating machines. With the impact of DG connections, thevenin impedance of the network decreases [13]. Thus, short circuit level and fault current magnitude will increase with presence of the DG machines in parallel. Fault currents are much higher than load currents so it is the basic tripping circumstance for the OC relays to discriminate the measured current magnitude easily. In condition of asymmetrical faults, fault contribution of DG sources is slightly different because of the transformer vector group and the generator grounding connection [14-15]. Growing short circuit currents due to DG connections have negative impacts on OC relays as listed below:

- Bidirectional fault currents require more than one switching operation to isolate the faulted part.
- Directional OC relay utilization requires new protection and switching devices.

- Some DG sources may have restricted short circuit contribution to protect their power electronic parts [16]. The restricted fault contribution flowing over the high impedance line makes relay level detection harder when fault contributions are measured near load currents.
- Changing fault currents may cause blinding of protection or false tripping [17].

2.2. Short circuit behavior and fault contribution of various DG generators

Generally, synchronous and asynchronous machines are modelled with voltage source and reactance, in short circuit calculations and fault behaviour depends on subtransient and transient reactance with its time constants. However, DG sources are generally connected via power converters so the fault behaviour depends on converter controller and its temperature endurance or design parameters [16]. In some short circuit cases, fault currents might be evaluated as normal operating conditions when restricted fault contribution and high impedance lines considered. Table 1 shows the short circuit contribution of DG sources as percentage of its nominal current [1]. The values given in the table is calculated for the faults at the output terminals of the sources. The more the faulted point moves away from generator terminals, fault contribution of DG decreases immediately.

Table 1. Fault current contribution for various generator types [1].

Generator Type	Fault current at generator terminals, percentage of rated output current
Inverter interface	100-400
Seperately excited Synchronous generator	First few cycles: 100-500 Permanent: 200-400
Induction generator or self excited Synchronous generator	First few cycles: 500-1000 Permanent: approximately 0

3. Overcurrent relay parameters in radial network with multiple sources

OC relay trip settings are determined with "pickup current (Ip)" and "time delay" parameters. Pickup current is the threshold current value and it is sensed with current level detection principle which has to be exceeded for relay trip. There are two approaches applied in common to determine the pickup current. First approach specifies pick up value as two times of maximum load current or it should be equal 1/3 times of minimum fault current at the nearest busbar [4]. Second approach suggests that it should be chosen between 1.25 times of maximum load current and 2/3 times of minimum fault current [18]. Time delay settings are defined depend on definite time and inverse time characteristics. Definite time characteristic has a smooth time delay for changing fault currents which are measured over the pickup value. On the other hand, inverse time characteristic has different time delays value for changing fault currents over the pickup value. Time delay setting is specified by "time multiplier settings (TMS) for inverse time relays.

Multiple sources radial network OC protection scheme is generally provided with directional OC relay (67) function [4]. Relays with the directional element can produce trip signal only for one direction of fault current. Figure 1, illustrates working scheme of directional OC relay on the simple test system. Figure 1(a) represents the downstream relays as R12, R23 and R34 and upstream relays as R21, R32 and R43. Downstream relays have coordination between each other and they can only trip for the fault contribution comes from utility side. Likewise, upstream relays have coordination between each other and they can only trip for the fault contribution comes from DG side. As seen in Figure 1, if a 3-phase f1 fault happens at the middle of L23 line, R12, R23, R32 and R43 relays will sense the fault currents. R21 and R34 relays cannot see the fault because they are looking different directions which are contrary to fault currents. Figure 1(b) shows the main and backup relays for the fault at L23 line. R23 is assigned as main protection relay and should trip first. If R23 do not isolate downstream fault current, backup relay R12 has to trip with a sufficient time delay which is called coordination time interval (CTI). This value is usually between 0.2-0.5s [19]. Similarly, R32 relay is assigned as main protection relay and its backup relay is R43 for the upstream fault current. If there are DG connections at BB2 and BB3 as shown in Figure 2, traditional radial protection scheme will result in curve intersections which are undesirable. Thus, it is common using instantaneous elements of OC relays to prevent such situations.

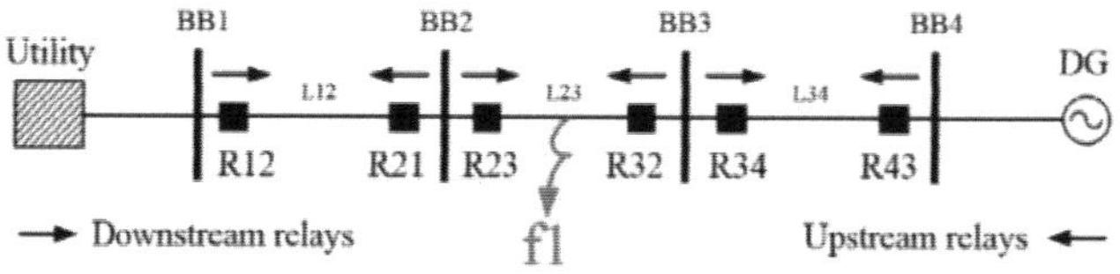

(a) Multiple sources radial network

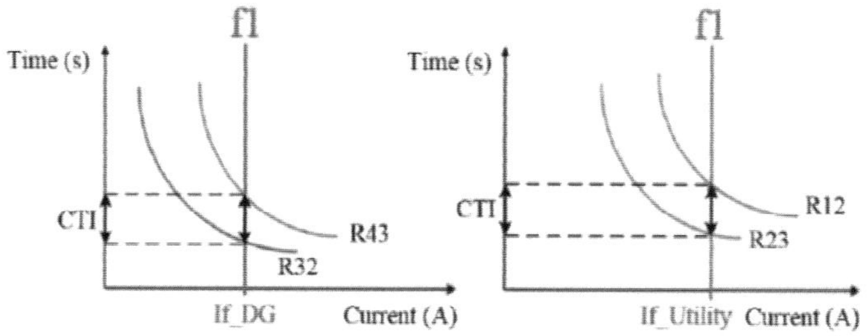

(b) Upstream and donwstream relay selectivity for f1 fault

Figure 1. Multiple sources radial network OC protection scheme

4. Analysis and simulation studies

Analysis and simulation studies are carried out in 4 busbars radial test system given in Figure 2 [20]. Electrical parameters of the system given in Table 2. DG sources are connected to busbars named BB2, BB3, BB4 and short circuit contributions of the sources are modelled considering generator type as inverter interface similarly given in Table 1. Each source has minimum short circuit contribution as twice of its nominal current and maximum fault contribution as three times of nominal current of a DG source. Directional OC relays are used for the protection coordination therefore downstream relays are R12, R23, R34 and upstream relays are R21, R32, R43. Relay R12 and R21 have 300/5 A and the other relays have 200/5 current transformer ratio.

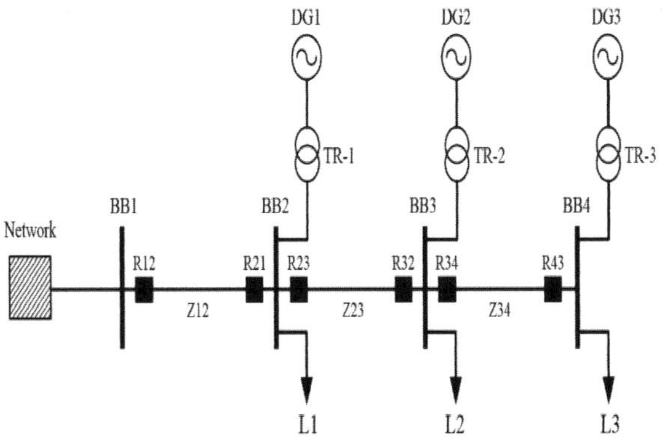

Figure 2. Radial test system with DG sources [20]

Table 2. Electrical parameters of test system [20].

Network	U	11 kV (phase to phase)
	S_k"	30.4841 MVA
	I_k"	1.6 kA
	x_0/x_1	1.2
Lines Z_{12}=Z_{23}=Z_{34}	Z_1	0.585+j2.9217 Ω
	Z_0	0.877+j4.382 Ω
	Z_0/Z_1	1.50
Transformers Uk_1=Uk_2=Uk_3	S	1 MVA
	U_1/U_2	11/0.4 kV
	U_k%	3.00
	U_{k0}%	3.30
Loads L1=L2=L3	S	1 MVA 0.8 (end.)
DG Sources DG1=DG2=DG3	S	1.111 MVA
	S_{kk}*	2.2 MVA

4.1. Scenario I

In this study case, all DG sources are connected. Before determining the relay parameters, it is necessary to look at minimum and maximum short circuit currents besides load currents at busbars. Minimum fault currents are calculated with 2-phase faults and maximum fault currents are calculated with 3-phase faults at the busbars. With the help of the calculated fault and load currents, downstream and upstream relay parameters are specified in Table 3 using the second approach mentioned in section 2.

Table 3. Relay parameters for scenario I.

Relay Parameters	Downstream Relays			Upstream Relays		
	R12	R23	R34	R43	R32	R21
Pickup current Ip (A)	4	4.25	2.5	1.55	2.75	2.92
Time multiplier settings	0.083	0.062	0.025	0.063	0.042	0.015

In order to make it clear how the relay parameters are set, downstream relay R34 settings will be explained. Each of the loads are given 1 MVA so it corresponds 52 A of current. Minimum fault current for R34 is calculated via 2-phase fault at BB4 without DG connections on test system as 427 A. Likewise, maximum fault current is calculated with 3-phase fault at BB4 as 582 A. Lower limit of the pickup current is calculated as 1.5 times of load current and higher limit of pickup current is calculated 2/3 times of minimum fault current respectively. Between the upper and lower values, pickup current of relay R34 is 100 A selected which corresponds 2.5 A when CT conversion considered. TMS value should be set as low as possible for the relay at far end so in this case TMS value is set 0.025 for R34 relay. Using the formula of IEC standard inverse curves, given in (1), (I_f represents maximum fault current and I_p is pickup current) with the parameters explained above, R34 pickup time is calculated 98 ms for the 3-phase fault at BB4.

$$t(s) = TMS \cdot \frac{0.14}{\left(\frac{I_f}{I_p}\right)^{0.02} - 1} \tag{1}$$

In case of R34 relay malfunction for fault at busbar BB4, R23 relay provide backup with a sufficient time delay of 300 ms. R23 will experience 503 A when fault occurs at BB4 so using (2) given below,

$$TMS_{R23} = \frac{(0.098+0.3)s}{\left(\dfrac{0.14}{\left(\dfrac{503}{170}\right)^{0.02}-1}\right)} \tag{2}$$

TMS value of R23 relay calculated as 0.062. By using the calculated TMS, pick up current and maximum fault current values of R23 relay from (1), tripping time for the fault at BB3 is obtained as 0.306 s. Same procedure is followed for the relay R12. This approach is applied for the upstream relays with some adaptations. It will be explained with the R21 relay parameters. When all DG sources online, R21 will measure the load current as the difference between the DG generation and load consumption of the whole system so it is hard to say exact value due to the changes of load and generation. Therefore it would be better look at the past load profile which illustrates the generation and consumption variation more precise. In our case, pickup currents of upstream relays will be selected near its higher limit value which corresponds 2/3 times of minimum fault current. Minimum fault current seen by R21 relay is the lowest fault contributions of three DG sources and it is calculated with 2-phase fault at BB1 as 263 A. Upper limit of the R12 pickup current should be 2/3 times minimum fault current which is calculated as 175 A. When considering the pickup current of R21, it would be better choosing it near the maximum fault current as 170 A due to the uncertainty of upstream load currents. TMS values are selected starting from 0.015 for the upstream relays. Using (1), R21 trip time for the 3-phase fault at BB1 calculated as 152 ms. If R21 relay has malfunction to clear the fault at BB1, R32 relay trip at 452 ms with the addition of CTI as 300 ms. When fault happens in BB1, R21 and R32 relays see different fault currents because of the presence of DG1 source so R32 pickup current should be set considering the DG1 fault contribution. However R21 measures 263 A for the 2-phase fault at BB1, R32 relay measures 165 A. Therefore R32 pickup current is calculated as 2/3 times of 165 A which is equal 110 A. When R32 provides backup for R21 relay, it measures 210 A for the 3-phase fault at BB1 so TMS value of R32 relay is calculated using (2) as 0.042. R32 relay measures its maximum current as 260 A and minimum current as 194 A for the 3-phase and 2-phase faults at BB2 respectively. Using the equation (1) with R32 relay trip time calculated as 339 ms. Same procedure is followed for R43 to calculate tripping times or TMS value. In order to prevent curve intersection to a certain degree, R21 R32 and R43 relays will use instantaneous elements with the pickup currents given; 415 A 300 A and 150 A respectively for the primary side of CT with 100 ms time delay. These values are determined as simulating the 3-phase fault near the upstream relays.

4.2. Scenario II

This scenario focuses on making a different approach for upstream relays so downstream relay parameters will stay same. Alternately, for the upstream relays, definite time curve characteristic can be used. Maximum load current can be measured flowing to utility side only if the loads are neglected. Upstream relays will experience the minimum fault currents in condition of 2-phase fault at BB1. In this case, coordination time interval is 200 ms specified. By using the information mentioned above, Table 4 shows the parameters for R21, R32 and R34 relays when all DG sources are connected to network. Pick up currents are selected as 1.2 times of load currents.

Table 4. Relay parameters for scenario II.

Relay Parameters	Upstream Relays		
	R43	R32	R21
Pickup current Ip(A)	1.56	3.12	3.12
Pickup time t(s)	0.6	0.4	0.2

Both scenarios prove that inverse or definite time curves can be used for the protection schemes when all DG sources connected radial network. On the other hand analyses indicate that coordination between the upstream relays might get lost according to connection combination of DG sources such as singular, dual groups, so relay coordination is kept sustained with updating upstream relay parameters.

4.3. Proposed algorithm for updating upstream relay parameters

A flow chart diagram is proposed in Figure 3. This diagram allows updating upstream relay parameters based on monitoring the network at upstream relay locations and common coupling points of DG sources. When there is a change in prevailing system condition such as connection or disconnection of DG sources or local loads, flow chart will update the upstream relay parameters. In the beginning, it is necessary to calculate load flow and short circuit analysis for the prevailing system conditions. At the upstream relay points; power flow direction may vary depend on consumption and DG productions in the system so each of the upstream relay should sense the magnitude and power flow directions wherever its location. If power flows to utility side (upstream), relay measures positive current magnitude. If upstream relay measures negative currents, it means that power flows to downstream or the local loads. "i" is an identifier for the 3 upstream relays. When "i" value is equal to 1, algorithm is calculating the parameter of R21 relay which is the far end upstream relay. If the measured current through the R21 relay is higher than zero for the certain working condition, pickup currents will be updated based on load flow analysis, as 1.2 times of maximum load current towards to utility side. If the measured current is negative, pickup current will be set to 2/3 times of minimum fault current, which is calculated in fault analysis for the R21 relay. For this case, it is essential to know which DG sources are grid connected and ready to contribute any fault events might happen so algorithm needs to monitor injected DG source currents to detect active current sources for both load flow and fault analyses. After choosing the pickup current (Ip) of R21 relay, tripping time can be calculated with the determined TMS value and stored fault currents calculated at BB1. Lowest TMS value for the upstream relay is given 0.015 for the far end relay R21. B(i,i) busbar fault current includes maximum and minimum short circuit currents calculated at busbar named BB1 for the relay R21 when "i" equals to 1. Then using the known values of CTI and R21 relay trip time, TMS value of relay R32 is calculated and stored for the next iteration of "i" value. Relay identifier is set its new value with an increment. Same procedure is followed for the R32 relay for choosing its pickup current. When calculating the trip time of R32, stored TMS value is used from the previous iteration. R43 relay will backup R32 relay so its TMS value is calculated and stored for the next iteration with a sufficient time delay. By using the proposed flow chart above in Digsilent Programing Language (DPL) script, relay coordination is tested in some different working conditions of the test system. Results are given in Table 5. Scenarios includes the working conditions which relays are able to sense both negative and positive currents. Load flow directions is varied depend on changing load scales and connecting different combinations of DG sources. Three phase fault events are simulated at the busbars for testing relay coordination. Each row shows the tripping time of relays for a 3-phase faulted busbar. For example, in scenario 1, for the case of faulted busbar is BB2, R12 and R32 relays will trip in 431 ms and 342 ms respectively. If R32 relay malfunctions, R43 backup relay will trip in 642 ms. In scenario 3, upstream R43 relay senses load flows to utility side so its pickup current is set based on upstream maximum load current. On the other hand, R32 and R21 relays sense downstream load flow directions so their pickup currents selected based on minimum fault current of DG source. However, there is a difference in way of selecting pickup currents, time grading margin is still satisfactory. Scenario 4 and 5 also diversify the load flow directions for each upstream relay. Coordination time interval is nearly 300 ms for the upstream relays so the algorithm determines the optimal parameters.

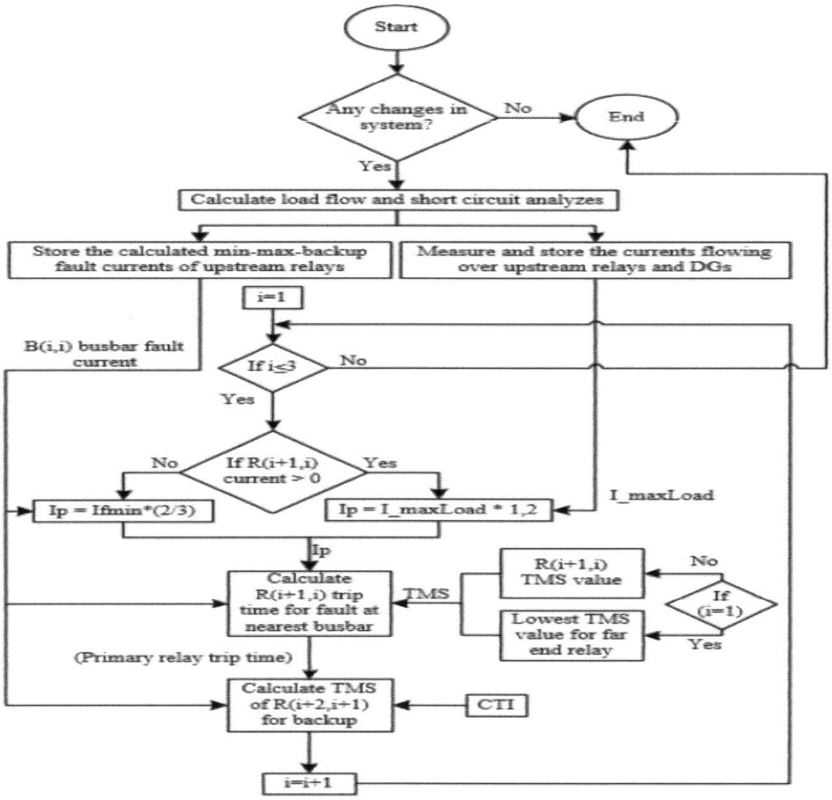

Figure 3. Proposed flow chart updating upstream relay parameters

Downstream relay trip times are slightly changed between the prevailing working conditions but time grading is in the acceptable range for the relay pairs although their relay parameters are constant. This algorithm could also be applied for the downstream relays but relay operating times are in the acceptable range as seen in Table 5.

Table 5. Relay trip times with proposed algorithm.

Scenarios	3-phase fault location	Relay Trip Time (s)					
		Downstream Relays			Upstream Relays		
		R12	R23	R34	R21	R32	R43
Scenario 1 (Inverse time Upstream Relays) Load scale 100%	BB1	-	-	-	0.160	0.460	0.944
	BB2	0.431	-	-	-	0.342	0.642
	BB3	0.625	0.308	-	-	-	0.517
	BB4	0.971	0.410	0.096	-	-	-

Scenarios	3-phase fault location	Relay Trip Time (s) Downstream Relays R12	R23	R34	Upstream Relays R21	R32	R43
Scenario 2 (Definite time Upstream Relays) Load scale 100%	BB1	-	-	-	0.2	0.4	0.6
	BB2	0.431	-	-	-	0.4	0.6
	BB3	0.625	0.308	-	-	-	0.6
	BB4	0.971	0.410	0.096	-	-	-
Scenario 3 Only DG3 connected Load scale 90%	BB1	-	-	-	0.291	0.591	0.897
	BB2	0.431	-	-	-	0.508	0.808
	BB3	0.591	0.334	-	-	-	0.727
	BB4	0.797	0.421	0.111	-	-	-
Scenario 4 Only DG2 connected Load scale 40%	BB1	-	-	-	0.240	0.540	-
	BB2	0.431	-	-	-	0.477	-
	BB3	0.592	0.333	-	-	-	-
	BB4	0.878	0.443	0.99	-	-	-
Scenario 5 DG1 and DG3 connected Load scale 70%	BB1	-	-	-	0.170	0.470	0.815
	BB2	0.431	-	-	-	0.349	0.649
	BB3	0.625	0.307	-	-	-	0.587
	BB4	0.901	0.388	0.105	-	-	-

5. Conclusion

In this study, such issues are investigated as; negative effects of DG sources on OC protection, DG fault contributions, regaining the lost relay coordination due to changing working conditions. Directional OC relay protection is applied to a radial test system with DG sources. Analysis show that upstream relay coordination might get lost due to connection variants of DG sources, because upstream relay coordination is suffered from changing fault contribution of DG sources. Therefore, a flow chart diagram is proposed to update upstream relay parameters, monitoring the system at upstream relay points and common coupling point of DG sources whether they are grid connected or not. Consequently, directional OC relay coordination is achieved for changing working conditions.

6. Further work

It seems that directional OC relays will continue to be used due to their simplicity and reliability. Future developments such as remote control capability, communication featured relays and network monitoring or scada systems will also favour the use of adaptive relay protection. This study focuses on protecting the DG connected system with adaptive relay parameters only for the upstream relays due to the prevailing working condition. Downstream relays may also require parameter update depending on DG fault contribution for the different test systems. Further studies will repeat the analysis process for earth fault protection and meshed distribution network.

References

[1] Ackermann T and Knyazkin V 2002 *IEEE/PES Transmission and Distribution Conf. and Exhibition* 2002 (Japan: Yokohama) pp 1357–62

[2] Hadjsaid N, Canard J F and Dumas F Dispersed generation impact on distribution networks 1999 *IEEE Comp. App.in Power* **12** pp 22–28

[3] IEEE Standard for Interconnecting Distributed Resources with Electric Power Systems IEEE Std. 1547–2003

[4] Horowitz S H and Phadke A G 2008 *Power System Relaying 3rd Edition* (Chichester: John Wiley & Sons)

[5] Mahat P, Chen Z, and Bak-Jensen B A simple adaptive overcurrent protection of distribution systems with distributed generation 2011 *IEEE Trans. on Smart Grid* vol **3** pp 428–437

[6] Javadian S, Haghifam M R and Barazandeh P 2008 *IEEE International Sym. on Industrial Electronics* (UK: Cambridge) pp 2520–25

[7] Singh M K and Reddy P N 2013 A fast adaptive protection scheme for distributed generation connected networks with necessary relay coordination *Students Conf. on Engineering and Systems* (India: Allahabad) pp 1-5

[8] Liu C, Chen Z and Liu Z A 2012 A Communication-less overcurrent protection for distribution system with distributed generation integrated *3rd IEEE Int. Sym. on Power Electronics for Distributed Generation Systems* (Denmark: Aalborg) pp 140-7

[9] Kumar D S, Radhakrishnan B M, Srinivasan D and Reindl T 2015 *IEEE Int. Conf. on Fuzzy Systems* (Turkey: Istanbul) pp 1-6

[10] Saha D, Datta A and Das P Optimal coordination of directional overcurrent relays in power systems using symbiotic organism search optimisation technique 2016 *IET Gen., Tran. & Dis.* **10** pp 2681-88

[11] Alex M and Josephine A. 2013 *Int. Conf. on Energy Efficient Technologies for Sustainability* (India: Nagercoil) 1225–29

[12] Lim S H and Kim J C Analysis on protection coordination of protective devices with a scfl due to the application location of a dispersed generation in a power distribution system 2012 *IEEE Trans. on App. Sup.* **22** p 5601104

[13] Jenkins N *et al* 2000 *Embedded Generation* vol 31 (London: Power & Energy Series 31)

[14] Blackburn J L and Domin T J 2006 *Protective Relaying Principles and Applications 3rd Edition*, (New York: CRC Press)

[15] Pradeep M V and Rama R 2008 *A rev. Power and Energy Society General Meeting-Conversion and Delivery of Electrical Energy in the 21st Cent.* (USA: Pittsburgh) pp 1-7

[16] Pan Y, Voloh I and Ren W 2013 *66th Annual Conf. for Protective Relay Engineers* (USA: College Station TX) pp 92-111

[17] Mohanty S R, Singh R K and Choudhary N K 2015 Coordination of overcurrent relay in distributed system for different network configuration *J. Power and Energy Eng.* **3** 1-9

[18] Soman S A 2016 PSM Setting and Phase Relay Coordination (Tutorial) URL: http://www.nptel.ac.in/courses/108101039/16

[19] Han Z 2014 *Protection coordination in networks with renewable energy sources* Master thesis, Fac. of Eng. and Phys. Sci. Univ. of Manchester

[20] M. Dewadasa, A Ghosh and G Ledwich 2011 *IECON 2011-37th Annual Conf. of the IEEE Industrial Electronics Society* (Australia: Melbourne) pp 924-9

IOP Conf. Series: Earth and Environmental Science **73** (2017) 012027 doi:10.1088/1755-1315/73/1/012027

Wisdom Appliance Control System

Hendrick, Jyun-Teng Jheng, Chen-Chai Tsai, Jia-Wei Liou, Zhi-Hao Wang, and Gwo-Jia Jong

Department of Electronic Engineering, National Kaohsiung University of Applied Sciences, Kaohsiung, Taiwan

Email: aa81330@gmail.com;

Abstract. Intelligent appliances wisdom involves security, home care, convenient and energy saving, but the home automation system is still one of the core unit, and also using micro-processing electronics technology to centralized and control the home electrical products and systems, such as: lighting, television, fan, air conditioning, stereo, it composed of front-controller systems and back-controller panels, user using front-controller to control command, and then through the back-controller to powered the device.

1. Introduction

The principle of this seminar will control all of electrical products in our home via smart phone or tablet and it can also monitor your home electrical usage. Using the Application to synchronize the Bluetooth and then transmit the signal to the relay station, Thorough the infrared rays signal to control appliances.

Currently in our society, wasteful consumption by our negligence, or often forget the remote control where we put, and now just using your smart phone, you can easily control all electrical appliances at home, the idea of this system is not to energy saving but also the reduce the loss caused by the waste. [1]

2. T Research method

Using the smart phone to combined the micro-controller, and via Bluetooth to transmit the data as a communication, another micro-controller the relay station of the signal, the relay station controls transmit infrared LED, and transmit infrared code (data length 32bit), to control the motor and the situations bulb.

On the use of electrical appliances it has a fixed or infrared code, so using this topic to retrieve the infrared receiver to control the electrical infrared code, and its receiving frequency is 38 kHz, in using a microcontroller to store the encoded data, after the completion of this action which can be used to control appliances. The control range is wisdom involves, as long as the infrared control of home electrical products can be controlled, such as television, electric water heaters, air conditioners, electric fans, and others electrical appliances. [2]

3. System Architecture

3.1. System Architecture Process

As shown in Fig.1, the system using Bluetooth to establish a wireless transmission module, user using a mobile phone connect to the Bluetooth module and complete the linear control, Bluetooth, infrared control the TV, stereo, lights and switch in our home, so if the person who go out instantly they don't need to worry about the appliances whether is turn off or not, to save time and home security to the win-win situation. [3]

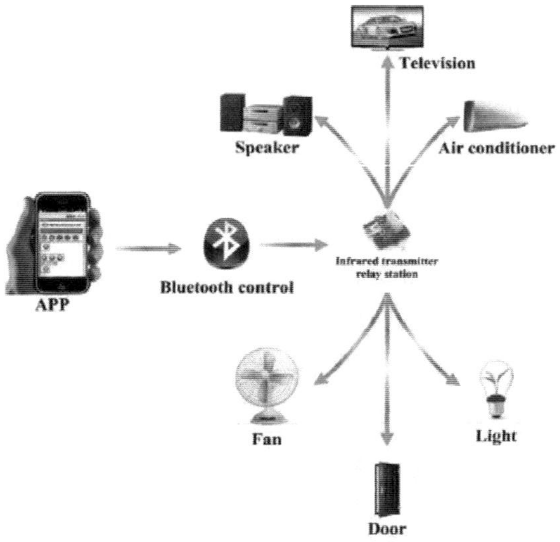

Figure 1. System Architecture flow chart

3.2. Infrared Transceiver System

As shown in Fig.2, the infrared signal using amplification Darlington circuit, in order to improve the emission from the shortage. Darlington amplifier is another form of direct coupling and the amplifier, between transistors connected in series in a straightforward manner, without adding any coupling element. The main effect is to provide the high-current amplification gain.

Darlington pair is the electronics by two (or more) bipolar transistor (or other similar integrated circuits or discrete components) to composed, first through this bipolar transistor amplified current can be further enlarged. Such a structure may provide the higher current gain than the others.

In the case of using the integrated current, the Darlington transistor can make the chip than two discrete transistor device the less space, because the two transistors can share a collector. Darlington transistor is usually packaged in a single chip, the look from the outside like a bipolar transistor [4].

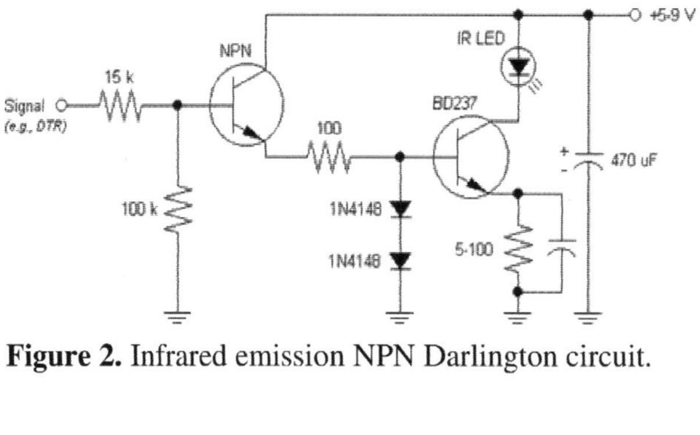

Figure 2. Infrared emission NPN Darlington circuit.

3.3. Infrared transceiver circuits

Receiving circuit will record the home appliance infrared code by it learning mode, after the record and then switch to remotely control mode, the signal will emit from the infrared transmitter circuit. As shown in Fig.3 and Fig.4.

Learning remote control base on the principle of demodulation, and then receive the square wave record completely, you can ignore the protocol by it advantage, to avoid additional unknown problems. Each button needs to re-learning; a complete record of the square wave multiplied the number of buttons, each data code is not the same every time after emitted, often use different coding instructions, or certain remote control have copy protection, deliberately let the signal different each time (for example, garage door remote control), in which will be converted into different coding remote control, also it might not be used.

Usually most of the remote control instruction within 0.1 to 0.3 seconds, and as in the use just need a few bytes of data to be transmitted for using the time savings, if the instruction time stretch, it will make the reaction rate becomes slow, so the learning remote control signal can be set in the range of 0.3 to 0.5 seconds when it is identifying.

In addition, learning remote control records the amount of data size will be based on the type of protocol as a modification, the number of multi-byte instruction, the amount of data square wave variation is varying, then the use of relatively is large. Use 256bytes to record amount every single button more appropriate, if we don't consider the use of space-saving, the best record of the amount of data is512~1Kbytes. [5]

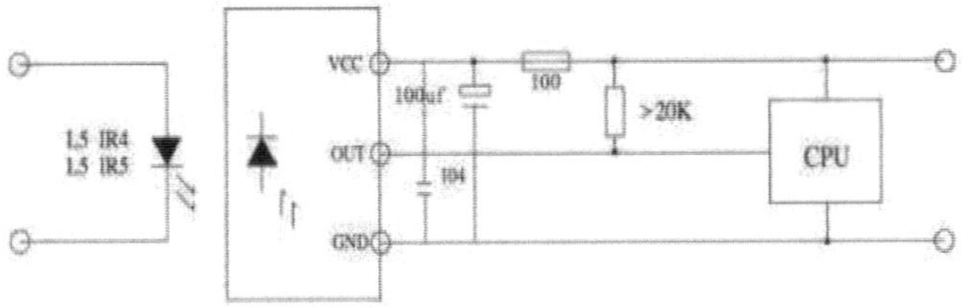

Figure 3. Infrared receiver circuit

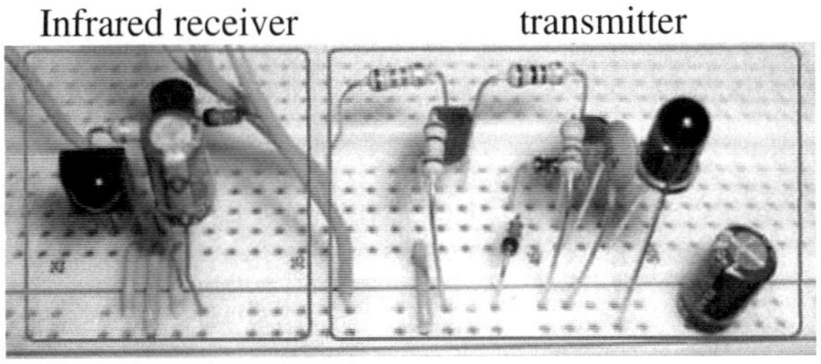

Figure 4. Infrared receiver and transmitter circuitry

3.4. Microcontroller Bluetooth module, protocol

FIG. 5 is a microcontroller Bluetooth module, this topic using Bluetooth as a medium of communication to hardware, data transmission and reception. It works in the 2.45GHz band. Each Bluetooth technology of connect devices has base on the IEEE 802 standards developed by the 48-bit address; can be connected to one or one more, Bluetooth transmission range within 10 meters (0dBm) to 100 meters (20 dBm) about 1600 times per second using frequency hopping spread spectrum technology.

IOP Conf. Series: Earth and Environmental Science **73** (2017) 012027 doi:10.1088/1755-1315/73/1/012027

In the data transmission, Bluetooth with ACL (Asynchronous Connection-Less) type of connection, to provide the highest download data 723.2kbps and asymmetric nature upload data transmission rate 57.6kbps or symmetrical nature of the transmission rate 433.9kbps. In the parts of speech to SCO (Synchronous Connection-Oriented) type of connection, providing audio transmission rate of 64kbps.
In addition, frequency hopping spread spectrum in the synchronous case, the transmitting and receiving via narrowband radio waves to transmit signals specific pattern, in order to avoid a specific frequency bands affected by other noise interference, sending and receiving transmit data in a short time, it will also switch to another frequency band, due to the constant switching frequency, and therefore more able to reduce a particular channel disruption. The launching of frequency hopping spread spectrum signals to follow specifically designed to avoid noise or duplicate channels and frequency hopping signals must follow the FCC (Federal Communications Commission) requirements, the use of more than 75 signal hopping and hopping a frequency to the maximum time interval of 400ms, in IEEE 802.11 the maximum time interval is typically set with 250ms, i.e. Frequencies 4 times per second. Moreover, the Bluetooth signal emitted must be less than 10 pico-W, it will emit one to three watts of transmit power when GSM phone starts, compared to mobile phone, Bluetooth damage caused is much lower. [6]

Figure 5. Microcontroller Bluetooth Module

Bluetooth protocol is divided into Radio, Baseband, Link Manager, L2CAP, HCI and Application Framework and other components, which is mainly responsible for the synthesis and Radio frequency noise filtering, Baseband main message handling code, and the code error resend hopping mechanism work, Link Manager is responsible for establishing the related Link, release, and even secrecy, etc., L2CAP is responsible for multi-tasking of different protocols, packet cutting and restructuring and quality of service and so on. HCI provides the interface between the Host and control Bluetooth, and as a hardware-independent standard control commands. As shown in Fig.6. [7]

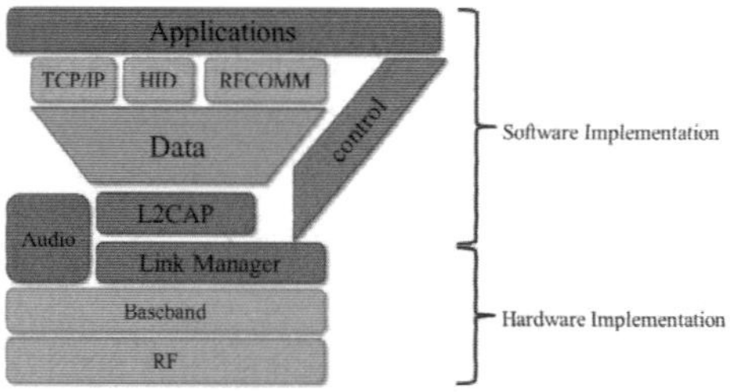

Figure 6. Bluetooth Module

3.5. Arduino microcontroller

Arduino, is an open source microcontroller, which uses Atmel AVR microchip is a low-cost microcontroller, this is the century we using smart phone very often, Android is the open-source software and hardware platform. It provides a I / O interface, and can be used with the high-level language similar to Java, C language development environment.

Arduino is an open authorization interactive development environment technology, interactive installations everywhere, such as air-conditioning sensors to detect the temperature, the indoor temperature is adjusted automatically; the radar using on car, if the car too closed to the object the alarm will warning the driver. These devices increase the number of life safety and convenience. [8]

In the past, we need engineer to deal with the related electronic equipment, each single piece together a small element of the whole circuit. Most of the tools are designed for engineers; in addition to external circuits also we require extensive knowledge to complete the circuit. Fortunately, the microprocessor has made substantial progress, the reducing prices makes learning easily.

Open Source (open-source) is one of Arduino's features, not only the open source software but also the hardware. Software development environment available for free download on the Internet, while the Arduino circuit design also available to download from the official website, but must follow with Creative CC license terms (Creative CC license terms). As shown in Fig.7. [9]

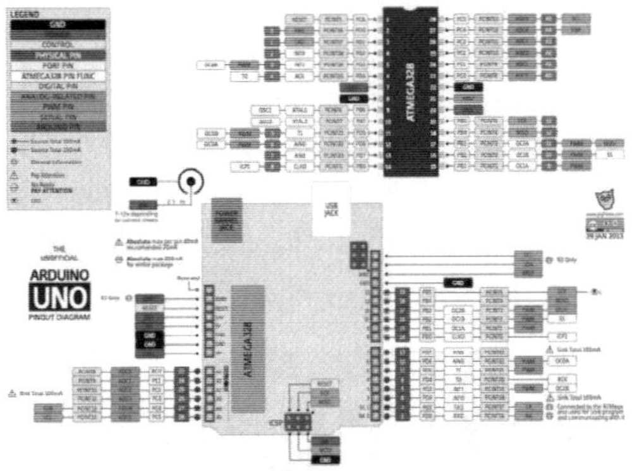

Figure 7. Arduino ATMEGA328

3.6. User interface

In the current ubiquitous intelligent application system, we need a powerful and efficient platform interoperability. Android's open source architecture that provides developers and users a very high flexibility, not only to attract a lot of smart developers have been put into the system, and there is a growing number of end-user devices are willing to import the Android platform.

Android is a Linux 2.6 kernel-based software platform and operating system, mainly developed by Google, after Google and semiconductors, mobile phones, telecommunications services and other manufacturers have established the Open Mobile Alliance (Open Handset Alliance), jointly for the subsequent development services. Because Android has been used as a complete, open and free mobile platform, so, Android mobile phone operating system by the community attention. As an open wisdom OS, Android system includes a number of features and a huge library of code. As shown in Fig.8. [9]

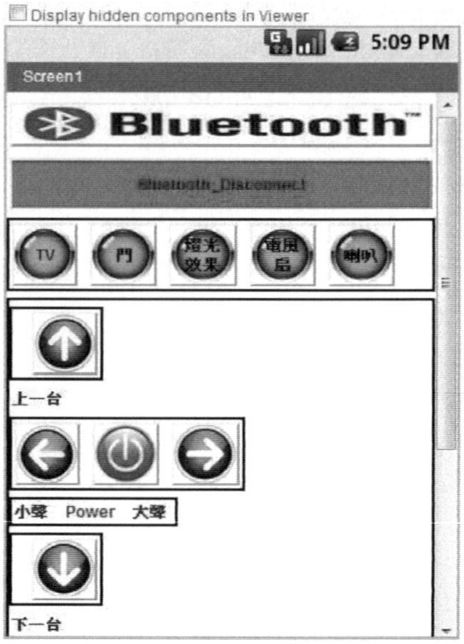

Figure 8. User interface for smart phone

Android is a complete system for mobile phone, can be divided into two layers, the upper use Java to write applications, while the lower system program uses in C language. If we subdivided into two layers, it can be divided into the application layer on top of the Java application framework (Application Framework) and applications (Applications), and then the C language system layer is divided into the system library (Libraries) and the Linux operating system layer, thus forming a four-layer architecture as shown below in Fig.9.

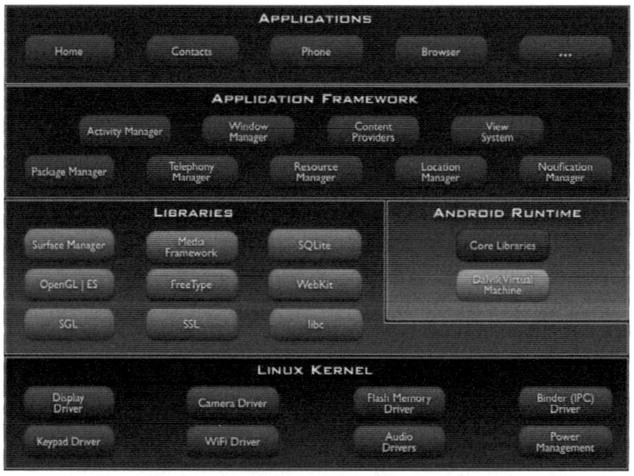

Figure 9. Android System Architecture

4. Conclusion

Smart home combined the computer, communications, consumer electronics products, our home life can completely control via the computer, such as air conditioning, lighting, energy equipment, health care, energy saving and carbon reduction, to make the life more comfortable. Although the digital home is not equivalent to the wisdom of the family, but based on the digital home past few years we established, so that in recent years, smart home can be implemented quickly. It is necessary for a smart home to integrate security, IT, networks, service-related industries, science and technology into the building, the future will be safe, convenient and comfortable.

The study designed a smart home control system; the results can be confirmed through the integration of a variety of wireless sensor modules to be customized production. And we developed a system can control appliances correctly, such a system can be designed and expand for the features we need to carry out development portfolio, but some electrical equipment we need to modify to control, such as air conditioners or dehumidifiers and other household electrical equipment, to achieve the most comfortable environment. The most difficult in modify is to realize the different types of component specifications, welding circuit design, coordination and communication between each other. [10]

References

[1] G. Eason, B. Noble, and I. N. Sneddon, "On certain integrals of Lipschitz-Hankel type involving products of Bessel functions," Phil. Trans. Roy. Soc. London, vol. A247, pp. 529–551, April 1955. *(references)*

[2] J. Clerk Maxwell, A Treatise on Electricity and Magnetism, 3rd ed., vol. 2. Oxford: Clarendon, 1892, pp.68–73.

[3] I. S. Jacobs and C. P. Bean, "Fine particles, thin films and exchange anisotropy," in Magnetism, vol. III, G. T. Rado and H. Suhl, Eds. New York: Academic, 1963, pp. 271–350.

[4] K. Elissa, "Title of paper if known," unpublished.

[5] R. Nicole, "Title of paper with only first word capitalized," J. Name Stand. Abbrev., in press.

[6] Y. Yorozu, M. Hirano, K. Oka, and Y. Tagawa, "Electron spectroscopy studies on magneto-optical media and plastic substrate interface," ASME Transl. J. Magn. Japan, vol. 2, pp. 740–741, August 1987 [Digests 9th Annual Conf. Magnetics Japan, p. 301, 1982].

[7] M. Young, The Technical Writer's Handbook. Mill Valley, CA: University Science, 1989.

[8] Electronic Publication: Digital Object Identifiers (DOIs):

[9] D. Kornack and P. Rakic, "Cell Proliferation without Neurogenesis in Adult Primate Neocortex," Science, vol. 294, Dec. 2001, pp. 2127-2130, doi:10.1126/science.1065467.

[10] H. Goto, Y. Hasegawa, and M. Tanaka, "Efficient Scheduling Focusing on the Duality of MPL Representatives," Proc. ASME Symp. Computational Intelligence in Scheduling (SCIS 07), ASME Press, Dec. 2007, pp. 57-64, doi:10.1109/SCIS.2007.357670

PSO Algorithm for an Optimal Power Controller in a Microgrid

W Al-Saedi[1, 2], S Lachowicz[2], D Habibi[2] and O Bass[2]
[1] University of Technology, Baghdad-Iraq
[2] School of Engineering, Edith Cowan University, 270 Joondalup Drive, Joondalup, WA 6027, Australia

Email: walsaedi@our.ecu.edu.au

Abstract. This paper presents the Particle Swarm Optimization (PSO) algorithm to improve the quality of the power supply in a microgrid. This algorithm is proposed for a real-time self-tuning method that used in a power controller for an inverter based Distributed Generation (DG) unit. In such system, the voltage and frequency are the main control objectives, particularly when the microgrid is islanded or during load change. In this work, the PSO algorithm is implemented to find the optimal controller parameters to satisfy the control objectives. The results show high performance of the applied PSO algorithm of regulating the microgrid voltage and frequency.

1. Introduction

The PSO algorithm was proposed by Kennedy and Eberhart in 1995. This algorithm simulates the social behaviour of the swarm such as schools of fish, flocks of birds, or swarm of bees where they find food together in a specific area. Therefore, this algorithm uses swarm intelligence concept which can be defined as a collective behavior of unsophisticated agents when they create coherent global functional patterns by interacting locally with their environment [1]. Therefore, the PSO can be accomplished based on the three main concepts, namely: social, intelligence, and the computational characteristics.

The social concept usually refers to the interaction and the collective coexistence between the members of group of humans or other animals. In other words, this concept describes the living characteristics of such groups in the environment, irrespective of whether they are aware or not of their interaction, and regardless of the interaction is voluntary or involuntary. Therefore, idea of the PSO algorithm is proposed based on two main theories as follows. First, "*human intelligence results from social interaction*". That means activities like evaluation, comparison, and learning from experience help humans to be familiar with the environment and establish optimal patterns of behaviour and attitudes. Second, "*culture and cognition are inseparable consequences of human sociality*", which means the mutual social learning leads individuals to become more similar [2]. Swarm intelligence is the second concept that provides integrated operation of the PSO technique when it consolidates the social behavior. This concept can be defined as a collective behavior system that simulates the cooperative work of the swarm when they interact locally in nature. In other words, swarm intelligence is a kind of ability that almost uses to solve an optimisation problem in the artificial intelligence applications. Additionally, it is important to explain that swarm intelligence includes two fundamental concepts, namely: the concept of a swarm that suggests multiplicity, randomness, stochasticity, and messiness, and the concept of intelligence that suggests a method for solving a problem which is somehow successful [3], [4]. A computational characteristic is another positive feature to the computational process of the PSO algorithm. That is because the PSO algorithm mainly uses swarm intelligence which provides sufficient computational characteristics.

To solve the optimization problems, many of the optimization techniques have been emerged to address the nonlinear problems, but their applications were with some disadvantages [5]. These techniques are classified based on the type of search space and the objective function, for instance the Linear Programming (LP), Nonlinear Programming (NLP) and Dynamic Programming (DP). The Genetic Algorithm (GA) and Particle Swarm Optimization (PSO) are computational-intelligence based techniques that proposed to solve an optimization problem. GA is a search method that emulates the evolutionary biology to find the approximate optimal solutions [6]. Although a good solution can be located rapidly, it also has some negative aspects, namely: (i) the convergence moves toward the local solution rather than the global solution unless the objective function is defined properly, (ii) it is difficult to run with sets of the dynamic data, and (iii) in a particular optimization problems and computation time, simple optimization technique may give better results than GA. In comparison, and as reported in [7], [8], the best results are achieved by the PSO algorithm compared to other optimization techniques. This is because it outperforms other methods, especially GA in some positive aspects namely:

- The PSO is easier to implement with less parameters for tuning.
- The memory capability of the PSO is more effective than the GA because of each particle is able to remember its own previous best position and the neighborhood best too.
- The PSO is more efficient to maintain the diversity of the swarm. This is because the swarm uses the most successful information to move toward the best which is similar to the community social behavior. While, the GA neglects the worse solution and passes only the good ones.

In this paper, the PSO algorithm is developed and implemented to find optimum power control parameters. This algorithm has been incorporated into the voltage-frequency control mode for a real-time self-tuning method, in order to regulate the microgrid voltage and frequency, especially when the microgrid transits to the islanding mode or during load change.

2. Developed PSO algorithm

The implemented PSO algorithm has been outlined based on the fundamental concepts described above. The essential steps of this algorithm are represented in a flowchart diagram shown in figure 1. These steps describe that this algorithm is an iterative technique that searches the space to determine the optimal solution for an objective function (fitness function). The PSO algorithm evaluates itself based on the movement of each particle as well as the swarm collaboration. Each particle starts to move randomly based on its own best knowledge and the swarm's experience. It is also attracted towards the location of the current global best position X_{gbest} and its own best position X_{pbest} [9]. Therefore, the basic rules of this algorithm can be explained in three main stages:

- Evaluating the fitness value of each particle.
- Updating local and global best fitness and positions.
- Updating the velocity and the position of each particle.

Mathematically, the search process can be expressed by simple equations, using the position vector $X_i = [x_{i1}, x_{i2}, ..., x_{in}]$ and the velocity vector $V_i = [v_{i1}, v_{i2}, ..., v_{in}]$ in the specific dimensional search space. In addition, the optimality of the solution in the PSO algorithm depends on each particle position and velocity update using the following equations [5]:

$$V_i^{k+1} = w.V_i^k + c_1.r_1[X_{pbest}^k - X_i^k] + c_2.r_2[X_{gbest}^k - X_i^k] \tag{1}$$

$$X_i^{k+1} = X_i^k + V_i^{k+1} \tag{2}$$

where i is the index of the particle; V_i^k, X_i^k are the velocity and position of particle i at iteration k, respectively; w is the inertia constant and is often in the range [0 1]; c_1 and c_2 are the cognitive coefficients which are usually between [0 2]; r_1 and r_2 are random values which are generated for each velocity update; X_{gbest} and X_{pbest} are the global best position that is achieved so far based on the swarm's experience, and the local best position of each particle that is achieved so far, based on its own best position, respectively. Moreover, each term in equation (1) can be defined according to its task as follows:

- The first term $w.V_i^k$ is called the inertia component; it is responsible for keeping the particles search in the same direction. The low value of the inertia constant w accelerates the swarm's convergence toward the optimum position, while the high value discovers the entire search space.
- The second term $c_1.r_1[X_{pbest}^k - X_i^k]$ is called the cognitive component; it represents the particle's memory. The particle tends to return to the field of search space in which it has high individual fitness and the cognitive coefficient c_1 affects the step size of the particle to move toward its local best position X_{pbest}.
- The third term $c_2.r_2[X_{gbest}^k - X_i^k]$ is called the social component; it is responsible to move the particle toward the best region found by the swarm so far. The social coefficient c_2 affects the step size of the particle to find the global best position X_{gbest}.

According to equation (2), the position of each particle updates itself by using the new velocity and its previous position. In this case, a new search process starts over the updated search space in order to find the global optimum solution. This process repeats itself until it meets the termination criterion such as the maximum number of iterations or the required fitness value, which are described as follows.

2.1. Fitness function

The fitness function is a particular criterion that is used to evaluate an automatic iterative search such as PSO or GA. In this case, regarding the control objectives, the minimization of error-integrating function is the most relevant function of the four error criteria techniques, namely: 1) *Integral Absolute Error (IAE)*, 2) *Integral Square Error (ISE)*, 3) *Integral Time Square Error (ITSE)*, and 4) *Integral Time Absolute Error (ITAE)*; which offered the best results in the previous study [10]. The *ISE* and *ITSE* are very aggressive criteria because squaring the error produces unrealistic evaluation for punishment. Also, the *IAE* is an inadequate technique compared with the *ITAE* which represents more realistic error index because the error multiplies by time. For these reasons, the controller's objective function is formulated based on *ITAE* in this work, which is calculated using *Simpson's 1/3 rule* that uses the area under the function ($y = A + Bx + Cx_2$, where A, B, and C are constants) between x_1 and x_2 (see figure 2), which is given by:

$$\int_{x_1}^{x_2}(A + Bx + Cx^2)dx \approx \frac{1}{6}(x_2 - x_1)[y_1 + 4y_m + y_2] \tag{3}$$

where:

$y_1 = A + Bx_1 + Cx_1^2$, $y_2 = A + Bx_2 + Cx_2^2$ and $y_m = A + B(x_1 + x_2)/2 + C\{(x_1 + x_2)/2\}^2$, which is located at the midpoint between x_1 and x_2, i.e. $(x_1 + x_2)/2$.

As shown in figure 3, assuming that the area under the curve is divided into an even number of strips *n*, then the area of each double strips can be approximated using equation (3), and the width of each double strips is *2(b-a)/n*. The reason for the double strip is to enable the central ordinate of each strip to give $y_m$value in equation (3). Thus, this method called *Simpson's 1/3 rule*, and the integration between *a* and *b* can be expressed as:

$$\int_a^b f(x)dx \approx \frac{1}{3}\frac{b-a}{n}[y_a + 4y_1 + 2y_2 + 4y_3 + 2y_4 + 4y_5 + y_b] \tag{4}$$

2.2. *Termination criteria*

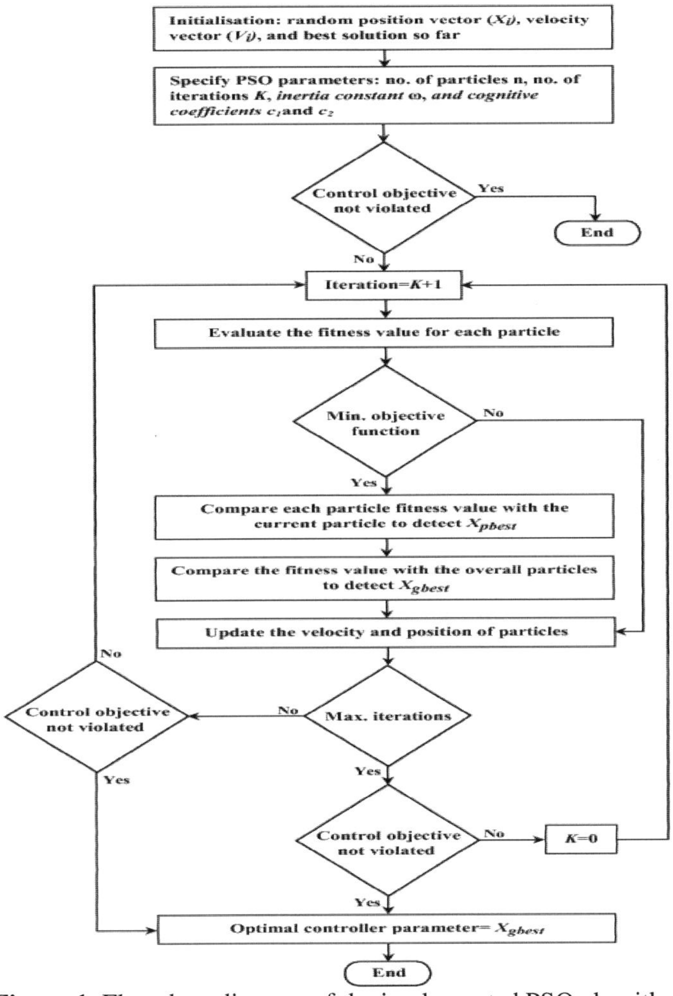

Figure 1. Flowchart diagram of the implemented PSO algorithm.

The termination criteria of a PSO algorithm can be either when the algorithm completes the maximum number of iterations or achieves an acceptable fitness value. In this work, the minimization of the objective function is considered with the maximum number of iterations to find optimum power control parameters. The implemented PSO algorithm and its objective function are individually constructed for each DG unit that allows dealing with more than one DG unit under the supervision by the Microgrid Control Centre (MGCC) unit.

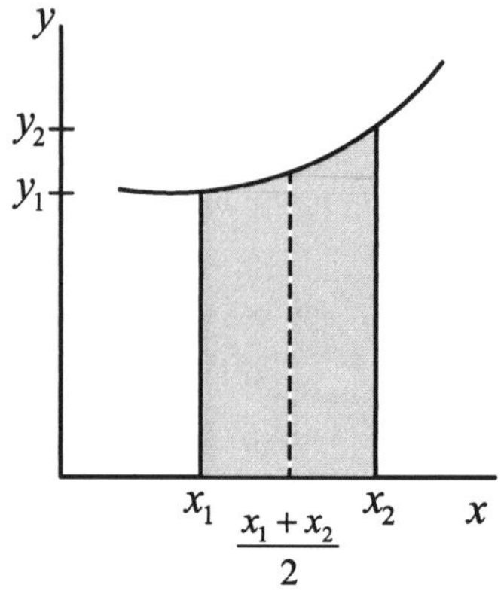

Figure 2. Numerical approximation integral: Simpson's rule.

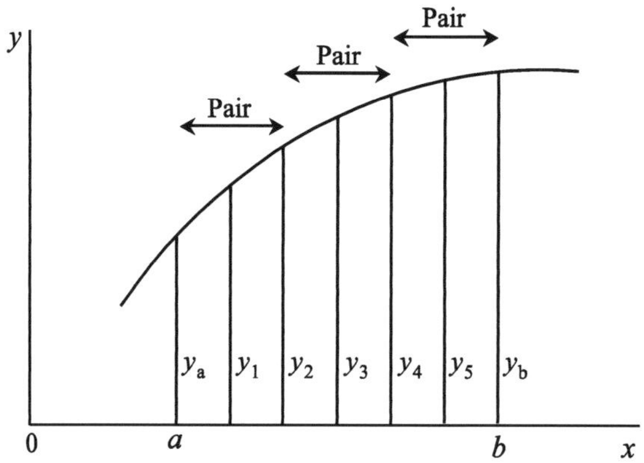

Figure 3. Simpson's rule: area under the curve.

3. Microgrid model

A microgrid is a recent innovation of the small-scale power generation network that aggregates a cluster of DG units using power electronic devices such as the VSI system [11]. This scenario can represent a complementary infrastructure to the utility grid due to the rapid change of the load demand. The high market penetration of the micro-sources such as wind, photovoltaic, hydro, and fuel cell emerge as alternatives which provide green energy and a flexible extension to the utility grid. These sources are usually connected to the power system by widely used Pulse-Width-Modulation (PWM)-VSI systems. While these systems offer flexible control and operation compared to the conventional power generators [12]. Figure 4 shows an example of the microgrid. In such system, a robust control strategy is required to provide acceptable power quality. Therefore, the power controller is usually used for better microgrid configuration. This controller needs to be quite adequate for the purpose of improving the quality of the power supply. In this paper, and as shown in figure 5, the voltage-frequency mode is proposed with the aim of maintaining the system voltage and frequency within acceptable limits.

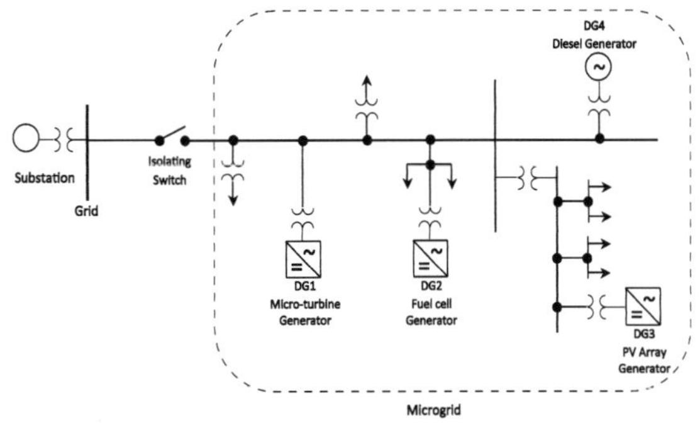

Figure 4. An example of microgrid: islanding mode.

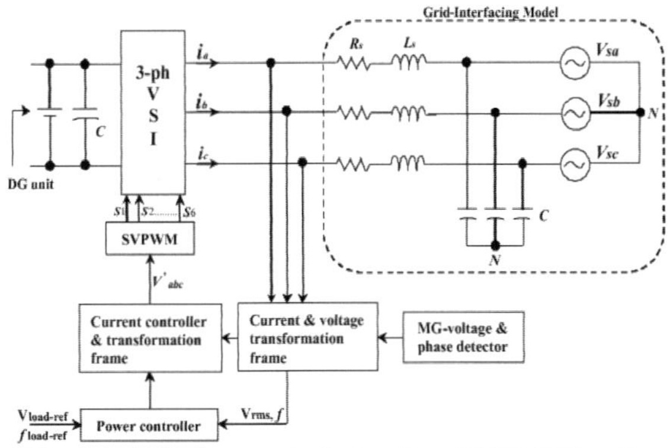

Figure 5. Power control circuits of VSI-DG unit

4. Simulation results

In this work, using MATLAB environment, the PSO algorithm and its objective function are individually constructed for each control objective for one DG unit, which allows dealing with more than one DG unit under the supervision by the MGCC. The voltage and frequency are two control objectives which are considered in this work. Table 1 shows the parameters of the applied PSO algorithm which sets to optimize 50 particles for each cycle of 50 iterations.

Table 1. The parameters of PSO algorithm.

PSO parameters	K_{pf}	K_{if}	K_{pv}	K_{iv}
Acceptable violation (p.u.)	±0.01	±0.01	±0.1	±0.1
Initial velocity (V)	0	0	0	0
Initial fitness value	800	800	800	800
Inertia constant(w)	0.05	0.5	0.05	0.5
Cognitive coefficients	0.09	0.1	0.09	0.1

The search spaces of the parameters of the voltage control loop K_{pv} and K_{iv} are limited to [0 -20] and [0 5e $^{-3}$], respectively. Similarly, the search boundaries of the parameters of the frequency control loop K_{pf} and K_{if} are set to [0 30] and [0 5e $^{-3}$], respectively. Figures 6, 7, 8, and 9 show an example of the search process of the candidate particles when the microgrid starts the islanding mode (at 0.6s) and during load change (at 1.8s). These particles select their trajectories based on their best fitness values, and the results show that the particles stop their movements at the best positions which are represent the power control parameters (see table 2). Figures 10 and 11 prove the high performance of the applied PSO algorithm, when it restored the microgrid voltage and frequency within the acceptable limits.

Table 2. Power control parameters.

Control parameters	Islanding mode	Load change
K_{pf}	3.010859	2.561841
K_{if}	0.000377	0.000778
K_{pv}	-0.99369	-1.01285
K_{iv}	0.003377	0.003196

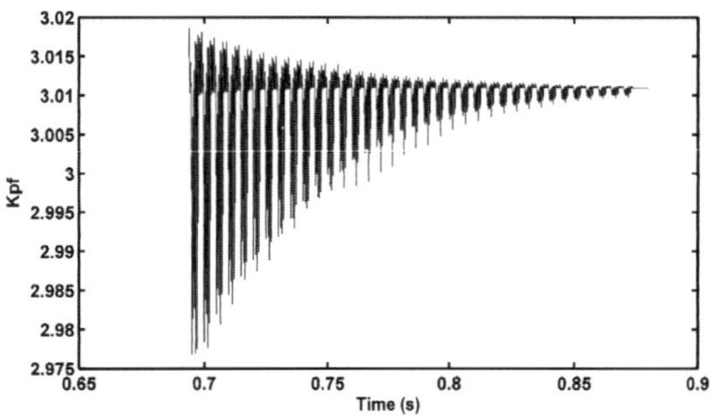

Figure 6. Search process when microgrid islanded (K_{pf}).

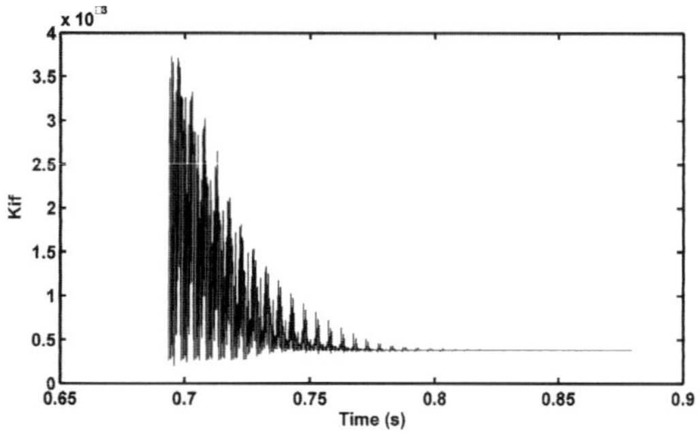

Figure 7. Search process when microgrid islanded (K_{if}).

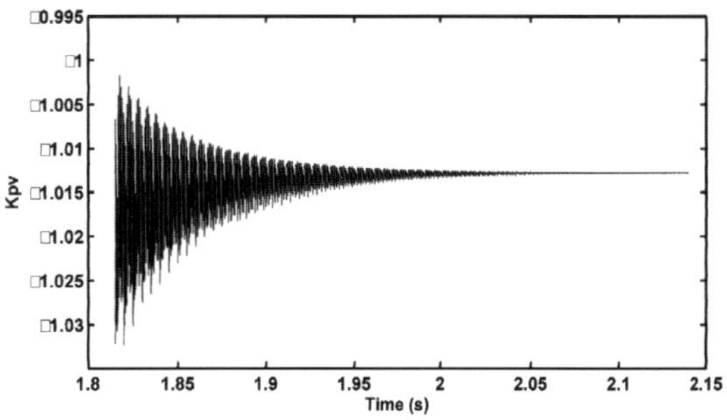

Figure 8. Search process at load change (K_{pv}).

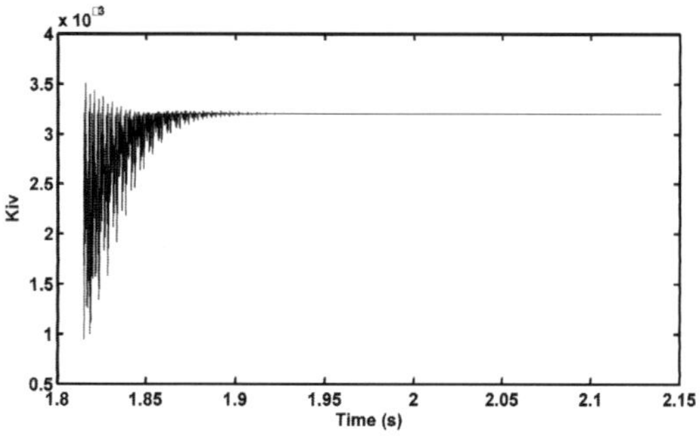

Figure 9. Search process at load change (K_{iv}).

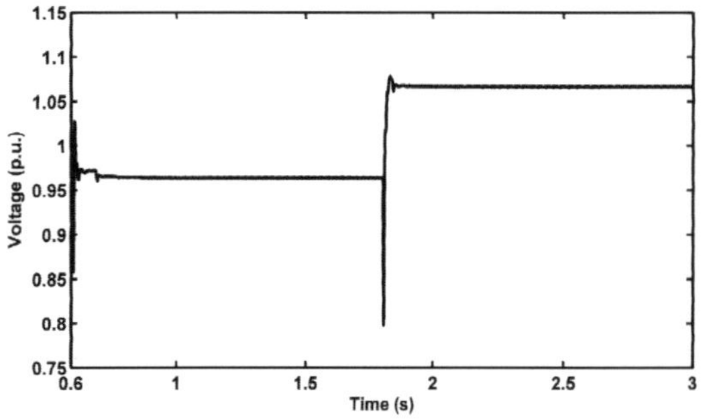

Figure 10. Microgrid voltage.

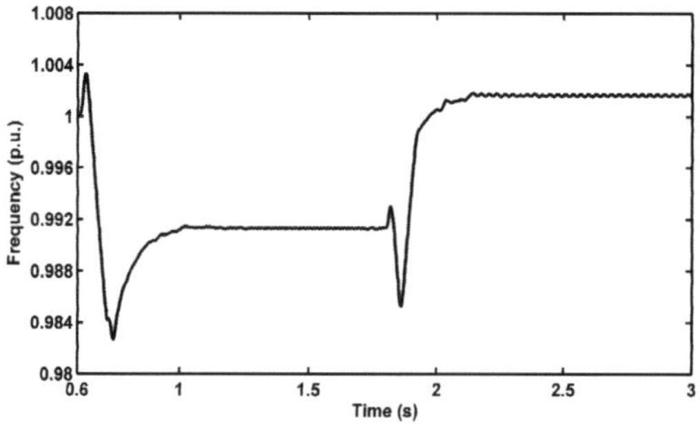

Figure 11. Microgrid frequency.

5. Conclusion

In this paper, the PSO algorithm has been proposed to improving the quality of the power supply in a microgrid. This algorithm is incorporated into the power controller to implement real-time self-tuning method. Thus, an optimization technique is embedded in the voltage-frequency power controller for the inverter based DG unit in a microgrid. The results showed that the proposed PSO algorithm offered high performance of regulating the microgrid voltage and frequency to be within the acceptable limits.

References

[1] Kennedy J 2006 *Handbook of Nature-Inspired and Innovative Computing of Swarm Intelligence* pp 187-219

[2] Eberhart R and Kennedy J 1995 A new Optimizer Using Particle Swarm Theory *Proc. of the 6th Int. Symp. on Micro Machine and Human Science (IEEE)* pp 39-43

[3] Shuyuan Y, Min W and Licheng J 2004 A quantum Particle Swarm Optimization *Proc. Congress on Evolutionary Computation (CEC2004)* 1 pp 320-324

[4] Millonas M 1992Swarms Phase Transitions and Collective Intelligence *Los Alamos National Laboratory Technical Report* (United States)

[5] Del Valle Y, Venayagamoorthy G K, Hernandez J C and Harley R G 2008 particle swarm optimization: basic concepts, variants and applications in power systems *IEEE Trans. on Evolutionary and computations* **12** pp 171-195

[6] Nocedal J and Wright S 1999 *Numerical Optimization* (Springer-Verlag)

[7] Arya L, Titare L and Kothari D 2010 improved particle swarm optimization applied to reactive power reserve maximization *International Journal of Electrical Power and Energy Systems* **5** pp 68-74

[8] Hasan R, Cohanim B, De Weck O and Venter G 2005 Acomparison of Particle Swarm Optimization and the Genetic Algorithm *Proc. Conf. of the 1st AIAA Multidisciplinary Design Optimization Specialist*

[9] Yang and Xin-She 2010 *Engineering Optimization: An Introduction with Metaheuristic Application* (Hoboken: John Wiley)

[10] Killingsworth K and Krstic M 2005 Auto-tuning of PID Controllers via Extremum Seeking *Proc. of the 2005 American Control Conf.* 1 pp 2251-2256

[11] Lasseter R 2002 microgrids *Power Engineering Society Winter Meeting IEEE* **1** pp 305-208

[12] Strzelecki R and Benysek G 2008 *Power Electronics in Smart Electrical Energy Networks* (London: Springer-Verlag)

IOP Conf. Series: Earth and Environmental Science **73** (2017) 012029 doi:10.1088/1755-1315/73/1/012029

Framework for Sustainability Performance Assessment for Manufacturing Processes- A Review

K Singh and I Sultan

School of Engineering and IT, Faculty of Science & Technology, Federation University, Mt Helen campus, Australia

Email: karmjitsingh@federation.edu.au;

Abstract. Manufacturing industries are facing tough competition due to increasing raw material cost and depleting natural resources. There is great pressure on the industry to produce environmental friendly products using environmental friendly processes. To address these issues modern manufacturing industries are focusing on sustainable manufacturing. To develop more sustainable societies, industries need to better understand how to respond to environmental, economic and social challenges. This paper proposed some framework and tools that accelerate the transition towards a sustainable system. The developed framework will be beneficial for sustainability assessment comparing different plans alongside material properties, ultimately helping the manufacturing industries to reduce the carbon emissions and material waste, besides improving energy efficiency. It is expected that this would be highly beneficial for determination of environmental impact of a process at early design stages. Therefore, it would greatly help the manufacturing industries for selection of process plan based on sustainable indices. Overall objective of this paper would have good impact on reducing air emissions and protecting environment. We expect this work to contribute to the development of a standard reference methodology to help further sustainability in the manufacturing sector.

1. Introduction

Sustainable manufacturing broadly implies the development of innovative manufacturing sciences and technologies that span the entire lifecycle of products and services to minimize negative environmental impacts, conserve energy and natural resources, are safe for employees, communities, and consumers, and are economically sound [1].

Sustainable manufacturing plays an important role in the manufacturing of products, materials and processes. For sustainable manufacturing development, enhanced modeling techniques are needed to understand and predict the sustainability aspects through design and manufacturing where technologies can be applied to transform materials with reduced energy consumption, reduced emissions, reduced generation of waste products, and reduced use of non-renewable or toxic materials. The scope of this paper is to propose some significant findings for sustainable manufacturing assessment.

However, every manufacturing industry has always affected the environment and will continue doing so in one or the other ways, demanding a never-ending research for sustainability. Our goal will be continuous improvement making constant advances in manufacturing company's overall sustainability performance. To develop more sustainable societies, industry needs to better understand how to respond to environmental, economic and social challenges. Our motive is to develop framework and tools that accelerate the transition towards a sustainable system.
This paper is organized as follows, Section 2 describe the literature reviews on sustainable manufacturing assessment and identify the research gap based on various authors work in the field of sustainable manufacturing processes followed by some useful findings in this area. Section 3 propose some basic steps of methodology for sustainable product development. Section 4 present conclusions and future scope of work.

2. Related work

Earlier, researchers have investigated performance evaluation and improvement for manufacturing process variable consumption. This section describes previous research efforts related to sustainable manufacturing assessment and sustainability indicators of various manufacturing processes. Lee et al. [2] introduces MAS2, which is an integrated modeling and simulation-based life cycle evaluation approach for sustainable manufacturing. This work has provided a way to assess the sustainability performance by combining sustainability concepts with engineering technologies using mathematical modeling. Zhang et al. [3] developed process-oriented Information Model (PIM) to integrate the relevant information regarding the sustainable manufacturing with the product design information. Seow et al. [4] developed an approach to model energy flows within a manufacturing system with an aim of representing the amount of energy attributed for manufacturing a product. They applied the concept of direct energy, indirect energy and auxiliary energy. Ciceri et al. [5] proposed a tool to estimate materials and manufacturing energy based on the bill of material for a product. The mechanism proposed by them determines the energy estimate by compiling available data from material embodied figures, empirical and bill of material. Kellens et al. [6] proposed a life-cycle analysis (LCA) oriented methodology for systematic inventory analysis of the use phase of manufacturing unit processes providing unit process datasets to be used in life-cycle inventory (LCI) databases and libraries. The methodology has been developed as a framework of the CO2PE! and comprises two approaches namely, screening approach and the in-depth approach with different levels of detail. Kellens et al. [7] presented two case studies, one for both the screening approach and the in-depth approach, demonstrating the application of the life cycle assessment-oriented methodology for systematic inventory analysis of the machine-tool use-phase of manufacturing unit processes. Methodology is explained with the help of machining and laser cutting processes. Jeswiet and Kara [8] proposed a method that connects the electrical energy used in manufacturing directly to the carbon emissions generated during production of that electrical energy. The developed methodology was implemented in machining operations, such as turning and milling. Taha et al. [9] presented a methodology that links design features of a product to the environmental impact. The study determines the specific energy consumption for the turning operation experimentally, which is in turn interpreted into CO_2 emissions. Jayal et al. [10] presented an overview of present tools and methodologies for evaluating sustainability contents at the product, process and system levels. The authors consider dry, near-dry and cryogenic machining as examples for optimization of sustainable manufacturing processes. Feng and Juong [11] proposed a framework for quantitative measurement of sustainability for machining operations. They take into consideration carbon emission and energy use in their study. Gutowski et al. [12] studied the environmental impact of different manufacturing processes, which accounts for the exergy used during the manufacturing process. Krajnc et al., [13] proposed a method to measure sustainability indicators and a strategic set of metrics for assessing the sustainability level of a company. The choice of the sustainability indicator for a company is left at the option of the user. Madan et al. [14] develop the needed measurement science, standards and methodologies to evaluate and improve sustainability of manufacturing processes. Authors present a science-based guideline to characterize energy consumption for a part manufactured using the injection molding process. Companies use these indicators to set targets and monitor progress for sustainability in manufacturing. A survey of the corporate sustainability reports demonstrated that environmental and social impact assessment has received less attention as a whole to evaluate sustainability of a manufacturing process. Furthermore, comparison and assessment of production systems demand quantitative measures for sustainability. Such quantitative measurement requires environmental indicators. For simplicity, environmental indicators are classified into input and output indicators. Input indicators for sustainability include energy use, material use, and water use indicators. Output indicators include product, solid waste, liquid waste, and air emission indicators. It is identified that some indicators do affect the performance of the other indicators. For example, electrical energy affects indicators like the Co^2 emission or renewable of energy. In the following section, we discuss the research gap followed by objectives of present work.

2.1. Research gaps

Despite very useful work presented in the literature, there are still gaps that need to be addressed, the prominent ones are mentioned below:

- Available approached do not satisfy the need for early estimation of sustainable manufacturing indicators that are useful for decision support to identify the most efficient manufacturing/process plans.
- Few researchers have made efforts in selecting indicators effecting environmental, economic and social aspects of manufacturing sustainability at the factory level. However, the use of specific methodologies and tools to measure and evaluate such indicators based on data collection from the industry shop floor or LCI database.
- Focus of most of the research papers is to collect life cycle inventory data for the database. This database is later used in the various inventory databases, such as Ecoinvent. The process of this data collection takes long time. Furthermore, such databases only represent average sustainability information and does not represent specific improvements made by the manufacturing.
- The manufacturing industry is more interested in knowing sustainability assessment of their product manufacturing in terms of
  - inter-process sustainable manufacturing comparisons
  - comparing alternative process plans for manufacturing a product
- Therefore, a deeper science based study to further carry on the sustainability assessment research to the process level is required.
- In demand-driven manufacturing scenario, estimation of sustainability assessment is very important and available systems do not support it.

- There has been a very limited work on sustainability information with integrating product design, which is essential for achieving overall sustainability. Therefore, further research is needed to address how to evaluate product sustainability using economic and environmental factors.
- Pre-and post-operations, such as joining, coating which are closely associated with the processes, have not been given much attention.
- Providing a computer-aided approach would help generate better sustainability evaluation and inter-process sustainability comparison for selected manufacturing processes.

2.2. Objectives of the present work

To address the research gaps, there is a need to develop a methodology, which can be used to evaluate and improve the sustainability of manufacturing processes. In this paper, we try to develop a guideline specify to several manufacturing processes. The objectives of this paper are mainly:

- Study the manufacturing processes with an objective to develop a sustainability assessment methodology for manufacturing processes
  - Step 1: Study taxonomies and structure of manufacturing processes
  - Step 2: Select representative manufacturing processes.
- Identify key performance indicators (KPIs) for manufacturing sustainability assessment
- Analyze the selected unit manufacturing processes and their structure/sub-processes.
  - For example, Heating and melting, casting, cooling, machining and finishing process.
- To propose a science based approach for sustainability assessment of the manufacturing processes with respect to identified KPIs.
- To develop a framework that helps assess sustainability for product manufacturing with selected manufacturing processes and process plans.
  - provide a basis for comparison and selection of the most sustainable process plan.
- Optimize product manufacturing for sustainability by selecting appropriate process parameters
- Verify the proposed methodology, compare the developed model by comparing results with the actual shop floor data obtained from the manufacturing industries.

3. Methodology for Sustainable manufacturing assessment

Methodology for proposed work is listed below.

- To review the literature related to Sustainable assessment of manufacturing processes, key performance indicators and various techniques to solve above said problem.
- After that this project proposes a structure for manufacturing processes and identification of representative manufacturing processes with an objective to develop a sustainability framework for manufacturing processes.
- Structured information models for seamless flow of information across design and manufacturing domains for selected manufacturing processes will then be defined.
- The proposed work also intends to identify key performance indicators (KPIs) for manufacturing sustainability assessment and perform analysis of selected unit manufacturing processes and their sub-processes aiming to propose a methodology for determining science based measurements of the manufacturing processes affecting the KPIs.

The below figure 1.1 chart shows the methodology of proposed system.

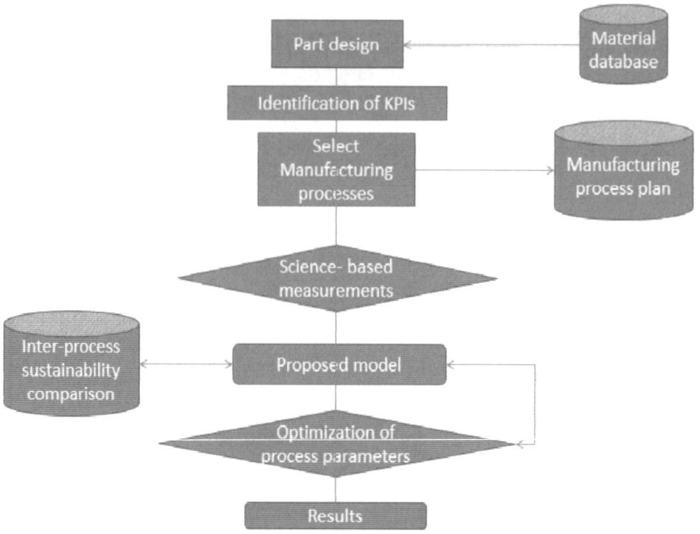

Figure 1. Methodology of proposed system

- The theoretical foundations determined will be used to develop a model that could evaluate sustainability of selected manufacturing processes and their respective process plans providing a basis for inter-process comparison and selection of the most sustainable process plan.
- The framework aims to consider different influencing factors such as product information, manufacturing process plan, part geometry, material related physical and processing properties and the manufacturing equipment information.
- Optimize product manufacturing for sustainability by selecting appropriate process parameters.
- Obtaining best compromising results.
- The proposed methodology will be verified with actual shop floor data obtained from the industry.

4. Conclusion
Since In recent years, the concepts of sustainability have gradually evolved and has begun receiving international attention. In addition, companies are under pressure to have environmental friendly manufacturing and service sectors, hence, this paper provides an attempt to have a brief overview towards the sustainability and related issues. Sustainability in manufacturing is considered as an important tool strategy similar to other manufacturing tools and systems for example lean production, supply chain management etc. This concept helps organization to have economically improvement but not only this, but also, environmentally and sociality. In the other words, environmentally friendly product and totally sustainable manufacturing system helps organization to reduce use of material, minimize environmental impact and make product economical sustainable. Although there are lots of researches which have been done on the area of sustainability, there are few works on sustainability of early product development and inter-process sustainable comparison, therefore, these issues need more investigation in future research.

Acknowledgements
The authors sincerely extend their appreciation to Federation University Australia for their support of this research project.

References

[1] Allwood, J 2005 *Sustainable Manufacturing Seminar Series Institute of Manufacturing, University of Cambridge*.

[2] Lee, J. Y., Kang, H. S., Noh, S.D 2014 *Journal of Cleaner production*. 66, pp. 146-163.

[3] Zhang, H., Zhu, B., Li, Y., Yaman, O., and Roy, U 2015 *J. Manuf. Syst.*, 37 (2), pp. 459–466.

[4] Seow, Y., Rahimifard, S 2011 *CIRP J. Manuf. Sci. Technol.* 4 (3), pp.258-264.

[5] Ciceri., Natalia., Gutowski, T., Garetti. M 2010 *IEEE/International Symposium on Sustainable Systems and Technology*.

[6] Kellens, K., Dewulf, W., Overcash, M., Hauschild, M., Duflou, J R 2012 *Int J Life Cycle Assess.* 17(1), pp 69-78.

[7] Kellens, K., Dewulf, W., Overcash, M., Hauschild, M., Duflou, J R 2012 *Int J Life Cycle Assess.* 17(2), pp 242-251.

[8] Jeswiet, J., Kara, S 2008 *CIRP Annals Manufacturing Technology*, 57(1), pp 17–20

[9] Taha, Z., Kurniati, H., Aoyama, H., Ghazilla R.G., Gonzales, S. N 2010 *International Conference on Robotics Control and Manufacturing Technology*.

[10] Jayal, A.D., Badurdeen, F., Dillon Jr., O.W., Jawahir, I.S 2010 *CIRP J. Manuf. Sci. Technol.* 2(3), pp 144-152.

[11] Feng, S.C., Joung, C.B., *Proceedings of the 7th Global Conference on Sustainable Manufacturing*, (2009)

[12] Gutowski, T., Dahmus, J., Thiriez, A 2006 *CIRP International Conference on Life Cycle Engineering*.

[13] Krajnc, D., Glavic, P 2003 *Clean Technologies and Environmental Policy*, 5(3-4), pp 279-288.

[14] Madan, J., Mani, M., Lee, J.H., Lyons, K.W 2014 *Journal of Cleaner production*, pp. 1-14.

IOP Conf. Series: Earth and Environmental Science **73** (2017) 012030 doi:10.1088/1755-1315/73/1/012030

Influence of Blasted Uranium Ore Heap on Radon Concentration in Confined Workspaces of Shrinkage Mining Stope

Y J Ye[1, 2], T Liang[2], D X Ding[1], B Lei[2], H Su[2] and Y F Zhang[2]
[1]Key Discipline Laboratory for National Defense for Biotechnology in Uranium Mining and Hydrometallurgy, University of South China, Hengyang, 421001, China
[2]School of Environmental Protection and Safety Engineering, University of South China, Hengyang, Hunan 421001, China

Email: yongjunye@163.com; 787346535@qq.com;

Abstract. A calculation model for radon concentration in shrinkage mining stopes under various ventilation conditions was established in this study. The model accounts for the influence of permeability and area of the blasted ore heap, ventilation air quantity, and airflow direction on radon concentration in a confined workspace; these factors work together to allow the engineer to optimize the ventilation design. The feasibility and effectiveness of the model was verified by applying it to mines with elevated radon radiation exposure. The model was found to accurately changes in radon concentration according to the array of influence factors in underground uranium mines.

1. Introduction

Radon concentration is higher in underground uranium mines and is known to be a severe health hazard for miners. Mining activity results in the release of radon gas and its daughter products into confined workspaces, where miners may be exposed to high levels of radon in ore heaps or from other sources. Poor ventilation condition can severely threaten the health and safety of miners due to exposure to radioactive materials. There is urgent demand for workable techniques to quantify radon concentration in confined workspaces in order to design ventilation systems for blasted uranium ore heaps in shrinkage mining stopes.

Radon measurement techniques were first investigated in coal mines in Pakistan [1] and Western Turkey [2]. EI-Fawal [3] established a calculation model for airflow, air pressure and radon/daughter concentration in mine ventilation networks. Sahu [4] evaluated the effect of various ventilation parameters on radon exposure to miners in underground uranium mines; airflow rate was considered the key parameter in controlling the radon/daughter concentrations in the mine. These researches have focused mostly on radon concentration measurement in assessing radiological hazards. Many have also explored the influence of ventilation parameters on radon concentration.

Data gathered in previous studies was utilized to build a calculation model for radon concentration in shrinkage mining stopes with various ventilation parameters. The calculation results may provide a workable reference for ventilation design in underground shrinkage uranium mines.

2. Mathematical Model

2.1. Calculating the radon release-diffusion-seepage migration in blasted uranium ore heap

The ore particles in blasted uranium ore heaps vary in size and can be roughly divided into n grades accordingly: $0 \sim r_1, r_1 \sim r_2, r_2 \sim r_3, \ldots\ldots, r_{n-1} \sim r_n$. The equivalent radius of the i_{th} grade ore particles can be calculated as follows:

$$r_i = \left(\frac{r_{i\min}^2 + r_{i\max}^2}{2} \right)^{\frac{1}{2}} \tag{1}$$

Where r_{min} and r_{max} is the minimum and maximum size of the i_{th} grade ore particles (m).
The mass fraction of the i_{th} grade ore particles is a function of the fractal dimension and particle size, it can be calculated as follows [5]:

$$w_i = \left(\frac{r_{i\max}}{r_{\max}} \right)^{3-D_f} - \left(\frac{r_{(i-1)\max}}{r_{\max}} \right)^{3-D_f} \tag{2}$$

Where D_f is the fractal dimension of the particle size distribution.
The bottom surface of uranium ore heap forms the origin of the coordinates and the top is the positive direction. Then the steady release-diffusion-seepage migration equation in the blasted uranium ore heap is as follows:

$$D_2 \frac{d^2 C}{dx^2} - \frac{v}{\eta_2} \frac{dC}{dx} - K_1 C + K_2 = 0 \tag{3}$$

$$K_1 = \lambda + \frac{3(1-\eta_2) D_1 \eta_1}{\eta_2} \sum_1^n \frac{w_i}{r_i} \left(\left(\frac{\lambda}{D_1} \right)^{\frac{1}{2}} cth\left(\left(\frac{\lambda}{D_1} \right)^{\frac{1}{2}} r_i \right) - \frac{1}{r_i} \right), \quad K_2 = \frac{3(1-\eta_2) D_1 \alpha_0}{\eta_2 \lambda} \sum_1^n \frac{w_i}{r_i} \left(\left(\frac{\lambda}{D_1} \right)^{\frac{1}{2}} cth\left(\left(\frac{\lambda}{D_1} \right)^{\frac{1}{2}} r_i \right) - \frac{1}{r_i} \right)$$

Where η_2 is ore heap porosity; K_1 is equivalent radon decay constant of the medium; K_2 is equivalent radon release rate of the medium (Bq m^{-3} s^{-1}).

2.2. Determination of radon exhalation rate of blasted uranium ore heap
Figure 1 and 2 show shrinkage stope with ascentional and descentional ventilation respectively. The pressure difference between the top and bottom surface of the ore heap is formed during the ventilation period. The value of said pressure difference is approximately equal to the ventilation resistance of the stull-supporting raise, which can be calculated as follows:

$$\Delta P = \frac{\alpha_f p L_0 Q^2}{s^3} \tag{4}$$

Where α_f is the frictional resistance coefficient of the stull-supporting raise (Pa s m^{-2}); p is the stull-supporting raise perimeter (m); s is the stull-supporting raise area, (m^2); Q is the ventilation air quantity of the stope (m^3 s^{-1}) and L_0 is the stull-supporting raise length (m).

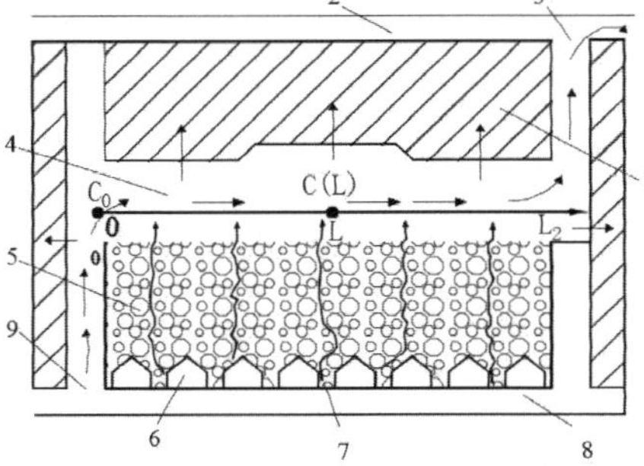

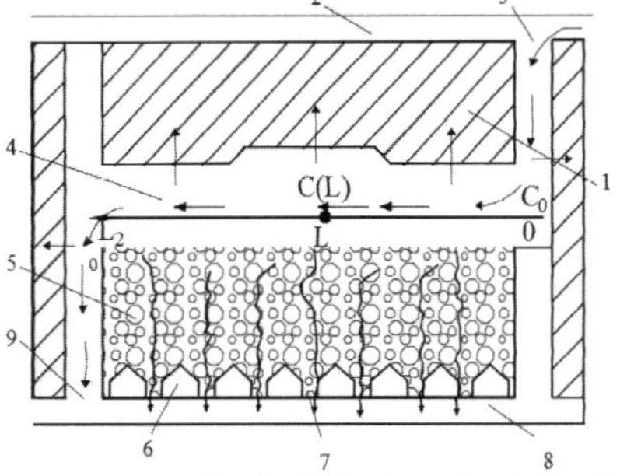

1--Unwrought ore body 2--Ventilation roadway 3--Return air shaft 4) Confined workplace
5--Blasted ore heap 6--Pedestal 7--Drawing chute 8--Haulage roadway 9--Stull-supporting raise

Figure 1. Schematic diagram of ascentional ventilation in shrinkage mining stope

Figure 2. Schematic diagram of decentional ventilation in shrinkage mining stope

The length of the stope's stull-supporting raise can be calculated as follows :

$$L_0 = \frac{H}{\sin\theta} = L_1 \tag{5}$$

Where H is the vertical height of the blasted uranium ore heap (m); θ is ore body obliquity; L_1 is the thickness of blasted uranium ore heap (m).

The air seepage velocity in the blasted uranium ore heap can be calculated through 'Darcy law'. In ascentional ventilation, then radon exhalation rate of the top surface of the blasted uranium ore heap is:

$$J_{ss} = \frac{\eta_2 D_2 K_2[b_1 e^{b_1 L_1} + b_2 e^{-b_2 L_1} - (b_1 + b_2)e^{(b_1 - b_2)L_1}]}{K_1(e^{b_1 L_1} - e^{-b_2 L_1})} \tag{6}$$

$$b_1 = \frac{\left(\left(\frac{\alpha_f kpQ^2}{\mu s^3}\right)^2 + 4K_1 D_2 \eta_2^2\right)^{\frac{1}{2}} + \frac{\alpha_f kpQ^2}{\mu s^3}}{2D_2\eta_2}, \quad b_2 = \frac{\left(\left(\frac{\alpha_f kpQ^2}{\mu s^3}\right)^2 + 4K_1 D_2 \eta_2^2\right)^{\frac{1}{2}} - \frac{\alpha_f kpQ^2}{\mu s^3}}{2D_2\eta_2}$$

Where D_2 is the radon diffusion coefficient of the blasted uranium ore heap ($m^2\ s^{-1}$). The radon diffusion coefficient can be calculated as follows [6]:

$$D = D_0\eta\exp(-6m\eta - 6m^{14\eta})(\frac{T}{273})^{0.75} \tag{7}$$

Where D_0 is the radon diffusion coefficient of air ($m^2\ s^{-1}$); T is absolute temperature (K); and m is water saturation in porous media.

The radon exhalation rate of the blasted uranium ore heap bottom surface is:

$$J_{sx} = \frac{\eta_2 D_2 K_2[b_1 e^{-b_2 L_1} + b_2 e^{b_1 L_1} - (b_1 + b_2)]}{K_1(e^{b_1 L_1} - e^{-b_2 L_1})} \tag{8}$$

During descentional ventilation, radon exhalation rate of the top surface of the blasted uranium ore heap is:

$$J_{xs} = \frac{\eta_2 D_2 K_2[b_1 e^{-b_2 L_1} + b_2 e^{b_1 L_1} - (b_1 + b_2)]}{K_1(e^{b_1 L_1} - e^{-b_2 L_1})} \tag{9}$$

The radon exhalation rate of the blasted uranium ore heap bottom surface is:

$$J_{xx} = \frac{\eta_2 D_2 K_2[b_1 e^{b_1 L_1} + b_2 e^{-b_2 L_1} - (b_1 + b_2)e^{(b_1 - b_2)L_1}]}{K_1(e^{b_1 L_1} - e^{-b_2 L_1})} \tag{10}$$

Under the effect of ventilation, radon diffusion is negligible if air seepage plays a dominant role in radon migration through the blasted uranium ore heap. The radon exhalation rate of the heap under the effect of ventilation can then be described as follows:

$$J_{\text{dui}} = \frac{\alpha_f kpQ^2 K_2}{K_1 \mu s^3}\left[1 - \exp\left(-\frac{K_1 \eta_2 \mu s^3}{\alpha_f kpQ^2} L_1\right)\right] \tag{11}$$

2.3. Radon concentration calculation model in confined shrinkage mining stope workspace

When calculating radon concentration in confined workspaces, as shown in Figures 2 and 3, the air inlet of the confined workspace sits at starting point 0 and the radon concentration at the point L meters from the air inlet is calculated as follows:

$$C(L) = C_0 e^{-\lambda\frac{LHW}{Q}} + \frac{\int_0^L (J_{dui}W + 2J_wH + J_dW)\, e^{-\lambda\frac{(L-x)HW}{Q}} dx}{Q}$$
$$= C_0 e^{-\lambda\frac{LHW}{Q}} + \frac{(J_{dui}W + 2J_wH + J_dW)}{\lambda HW}(1 - e^{-\lambda\frac{LHW}{Q}}) \quad (12)$$

Where C_0 is the air inlet radon concentration (Bq m^{-3}); J_w *and* J_d is the radon exhalation rate of surrounding rock and the top ore body(Bq m^{-2} s^{-1}); W is the width of the workspace (m); H is the height of workspace (m); λ is the radon decay constant (s^{-1}); Q is air quantity (m^3 s^{-1}).

The radon decay constant is 2.1×10^{-6} (s^{-1}), the length of the workspace does not exceed 100 m, and air velocity in the workspace is above 0.25 (m s^{-1}), so Equation (12) can thus be simplified as follows:

$$C = C_0 + \frac{(J_{dui}W + 2J_wH + J_dW)L}{Q} \le C_{\lim} \quad (13)$$

Again, the length of the confined workspace is L_2. If in ascentional ventilation -J_{dui}=J_{ss}. The exhaust radon concentration increment in the confined workspace-caused by the blasted uranium ore heap can be calculated as follows:

$$\Delta C_{ss} = \frac{J_{ss}WL_2}{Q} \quad (14)$$

Under descentional ventilation, J_{dui}=J_{xs}, so the exhaust radon concentration increment of workspace caused by the blasted uranium ore heap is calculated as follows:

$$\Delta C_{xs} = \frac{J_{xs}WL_2}{Q} \quad (15)$$

3. Results and Discussion

3.1. Determining parameters

Here, we assume that the uranium ore grade is 0.5%, the value of α_0 is 9 (Bq m^{-3} s^{-1}), ore particle size is 0-0.3 m, particle size distribution fractal dimension is 2.0, ore porosity is 0.035, ore heap porosity is 0.333, ore humidity is 0, air temperature is 20 (℃), α_f is 0.05 (Pa s m^{-2}) , workspace height is 2 (m). The top and bottom surfaces of the blasted uranium ore heap have the same area S_S is 200 (m^2).

3.2. Influence of ventilation airflow directions on radon concentration increment in confined workspaces

For simplicity, we assume that the height of blasted ore heap is 20 (m) regardless of changes in other parameters.

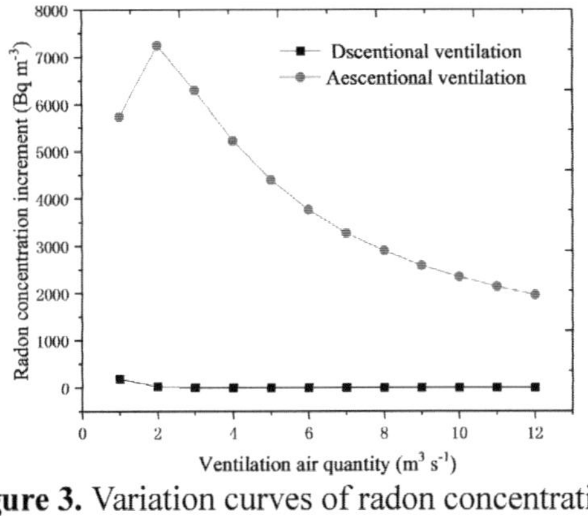

Figure 3. Variation curves of radon concentration increment in confined workspaces of shrinkage mining stope with different ventilation airflow

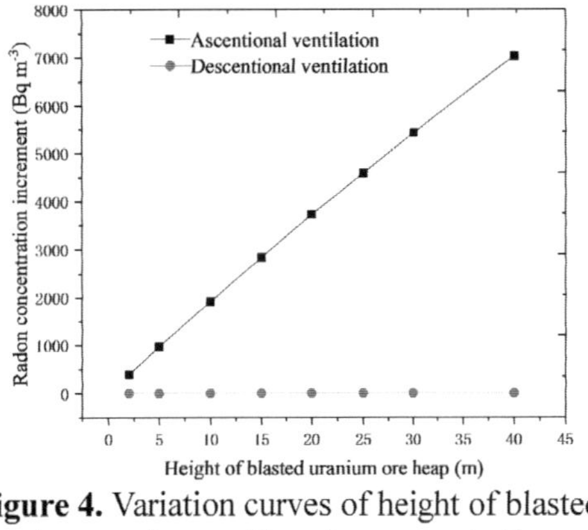

Figure 4. Variation curves of height of blasted uranium ore heap with radon concentration increment caused by ore heap exhalation

Figure 3 shows that in ascentional ventilation, as ventilation air quantity increases from 1 to 2 (m^3 s^{-1}), ventilation increases the velocity of radon exhalation to the point where the amount of radon emitted is less than the amount of radon exhaled. When the air quantity is over 2 (m^3 s^{-1}), however, the amount of radon emitted is greater than the amount of radon exhaled. With descentional ventilation, when ventilation air quantity is from 1 to 2 (m^3 s^{-1}), radon concentration increment in the workspace gradually decreases. When ventilation air quantity is from 2 to 12 (m^3 s^{-1}), radon concentration increment is close to 0. When permeability is invariable, the variations in radon concentration with ascentional ventilation are greater than with descentional ventilation.

3.3. Height of blasted uranium ore heap affects radon concentration in confined workspaces

Figure 4 shows that when air quantity is 6 (m^3 s^{-1}) and permeability is 1×10^{-8} (m^2), the radon concentration increment with ascentional ventilation increases as ore heap height increases. With descentional ventilation, radon concentration increment stays close to 0 regardless of heap height.

3.4. Blasted uranium ore heap area affects radon concentration increment in the confined workspaces

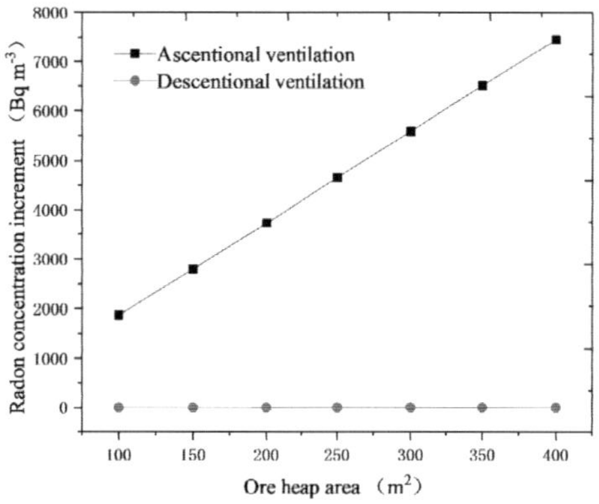

Figure 5. Variation curves of ore heap area with radon concentration increment caused by ore heap exhalation

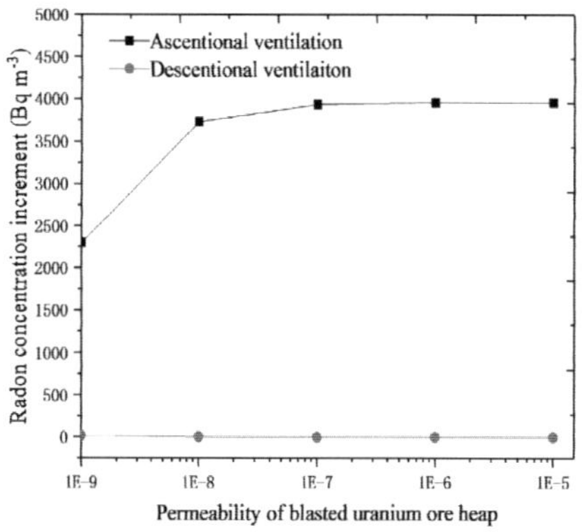

Figure 6. Variation curves of radon concentration increment in confined workspace of shrinkage stope with different permeability

Figure 5 shows that when air quantity is 6 (m^3 s^{-1}) and permeability is 1×10^{-8} (m^2), the radon concentration changes as the heap area increases. As area increases, however, the variation tendency of radon concentration increment in the workspace gradually decreases. Although the area of blasted uranium ore heap has nothing to do with the exhalation rate, it does alter the radon exhalation area.

3.5. Influence of permeability on radon concentration increment in confined workspaces

Figure 6 shows that radon concentration in the workspace under ascentional ventilation gradually increases until leveling off as heap permeability increases. With descentional ventilation, radon concentration decreases as permeability increases until it nears 0. Under ascentional ventilation, the airflow forces radon exhalation from the ore heap into the workspace. The higher the heap permeability, the more radon is exhaled. The opposite effects occur with descentional ventilation.

3.6. Frictional resistance coefficient affects radon concentration increment in confined workspace

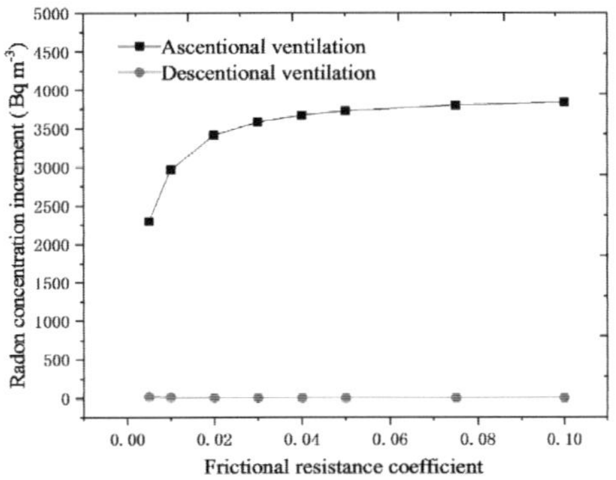

Figure 7. Variation curves of different frictional resistance coefficient with radon concentration increment caused by ore heap exhalation

Figure 7 shows that when air quantity is 6 (m^3 s^{-1}) and permeability is 1×10^{-8} (m^2), radon concentration increment continuously increases under ascentional ventilation as frictional resistance coefficient increases; the growth rate gradually decreases over the curve. Under descentional ventilation, the radon concentration is less influenced by frictional resistance coefficient.

4. Conclusions

A calculation model for radon concentration as-affected by source and ventilation conditions was established in this study for application in shrinkage mining stopes. The radon exhalation rate with varying airflow parameters was determined accordingly. The model accounts for the source of radon in the underground uranium mine, and allows the engineer to predict changes in radon concentration based on an array of influence factors.

The height, area, and permeability of the heap are positively correlated with radon concentration increment under ascentional ventilation in the workspace; ventilation air quantity has the opposite effect. Radon concentration increases to greater extent under ascentional ventilation than descentional ventilation. These observations may be used as a reference for ventilation design to control radon in shrinkage mining stopes.

5. Acknowledgements

This work was financially supported by Chinese National Natural Science Fund Projects (No.11105069) and (No.11575080), the Doctoral Scientific Research Foundation of University of South China (2015XQD38).

References

[1] Qureshi A A, Kakar D M, Akram M, et al. Radon concentrations in coal mines of Baluchistan, Pakistan. *Journal of Environmental Radioactivity*, 2000, 48(2):203-209.

[2] Yener G, Küçüktaş E. Concentrations of radon and decay products in various underground mines in western Turkey and total effective dose equivalents. *Analyst*, 1998, 123(1):31-4.

[3] El-Fawal M M. Mathematical modeling for radon prediction and ventilation air cleaning system requirements in underground mines. *Journal of American Science*, 2011; 7(2): 389–402.

[4] Sahu P. Evaluation of effect of ventilation on radon concentration and occupational exposure to radon daughters in low ore grade underground uranium mine. *Journal of Radioanalytical & Nuclear Chemistry*, 2015, 303:1933-1941.

[5] Tyler S W, Wheatcraft S W. Fractal Scaling of Soil Particle-Size Distributions: Analysis and Limitations. *Soil Science Society of America Journal*, 1992, 56(2):362-369.

[6] Rogers V C, Nielson K K. Correlations for predicting air permeabilities and 222Rn diffusion coefficients of soils. *Health Physics*, 1991; 61(2): 225-230.

To Investigate the Influence of Building Envelope and Natural Ventilation on Thermal Heat Balance in Office Buildings in Warm and Humid Climate

Pradeep G. Kini[1], Naresh Kumar Garg[2], and Kiran Kamath[3]
[1] Faculty of Architecture, Manipal University, Manipal, Karnataka, India
[2] School of Architecture and Design, Manipal University, Jaipur, Rajasthan, India
[3] Dept of Civil Engineering, Manipal University, Manipal, Karnataka, India

Email: kininet1@gmail.com; nkgcbri@rediffmail.com; kiran.kamath@manipal.edu;

Abstract. India's commercial building sector is witnessing robust growth. India continues to be a key growth market among global corporates and this is reflective in the steady growth in demand for prime office space. A recent trend that has been noted is the increase in demand for office spaces not just in major cities but also in smaller tier II and Tier III cities. Growth in the commercial building sector projects a rising trend of energy intensive mechanical systems in office buildings in India. The air conditioning market in India is growing at 25% annually. This is due to the ever increasing demand to maintain thermal comfort in tropical regions. Air conditioning is one of the most energy intensive technologies which are used in buildings. As a result India is witnessing significant spike in energy demand and further widening the demand supply gap. Challenge in India is to identify passive measures in building envelope design in office buildings to reduce the cooling loads and conserve energy. This paper investigates the overall heat gain through building envelope components and natural ventilation in warm and humid climate region through experimental and simulation methods towards improved thermal environmental performance.

1. Introduction
The building sector during its construction and operations consume more than 40% of the total energy end use worldwide [1]. The commercial building sector is expected to increase fivefold between now and 2030 in India. It is estimated that about 70% of the commercial floor space that will exist in India in 2030 is yet to be constructed [2]. It is important for India to promote research in energy conservation during the early stages of growth in the building sector [3]. The building envelope contributes to a large extent in heat transfer in to a building and makes the indoor uncomfortable in the naturally ventilated buildings and adds to energy load in the conditioned buildings. The thermal performance of building depends on the design variables such as the dimensions of building elements and their orientation, the thermo physical properties of the building materials and also the climatic data of the reference site location. Out of the total solar radiation incident on the external layer of the roof, a part of it is absorbed by the roof and some of it is reflected back to the environment. Heat is also lost to the environment through convection and radiation from the roof's external surface. The remaining heat is conducted into the roof raising the roof surface temperature. The inner surface of the roof transfers heat by convection and radiation to the room air, raising its temperature. Heat exchanges like

these take place through opaque building elements such as walls and roofs. Such heat transfer processes affect the indoor comfort temperature experienced by its occupants.
Providing a comfortable environment for occupants in office buildings should be a high priority without which the building will have decreased worker productivity. It is important to understand the contribution of building envelope components on thermal heat gain in buildings so that appropriate building envelope strategies can be undertaken to meet the adaptive comfort standards. Air-conditioning systems consumes the most energy of all building services, which comprises 55% of the total energy consumption in commercial buildings and has the largest potential in energy savings [2].This paper investigates the contribution of thermal heat gain of various building envelope components and natural ventilation for an office space during a peak summer day in warm and humid climate towards prioritizing the passive strategies to be adopted. The paper also investigates the effectiveness of ventilation and its effect on cooling the thermal mass during the day. The research objectives are to use simulation methods to analyze the contribution of building envelope design and natural ventilation towards thermal heat balance in office buildings in warm and humid climate.

2. Methodology

A study was carried out in 2012-2013 on office buildings in Coastal Karnataka to understand the construction practices adopted and energy consumption scenarios. The study revealed that the majority of the buildings surveyed the building roof consisted of RCC slab and concrete block was used for masonry walls. The mean radiant temperatures remains very high during the day because of heat gains through the building envelope and to maintain thermal comfort the air conditioning systems were extensively used by the occupants. To understand heat gain through the building envelope, an office building in coastal Karnataka, India is taken for detailed analysis and demonstration. The building is six floors high and consists of various naturally ventilated and air conditioned office spaces. The study area falls under identified warm and humid region and the location of the case study is a broad representative sample of this region. The building has a narrow floor plate of 11m and the effects of natural cross ventilation can be better understood in a narrow layout. The longer axis of the building is exposed to the east and west incident solar radiation. This layout can help to determine the thermal heat gain during the heat gain conditions on a peak summer day. The reference baseline model was modelled as per the as built documentation of the reference building. Simulation software design builder has been used for the study as it has the Energyplus simulation engine developed by department of energy, USA [4], [5]. Energyplus has been validated under the comparative standard method of test for the evaluation of building energy analysis computer program BESTEST and ASHRAE Std 140.A thermal comfort survey was conducted in the reference building. As per the survey, the occupants of the fifth floor office space were found to have the maximum thermal discomfort and that office space was chosen as the baseline case to analyse thermal heat gains. Climatic analysis of warm and humid climate along coastal Karnataka revealed that March, April, and May to be warmest months. May being the most critical month in terms of percentage of office hours outside the thermal comfort zone, detailed simulations were carried on the peak summer day of that month. The results of field studies and simulation results have been analysed to study the impacts of building envelope and natural ventilation on thermal heat gain.

3. Climate Study

Coastal Karnataka has warm and humid climate with intense radiation on clear days. In summer, temperatures can reach as high as 30-35°C during the day and 25-30°C at night. In winter, the maximum temperature is between 25°C to 30°C during the day and 20°C to 25°C at night. The relative humidity is generally very high, about 70-90% throughout the year. Precipitation is also high, being about 1200 mm per year [6]. The wind is generally along southwest to northeast direction with speeds ranging from extremely low to very high. The typical Meteorological weather data for the reference site location is shown in Fig. 1.

Month	Jan	Feb	Mar	Apr	May	Jun	Jul	Aug	Sep	Oct	Nov	Dec
Outside Dry-Bulb Temperature (°C)	26.24	26.86	27.93	29.14	29.31	26.12	25.44	25.26	26.08	26.59	26.87	26.57
Outside Dew-Point Temperature (°C)	18.62	20.01	21.43	23.06	22.36	23.71	23.76	23.72	23.69	23.21	21.80	18.49
Wind Speed (m/s)	2.64	2.52	2.54	2.64	3.37	3.23	3.21	2.42	2.80	2.20	2.07	2.93
Wind Direction (°)	133.25	132.15	155.93	161.20	162.63	144.00	185.20	145.43	161.71	116.02	106.68	130.06
Solar Altitude (°)	-6.83	-4.82	-0.80	3.70	6.36	7.29	6.92	5.08	1.23	-3.32	-6.24	-7.28
Solar Azimuth (°)	179.22	176.41	174.27	176.07	179.99	173.54	172.63	173.63	176.95	179.28	179.99	179.96
Atmospheric Pressure (Pa)x10^3	100.69	100.89	100.49	100.42	100.54	100.16	100.16	100.54	100.16	100.62	100.87	100.91
Direct Normal Solar (kWh)	133.07	154.86	140.31	118.01	133.04	54.71	52.73	65.73	84.09	116.32	128.13	185.63
Diffuse Horizontal Solar (kWh)	69.71	68.33	82.94	85.91	89.43	86.08	89.21	90.41	87.15	85.63	73.96	70.48

Figure 1. Climate Data of the reference site location

4. Reference Building Data

The reference building is located in Manipal in Coastal Karnataka, India as indicated in Fig.2. The details of the reference building details and also the building envelope details is listed in Table 1 and Table 2. The Building floor to floor height is 3.2 m height The window wall ratio of the simulated fifth floor office space is 40% on east side, 22.5% on south, 30% on the west side. The north side of the office space is blocked because of another office space. The window sill height is 0.9m from the floor.

Figure 2. View of the reference building showing the longer east facade and south facade.

Table 1. Reference building details & operation schedule

Building Information	Details
Location	13°20'58.5"N 74°47'14.8"E
Building Floor Plate	40m long, 11m wide
Building Orientation	East
Office Schedule	8 AM to 6 PM (6 days a week)
Occupancy	1 person per 9 m^2
Lighting Power Density	9.5 W/m^2

Table 2. Building Envelope Details

Building Component	Details
Wall	200mm thick masonry wall + 15mm cement plaster on the internal and external surfaces
Roof	150mm thick reinforced concrete slab
Floor	150mm thick reinforced concrete slab

Surface	Cream color interior paint, light brown color exterior paint
Window type	Clear Glass with 4 mm thickness
Window Frames	100 mm steel

5. Simulation and Data Analysis

The roof of the buildings is constructed using 150 mm Reinforced cement concrete roof with 12mm plaster on both sides. The calculated u-value from the thermo physical properties of the roof is listed in table 3. The Masonry wall is 200mm concrete block with 15mm cement plaster on both sides. These data as per the as built documentation has been used as a input condition for the energyplus model.

Table 3. Calculated thermal Transmittance Values for Roofs

Description of the roof	Conductivity W/mK	Density K/m^3	Specific heat J/Kg^{o}K	U Value W/m^{2}K
Baseline Case 1 : 150 mm RCC slab with 12mm plaster on both sides				2.864
1. 12 mm Cement Plaster	0.721	1762	0.84	
2. 150mm RC slab	1.580	2288	0.88	
3. 12 mm Cement Plaster	0.721	1762	0.84	

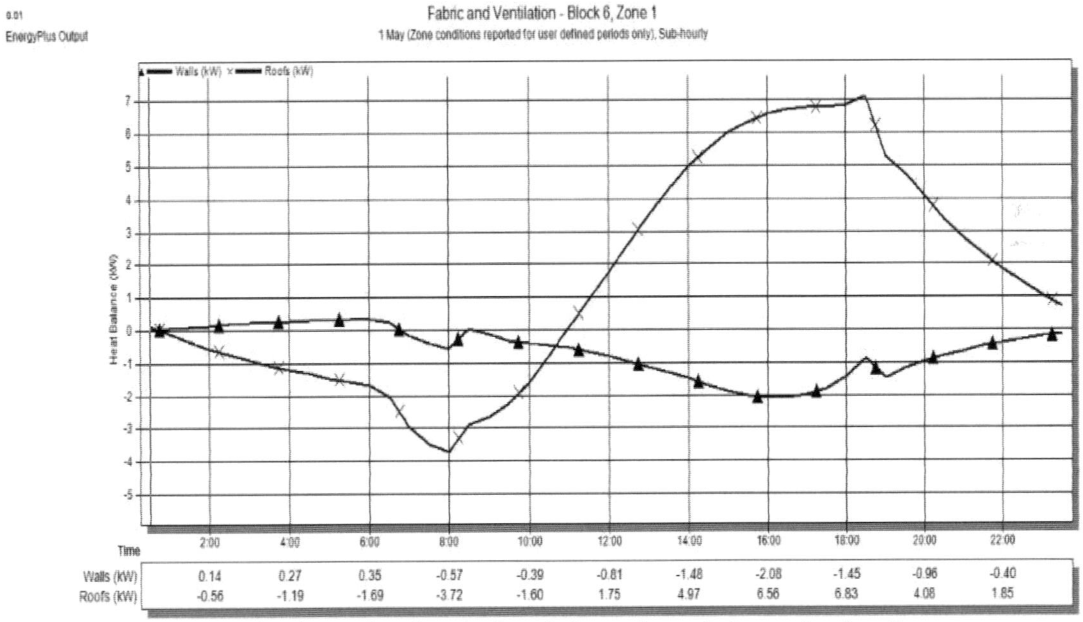

Figure 3. Figure showing the heat balance through the roof and walls

The heat balance through the roofs and walls is shown in Fig. 3. The surface temperature of the inside and the outside surface of the roof is shown in Fig. 4. Solar heat gains through the windows and heat losses due to natural ventilation are shown in Fig. 5.

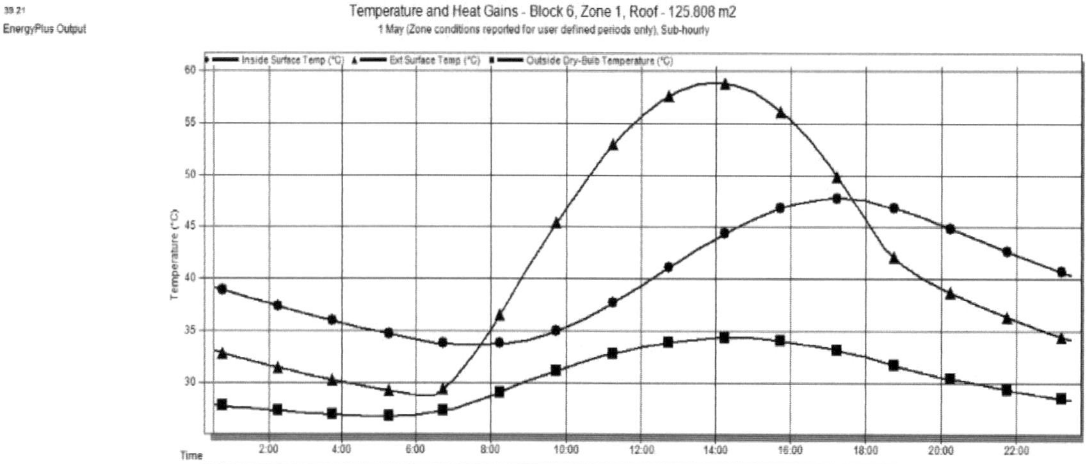

Time	2:00	4:00	6:00	8:00	10:00	12:00	14:00	16:00	18:00	20:00	22:00
Inside Surface Temp (°C)	37.61	35.73	34.13	33.70	35.37	39.38	43.86	47.11	47.57	45.22	42.33
Ext Surface Temp (°C)	31.63	30.05	28.81	35.10	46.81	55.71	58.90	55.32	45.64	39.14	36.01
Outside Dry-Bulb Temperature (°C)	27.32	26.85	26.85	28.70	31.47	33.40	34.40	33.94	32.55	30.55	29.16

Figure 4. Figure surface temperature of roofs

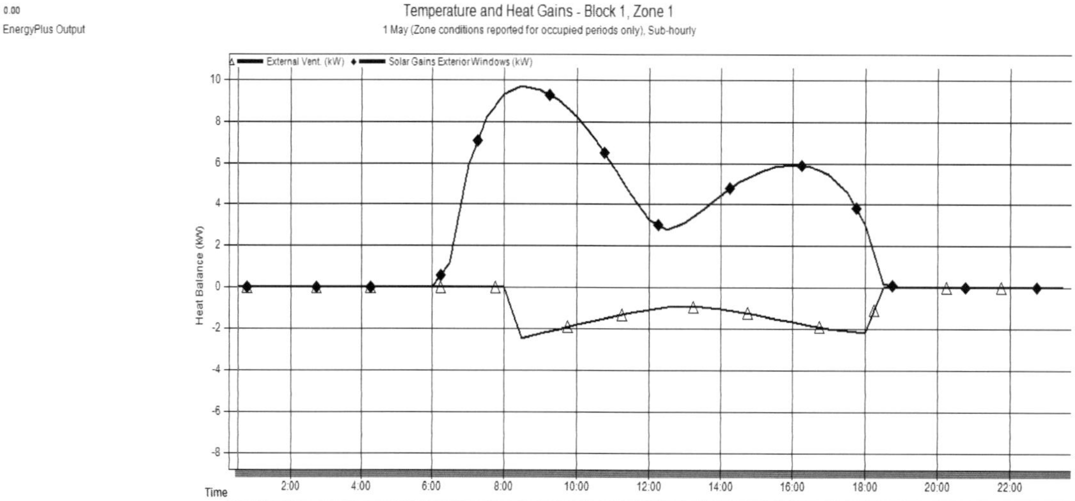

Time	2:00	4:00	6:00	8:00	10:00	12:00	14:00	16:00	18:00	20:00	22:00
External Vent. (kW)	0.00	0.00	0.00	0.00	-1.82	-1.08	-1.04	-1.67	-2.17	0.00	0.00
Solar Gains Exterior Windows (kW)	0.00	0.00	0.00	9.33	8.22	3.26	4.45	5.95	3.03	0.00	0.00

Figure 5. Solar gains through the windows and heat losses through natural ventilation

6. Discussions on simulation results

The simulation results show that the heat gain through the roof increases the most between 2pm and 6 pm during office hours due to the high U value of 2.864 W/m^{2}K of the widely used roof construction practices in the region. For the reference office space the heat balance through the roof is 4.97 KW at 2pm and 6.83 KW at 6 pm during the peak summer day. Heat gain through the roof elevates ceiling surface temperature and causes radiant heat load on the occupants. The solar gains through the exterior windows are the most during the morning hours because of exposed east openings. The solar gains through the window openings is 9.33 kW at 8 AM , reduces to 3.26 kW during noon and increases again to 5.95 kW at 4 PM due to west openings. The day time ventilation has a beneficial cooling effect during the office hours. The walls do not contribute significantly to heat gain during the office hours because of the time lag and the thermal mass providing a heat sink during the working day. In the conventional roof the temperature amplitude ratio of the outside surface temperature and the

IOP Conf. Series: Earth and Environmental Science **73** (2017) 012031 doi:10.1088/1755-1315/73/1/012031

inside surface temperature is low contributing to internal heat gains. The roof is the most exposed to impacts of solar radiation, as it receives sunlight for practically the whole of the day and in the tropics the angle of incidence is close to the normal in the hotter parts of the day. The results can be summarized as follows:

- The roof contributes the most to heat gain through the building envelope due to exposed solar radiation. It is very essential to construct roof with good insulation and layers of different thermo physical properties.
- The opaque parts of the building envelope should be constructed with an improved temperature amplitude ratio which specifies the time it takes for a temperature rise on the outside of the building component to pass on the inside.
- A building envelope optimized with low U-value and openings to allow natural ventilation provides good protection against overheating in summer.

7. Conclusions

India's commercial building sector is expected to grow fivefold between now and 2030. To reduce the dependence on fossil fuels it is essential to reduce the use of energy intensive systems in the building sector. Analysis of the building thermal performance is investigated in this paper with simulation approach to understand the source of the most significant heat gain in warm and humid climatic region which contributes to energy loads in buildings. This can help in developing specific strategies in the design of building envelope components to improve heat balance in buildings, conserve energy and contribute towards mitigation of climate change. Further research can be conducted on the various types roof and wall construction assemblies to optimize thermal heat balance in warm and humid climatic regions.

Acknowledgements

We thank Manipal University for all the support towards this research. We also thank Prof. Manohar Pai, Prof. Satish Shenoy, Prof. Nandineni Rama Devi, Prof. Udayashankar H.N and Prof. Pranav Kishore for providing insights that greatly assisted the research.

References

[1] International Energy Agency. 2015. *India Energy Outlook*. Paris (France) OECD/IEA. p 187

[2] The Energy and Resources Institute. 2015. *Green growth and buildings sector in India*. New Delhi

[3] Kini P. G., 2013. *International Journal of Earth Sciences Engineering*. Exploring strategies for sustainable site planning of built environment to mitigate the effects of rapid urbanization in warm and humid climate. Vol 06(4) p 697-701

[4] Design Builder Software Ltd. Available from http://www.designbuilder.co.uk/

[5] United States Department of Energy, 2016. *EnergyPlus™ Version 8.5 Documentation*, Input Output Reference.

[6] Nayak J. K. and, Prajapati J. A. 2006. *Handbook on energy conscious buildings*.Ministry of Non-conventional Energy Sources, Government of India.

IOP Conf. Series: Earth and Environmental Science **73** (2017) 012032 doi:10.1088/1755-1315/73/1/012032

Numerical Modelling of a Fast Pyrolysis Process in a Bubbling Fluidized Bed Reactor

S Jalalifar[1], M Ghiji[1], R Abbassi[1], V Garaniya[1], and K Hawboldt[2]
[1] Australian Maritime College, National Centre of Maritime Engineering and Hydrodynamics, University of Tasmania, Launceston, Tasmania, Australia.
[2] Faculty of Engineering and applied Science, Memorial University, St. John's, NL, Canada.

E-mails: salman.jalalifar@utas.edu.au;

Abstract. In this study, the Eulerian-Granular approach is applied to simulate a fast pyrolysis bubbling fluidized bed reactor. Fast pyrolysis converts biomass to bio-products through thermochemical conversion in absence of oxygen. The aim of this study is to employ a numerical framework for simulation of the fast pyrolysis process and extend this to more complex reactor geometries. The framework first needs to be validated and this was accomplished by modelling a lab-scale pyrolysis fluidized bed reactor in 2-D and comparing with published data. A multi-phase CFD model has been employed to obtain clearer insights into the physical phenomena associated with flow dynamics and heat transfer, and by extension the impact on reaction rates. Biomass thermally decomposes to solid, condensable and non-condensable and therefore a multi-fluid model is used. A simplified reaction model is sued where the many components are grouped into a solid reacting phase, condensable/non-condensable phase, and non-reacting solid phase (the heat carrier). The biomass decomposition is simplified to four reaction mechanisms based on the thermal decomposition of cellulose. A time-splitting method is used for coupling of multi-fluid model and reaction rates. A good agreement is witnessed in the products yield between the CFD simulation and the experiment.

1. Introduction

Concern about limited supply of fossil fuels and environmental risks such as global warming and climate change is a strong motivation for development of environmentally friendly technologies and renewable energy sources. Biomass is abundant and therefore an alternative energy source. All organic materials such as plants (byproduct and residue of agricultural products) and animal resources (available in land and water environments) can be classified as biomass. Agricultural waste such as corn stover, bagasse and residues associated with harvesting crops contain high-energy organics in the form of cellulose, hemicellulose, and lignin. Compared to conventional fossil fuels, energy derived from biomass has lower emissions of sulfur dioxides (SO_2) and particulate matter (PM). In addition, energy from biomass is considered carbon neutral because the released carbon dioxide (CO_2) into the environment is recycled by the photosynthesis process. The quality of energy derived from biomass can be improved by conversion to a liquid or more homogenous solid. Thermochemical conversions are mainly: direct combustion of biomass to provide heat; gasification to produce gas as a synthesis gas, heat, and/or power; pyrolysis to generate bio-oil, char, and gas [1].

Pyrolysis offers the advantage of producing three bioproducts; a bio-oil for liquid fuel or source of high value chemicals; a solid biochar (e.g. sustainable source for adsorbent, soil amendment, or catalyst); and a biogas for energy recovery. Pyrolysis is loosely divided into three different categories; slow, fast, and flash pyrolysis. The primary product of slow pyrolysis is the char, and for fast and flash pyrolysis, the primary product is the liquids. Extracted bio-oil from biomass and can be consumed directly for the generation of heat and power in boilers, gas turbines, and diesel engines, or it can be refined to produce a higher quality fuel [2, 3].

Fast pyrolysis has been studied for a variety of biomass feedstock, types of reactors, and operating conditions. The reaction mechanisms are complex due to the heterogeneity of the biomass and the number of phases, however some researchers have studied global and intrinsic reaction rates [4, 5]. The drawback to detailed reaction mechanisms is that they are specific to a feedstock and operating conditions and difficult to determine completely due to the multiple reactions occurring in the various phases and the ability to accurately measure the products. Typically, series of global reactions are assumed where the biomass is converted through a series of primary and secondary reactions [6]. The reaction rates are typically derived in reactors where heat and mass transfer resistances are minimized. To properly model a pilot or commercial scale reactor these resistances must be included in the form of transport equations. Physical phenomena such as flow dynamics (multi-phase flow and mechanically moving parts), phase equilibrium, and heat and mass transfer, can be modelled and predicted by means of Computational Fluid Dynamics (CFD) modelling techniques. The reaction rates can be included in the CFD model.

The widespread applications of fluidized bed reactors have motivated many researchers to use CFD techniques to obtain a better understanding of the reactor transport phenomena. Parameters involved in the simulation include; pyrolysis temperature, the size of biomass particles, biomass feed rate, residence time and carrier gas velocity. Temperature and residence time of the condensable and non-condensable gases determine largely the amount of tar (bio-oil) produced. Temperatures exceeding 550-600°C and longer vapour residence times (e.g. lower nitrogen flow rates) will result in secondary cracking reactions, which decrease yield of the tar and increase the water content [6]. One method to vary the residence times in fluidized beds is to vary the location of the biomass feed relative to the reactor height [7]. The biomass feed rate is a factor in solid particle residence time, there is a balance between minimizing secondary reactions and ensuring the biomass particles are well mixed and reach thermal equilibrium [8]. The carrier gas (typically nitrogen) velocity must be higher than the minimum fluidization velocity to maintain fluidization conditions and ensure the biomass and sand are well mixed to maximize heat transfer. However, if nitrogen flow rates become too high sand may become entrained in the exit gases/char. The biomass particle size determines the heating rate of the particle and ideally, fine particles are used to minimize intraparticle heat and mass transfer resistances. However, there is a balance between minimum particle size and costs to grind to this size.

2. Methodology

2.1. Chemical kinetics of a single particle

In this study, a combination of the Bradbury et al. [9] model of cellulose to "active material" which then reacts to tar and char is employed. The tar then further reacts to form gas. As illustrated in figure 1, three different phases including biomass phase, sand phase and gas phase are taken into account. Each phase involved in the process has a number of species. The biomass phase includes unreacted biomass and char, while the gas phase include condensable (tar), non-condensable and the nitrogen. The sand phase and nitrogen are inert and do not participate in the chemical reactions. The reaction rate constants are calculated as below:

$$k_i = A_i \exp[-Ea_i / (RT)] \quad (1)$$

Where k_i is the rate constant for reaction "i", and A_i and Ea_i are the associated Arrhenius constant and activation energies. "T" is the temperature in kelvin and "R" is the gas constant. As indicated previously, overall seven different groups are included in this reaction scheme the solid reaction phase (virgin cellulose, active cellulose, and biochar), the condensable/non-condensable phase, and the non-reacting sand phase. The nitrogen is included in the gas phase as it contributes to partial pressures but does not react. The values of the kinetic parameters for the reaction scheme and the obtained values for heats of reaction of cellulose are outlined in Table 1. Y is the formation ratio for the char component, which is 0.35 for pyrolysis of pure cellulose [10].

2.2. Mathematical methods/models

The fundamental governing equations are the conservation of mass, momentum, energy, and species. Thus, a combination of Multi Fluid Model (MFM) and a chemical solver is the most appropriate method to simulate biomass fast pyrolysis. In multi-fluid model, all phases are treated as continua. For the solution of the governing equations, the Kinetic Theory of Granular Flow (KTFG) [11] is applied to for the solid phases. Further details about the description of the models can be found in the literature [7]. Numerical models and settings.

The multiphase fluid flow, heat transfer, chemical reactions, and species mass transfer in the lab-scale fluidized bed reactor is simulated using the solver FLUENT16.0. The dimensions of the reactor are outlined in figure 2. The laminar multiphase Eulerian-Granular flow is used and the energy equations and species transport are activated. The first fractional step considers merely the spatial solution of the multiphase species with the reaction terms set to zero. In the next fractional step, the reaction terms are integrated in each cell using a stiff ODE solver. Time discretization was done using a second-order implicit method. The pressure-based solver used a least-square cell based phase coupled SIMPLE-algorithm for the pressure velocity coupling and second order (upwind) discretization was applied for the momentum, energy and species modelling parameters. QUICK algorithm is applied for volume fraction calculations. Finally, hybrid initialization is applied and each phases; especially the sand phase; are patched based on the initial packing limit.

2.2.1. Computational domain, initial and boundary conditions

The 2-D computational domain is shown in figure 2. Biomass with diameter of 0.4 mm at an inlet temperature of 300 K is fed at a rate of 100 g/h. Nitrogen flows from the bottom of the bed at a velocity of 0.36 m/s and temperature 773 K. The sand with diameter of 0.52 mm is initially packed to the height of 5.5 cm with the volume fraction of 0.59. The outflow boundary condition is used at the outlet. No-slip wall condition is applied for the solid walls. To simulate external heating, the wall temperature is kept constant at of 773 K. The bed temperature is initially set to 773 K. The biomass feedstock of pure cellulose. A study conducted for the grid independence shows that a 2-D grid with 2055 quadrilateral cells is capable of predicting all parameters to an acceptable level of accuracy. To avoid numerical instability small time step size of 1×10^{-4} s is used at the commencement of the simulation. However the grid resolution adopted here allow us to opt for the maximum time step size of 5×10^{-4} s without any numerical instability. In the solution of species transport equations, mass fraction of each species must sum to unity. Therefore, the N_{th} mass fraction is determined as one minus the sum of the N-1 solved mass fractions. To minimize numerical error, the N_{th} species should be selected as that species with the overall largest mass fraction (nitrogen in the gas phase and cellulose in the biomass phase). The thermo-physical properties of species involved in the reactions of the biomass are given in Table 2. It is worth noting that incompressible ideal gas model calculates the density of gaseous species and the viscosity of the solid species are calculated based on granular models.

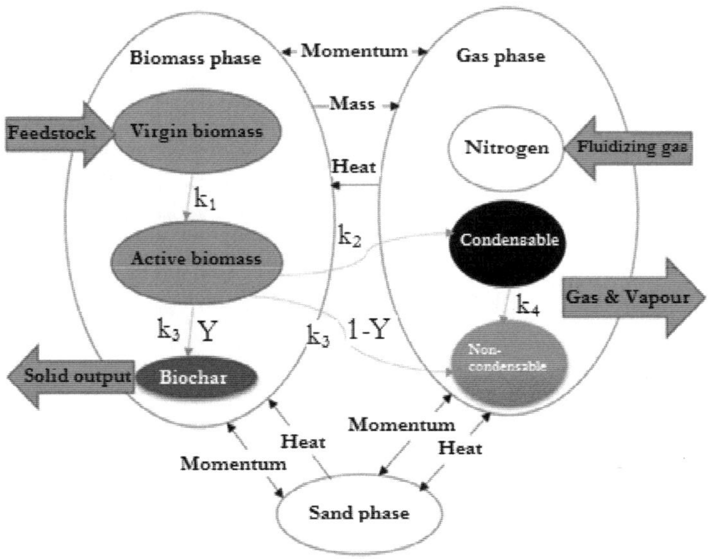

Figure 1. Chemical reactions and exchange of mass, momentum, and heat between phases.

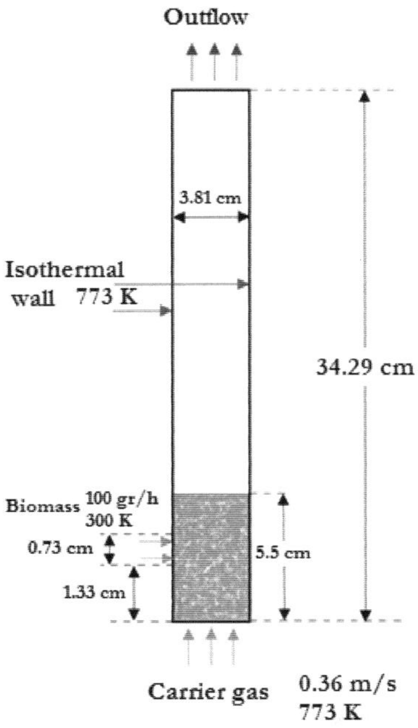

Figure 2. Schematic geometry of 2-D model of bubbling fluidized bed for simulations.

Table 1. Pre-exponential factor and activation energies for the biomass component [10].

Components	Reaction	A(s-1)	E(MJ/kmol)	Heat release, ΔH (kJ/kg)
	k_1	2.8×10^{19}	242.4	0
Cellulose	k_2	3.28×10^{14}	196.5	255
	k_3	1.3×10^{10}	150.5	-20
Tar	k_4	4.25×10^{6}	108.0	-42

Table 2. Thermo-physical properties of species [12]

Species	Density ρ (kg/m^3)	Particle diameter d_s (m)	Molecular weight (g/mol)	Heat capacity C_p (J/kg K)	Dynamic viscosity μ (kg/ms)	Thermal conductivity k (J/kg K)
Tar	-	-	100	2500	3×10^{-5}	2.577×10^{-2}
Gas	-	-	30	1100	3×10^{-5}	5.63×10^{-2}
N_2	-	-	28	1121	3.58×10^{-5}	2.577×10^{-2}
Cellulose	400	4×10^{-4}	12.14	2300	-	0.3
Biochar	2333	4×10^{-4}	12.01	1100	-	0.1
Sand	2649	5.2×10^{-4}	60.08	800	-	0.27

3. Results and discussions

A grid dependence study is carried out using 2-D mesh with four different mesh resolutions. The number of meshes for cases 1-4 are 225, 910, 2055, and 3640, respectively The temperature distribution from the centreline of the reactor is shown in figure 3. It can be seen that there is no substantial difference between cases 3 and 4. Thus, case 3 with 2055 meshes is selected in this study to save computational costs. The chemical reactions are not activated at this stage since this study focuses firstly on the capability of CFD model to simulate multiphase flow dynamics in the fluidized bed reactor. Figure 4 outlines the volume fraction of biomass inside the reactor to 160 s. These results are mapped and initialized for a more advanced simulation with the chemical reactions.

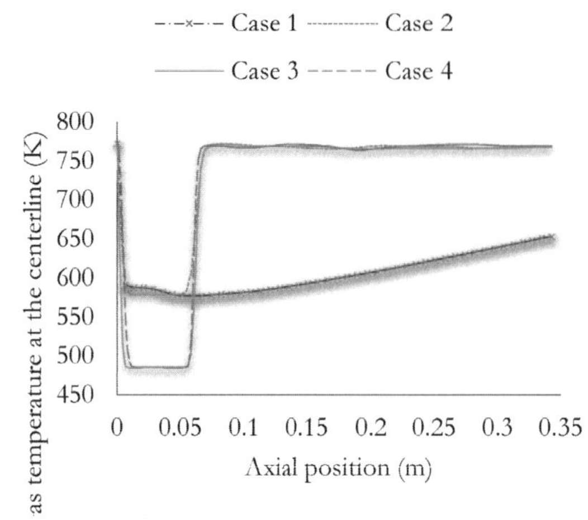

Figure 3. Axial distribution of gas temperature at statistically steady state condition using different grid sizes (no chemical reactions).

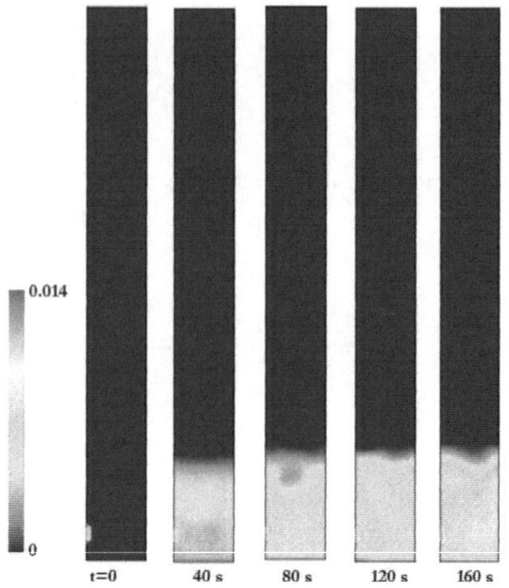

Figure 4. Snapshots of biomass volume fraction (no chemical reactions)

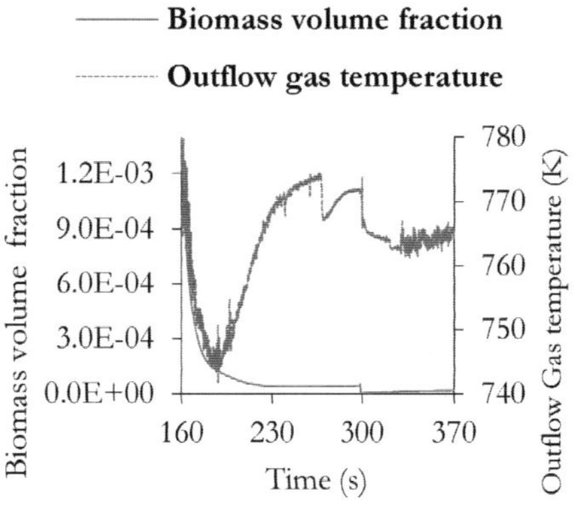

Figure 5. Temporal evolution of area-weighted average of biomass volume fraction and gas temperature at the reactor outlet

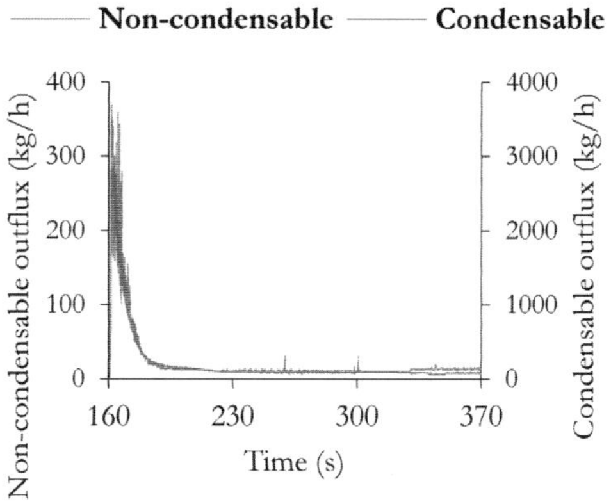

Figure 6. Temporal evolution of non-condensable and condensable outflux

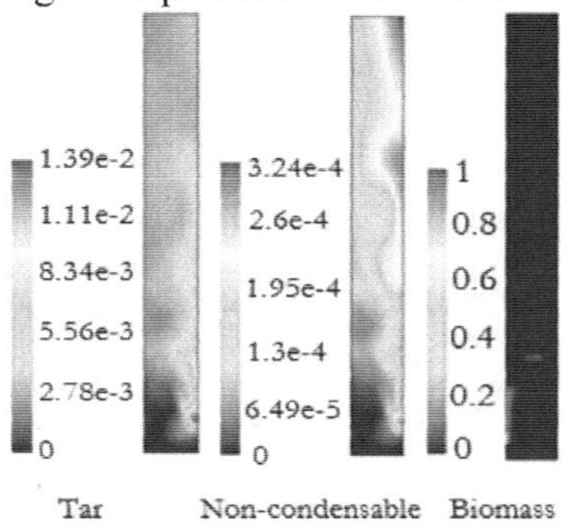

Figure 7. Mass fraction distribution at steady state condition

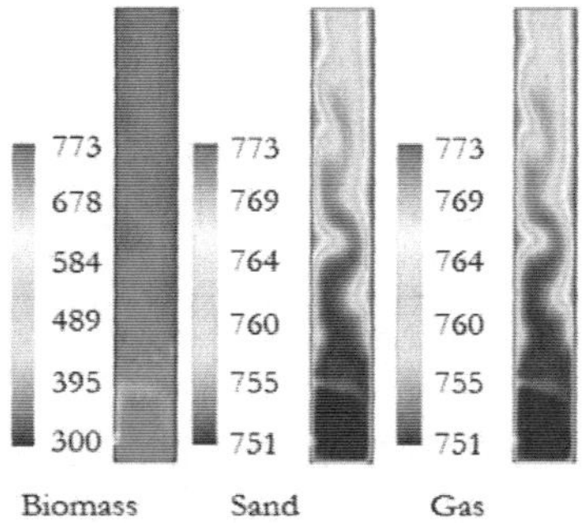

Figure 8. Contour plots of temperature for biomass, gas, and sand at statistically steady state.

In order to obtain a statistically steady state condition, area-averaged biomass volume fraction, outflow gas temperature, tar and non-condensable flows are monitored. Figures 5 and 6 show the temporal evolution of biomass volume fraction, the outflow gas temperature, non-condensable and tar outflux, respectively. It can be seen that all of the values reach statistically steady state condition after t=330 s. By monitoring the outflux of the products, the yield of tar, non-condensable, and bio-char in the last 40 seconds, are obtained. By integrating the flow of the product at the reactor outlet, the total mass flow rate of each species during the pyrolysis process is obtained. During the process, tar and non-condensable are added to the gas phase whereas biochar remains in the biomass phase. The mass fractions of products are shown in figure 7 at statistically steady state condition. Tar will be cracked into non-condensable as the temperature increases. Snapshots of temperature for the different phases are illustrated in figure 8. It is observed that except for lower temperatures at the biomass inlet due to cold virgin biomass, temperatures in other regions are in the range of 751-773 K.

Figure 9 illustrates the volume fraction of sand phase after reaching statistically steady state condition. The formation of bubbles in the mixing zone of the reactor is due to the flow of nitrogen as a fluidizing gas from the bottom of the reactor and entrained biomass from the left side of the reactor. These formations of bubbles enhance the gas-solid mixing and consequently facilitate the heat transfer from hot sand and gas and heated wall to the biomass particles. Hence, the required heat for the process is conveniently supplied.

The comparison of product yield against published experimental data [10] is shown in Table 3. As it can be seen, the predicted products yield match the experimental data very well.

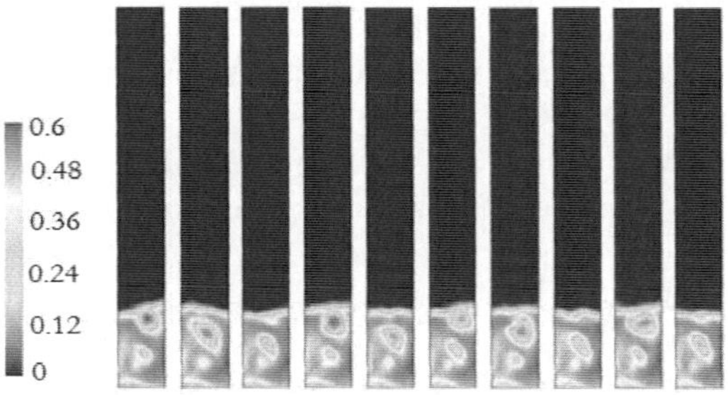

Figure 9. Snapshot of sand volume fraction at statistically steady state condition (t>340s). The first snapshot on the left is at t=340.1 s and the interval between the snapshots is 0.1 s.

Table 3. Comparison of product yield (wt %) between simulation and experiment

Components	Tar	Non-condensable	Biochar
Experiment [10]	82.1	12.4	2.2
Current study	77.8	13.8	2.4

4. Conclusion

In this study, a numerical framework for simulation of fast pyrolysis process is developed. In order to verify the capability of this numerical framework, a standard 2-D lab scale bubbling fluidized bed is selected due to its simplicity and applications. In this simulation, the multi-fluid model with a gas phase and two solid phases are studied. Multiple species are available in each phase. To implement the chemical reactions, the global reaction mechanisms are considered. A list of mathematical formulations for the governing equations is tabulated. The simulated results show an acceptable level of agreement with the published experimental data. It is concluded that the provided numerical framework can predict different properties such as volume fractions, temperatures, velocity, and above all, the product yield of the process accurately. This numerical framework can be applied to more complex geometries for further applications.

References

[1] Panwar N, Kothari R, Tyagi V 2012 Thermo chemical conversion of biomass–Eco friendly energy routes, *Renewable and Sustainable Energy Reviews*, **16** 1801-1816.

[2] Goyal H, Seal D, Saxena R 2008 Bio-fuels from thermochemical conversion of renewable resources: a review, *Renewable and sustainable energy reviews*, **12** 504-517.

[3] Balat M, Kırtay E, Balat H 2009 Main routes for the thermo-conversion of biomass into fuels and chemicals. Part 1: Pyrolysis systems, *Energy Conversion and Management*, **50** 3147-3157.

[4] Xiong Q, Aramideh S, Kong S C 2014 Assessment of devolatilization schemes in predicting product yields of biomass fast pyrolysis, *Environmental Progress & Sustainable Energy*, **33** 756-761.

[5] Mellin P, Kantarelis E, Yang W 2014 Computational fluid dynamics modeling of biomass fast pyrolysis in a fluidized bed reactor, using a comprehensive chemistry scheme, *Fuel*, **117** 704-715.

[6] Papari S, Hawboldt K, Helleur R 2017 Production and Characterization of Pyrolysis Oil from Sawmill Residues in an Auger Reactor, *Industrial & Engineering Chemistry Research*, **56** 1920-1925.

[7] Xiong Q, Kong S C, Passalacqua A 2013 Development of a generalized numerical framework for simulating biomass fast pyrolysis in fluidized-bed reactors, *Chemical Engineering Science*, **99** 305-313.

[8] Xiong Q, Aramideh S, Kong S C 2013 Modeling effects of operating conditions on biomass fast pyrolysis in bubbling fluidized bed reactors, *Energy & Fuels*, **27** 5948-5956.

[9] Bradbury A. G., Sakai Y., Shafizadeh F. 1979 A kinetic model for pyrolysis of cellulose, *Journal of Applied Polymer Science*, **23** 3271-3280.

[10] Xue Q, Heindel T, Fox R 2011 A CFD model for biomass fast pyrolysis in fluidized-bed reactors, *Chemical Engineering Science*, **66** 2440-2452.

[11] Gidaspow D 1994 *Multiphase flow and fluidization: continuum and kinetic theory descriptions*, Academic press, Department of Chemical Engineering, Illinois Institute of Technology, Chicago, Illinois.

[12] Lathouwers D, Bellan J 2001 Yield optimization and scaling of fluidized beds for tar production from biomass, *Energy & Fuels*, **15** 1247-1262.